高等职业教育“职教出海”特色教材

数控机床故障诊断与维修

（双语）

主　编　韩志国　李文强　王同庆

主　审　李云梅

北京理工大学出版社

BEIJING INSTITUTE OF TECHNOLOGY PRESS

内容简介

本书依据亚龙 YL-569 型 0i MF 数控设备能够开展的实训项目及数控机床装调维修工相关技能的要求并结合鲁班工坊的实际情况编写而成。本书内容主要包括数控机床故障诊断与维修概述、亚龙 YL-569 型 0i MF 数控设备整体认知、数控机床电气装调、FANUC 0-F 数控系统介绍、伺服系统故障诊断与维修、数控机床机械装调与检测这六个模块内容。每个模块开始都有学习目标,末尾都有复习思考题。本书作为鲁班工坊的配套教材,不仅可以满足国内学生对数控设备应用与维护类相关专业的学习需求,而且可以满足海外相关技能的培训工作和国外学生的自学需求。

本书可作为高等学校应用型本科、高职高专院校智能制造装备技术、机电一体化等相关专业的教学用书,也可以作为相关技术人员的参考书或技能培训用书。

版权专有　侵权必究

图书在版编目(CIP)数据

数控机床故障诊断与维修 / 韩志国, 李文强, 王同庆主编. -- 北京 : 北京理工大学出版社, 2023.10(2025.2 重印)

ISBN 978-7-5763-2997-1

Ⅰ. ①数… Ⅱ. ①韩… ②李… ③王… Ⅲ. ①数控机床-故障诊断-职业教育-教材 ②数控机床-维修-职业教育-教材 Ⅳ. ①TG659

中国国家版本馆 CIP 数据核字(2023)第 202082 号

责任编辑：赵　岩　　**文案编辑**：辛丽莉
责任校对：周瑞红　　**责任印制**：李志强

出版发行 / 北京理工大学出版社有限责任公司
社　　址 / 北京市丰台区四合庄路 6 号
邮　　编 / 100070
电　　话 / (010) 68914026 (教材售后服务热线)
(010) 63726648 (课件资源服务热线)
网　　址 / http://www.bitpress.com.cn

版 印 次 / 2025 年 2 月第 1 版第 2 次印刷
印　　刷 / 涿州市新华印刷有限公司
开　　本 / 787 mm×1092 mm　1/16
印　　张 / 21.75
字　　数 / 508 千字
定　　价 / 58.00 元

图书出现印装质量问题，请拨打售后服务热线，负责调换

前　言

为了更好地服务我国“一带一路”倡议，服务鲁班工坊，为国家输出优质的技术技能人才，鲁班工坊项目组的教师秉持教育现代化的基本理念，结合我国职业技能标准，依托先进的培训设备编写了本书。本书基于数控机床装调与维修工国家职业标准的技术技能及理论知识的要求，结合目前国内职业教育领域最先进的实训设备，灵活运用了工作过程系统化的科学教育理论，将学生所需掌握的知识和技能充分体现在了每一个模块和具体的工作任务中。本书以能力为本，以实际工作任务为导向，按照项目化教学模式的原则进行编写，在内容和结构上进行了科学合理的创新和突破。本书将理论和实践相结合，学生可根据实际任务需求边学边练，真正做到知识和技能的学以致用。

本书所设计的模块及任务内容满足钳工国家职业标准对初级理论知识及相关技能的所有要求及规定。本书采用中英双语进行编写，内容与国际化教材接轨，不仅可以满足国内学生对数控设备应用与维护类相关专业的学习需求，而且也可以满足海外相关技能的培训工作和国外学生自学需求。

韩志国、李文强、王同庆担任本书主编，杨国星、周京、商丹丹、王世委（企业）担任副主编，周树清、李洪、张川参编。具体分工：韩志国编写模块一、模块五，李文强编写模块六，王文庆编写模块三，杨国星、周树清编写模块二，周京编写模块四，王世委提供相关企业的生产项目和工作任务案例，李洪、张川负责全书素材的整理，李文强负责整体统稿。天津轻工职业技术学院李云梅院长担任本书主审。

由于编者能力有限，书中难免有不妥和错误之处，希望广大读者和同行批评指正。

编　者

目 录

模块一　数控机床故障诊断与维修概述

数控机床
维修的概述

随着电子技术和自动化技术的发展，计算机数控技术（computer numerical control，CNC）的应用越来越广泛。数控设备是一种自动化程度高、结构复杂的先进加工设备，是企业的关键设备。要发挥数控设备的高效益，就必须正确地操作和精心地维护，才能保证设备的利用率。正确的操作能够防止机床非正常磨损，避免突发故障；做好日常维护保养，可使设备保持良好的技术状态，延缓劣化进程，及时发现和消灭故障隐患，从而保证设备的安全运行。故障诊断是进行数控机床维修的第一步，它不仅可以迅速查明故障原因，排除故障，也可以起到预防故障发生与扩大的作用。故障诊断的实质是在诊断对象出现故障的前提下，通过来自外界或系统本身的信息输入，经过处理，判断出故障种类，定位故障部位（元部件），进而估计故障可能发生的时间、严重程度、故障原因等，甚至还可以提供评价、决策以及进行维修的建议。

现代故障诊断的主要内容应包括实时监测技术、故障分析（诊断）技术和故障修复方法三个部分。从信息获取到故障定位，再到故障排除，数控机床故障诊断与维修作为单独技术领域发展的同时，又与故障诊断的技术共同协调发展。

任务一　数控机床故障诊断与维修的工作任务

【学习目标】

（1）了解数控机床故障的分类。

（2）了解数控机床故障诊断与维修的工作任务。

（3）掌握数控机床故障诊断与维修具体工作要求的知识与技能。

（4）具有自主分析问题和解决问题的能力。

（5）掌握工程工作方法，培养严谨的工作作风，遵守5S管理制度。

【任务描述】

以小组为单位，学习本任务，并将数控机床故障诊断与维修的故障分类、故障现象及故障诊断方法等填写到任务工单中。要求在学习过程中务必注意安全，遵守实训场所相关操作规范。

【知识链接】

一、故障的分类

1. 按故障发生的部位分类

（1）主机故障。数控机床的主机通常指组成数控机床的机械、润滑、冷却、排屑、液压、气动与防护等部分。常见的主机故障主要有以下三个。

1）因机械部件安装、调试、操作使用不当等原因引起的机械传动故障。

2）因导轨、主轴等运动部件的干涉、摩擦过大等原因引起的故障。

3）因机械零件的损坏、连接不良等原因引起的故障等。

主机故障主要表现为传动噪声大、加工精度差、运行阻力大、机械部件动作不进行、机械部件损坏等。润滑不良，液压、气动系统的管路堵塞和密封不良，是主机发生故障的常见原因。数控机床的定期维护保养，控制和根除“三漏”（漏油、漏水、漏气）现象的发生是减少主机部分故障的重要措施。

（2）电气控制系统故障。从所使用的元器件类型上，根据通常习惯，电气控制系统故障通常分为“弱电”故障和“强电”故障两大类。

“弱电”部分是指控制系统中以电子元器件、集成电路为主的控制部分。数控机床的弱电部分包括 CNC、PMC、MDI/CRT 以及伺服驱动单元、输入/输出单元等。

“弱电”故障又有硬件故障与软件故障之分。硬件故障是指上述各部分的集成电路芯片、分立电子元器件、接插件以及外部连接组件等发生的故障。软件故障是指在硬件正常情况下所出现的动作出错、数据丢失等故障，常见的有加工程序出错、系统程序和参数的改变或丢失、计算机运算出错等。

“强电”部分是指控制系统中的主回路或高压、大功率回路中的继电器、接触器、开关、熔断器、电源变压器、电机、电磁铁、行程开关等元器件及其所组成的控制电路。

“强电”故障虽然维修、诊断较为方便，但由于元器处于高压、大电流工作状态，发生故障的概率要高于“弱电”部分，必须引起维修人员足够的重视。

2. 按故障的性质分类

（1）确定性故障。确定性故障是指控制系统主机中的硬件损坏或只要满足一定的条件，数控机床必然会发生的故障。这一类故障现象在数控机床上最为常见，但由于它具有一定的规律，因此维修也方便。

确定性故障具有不可恢复性，故障一旦发生，如不对其进行维修处理，机床不会自动恢复正常，但只要找出发生故障的根本原因，维修完成后机床立即可以恢复正常。正确的使用与精心的维护是杜绝或避免故障发生的重要措施。

（2）随机性故障。随机性故障是指数控机床在工作过程中偶然发生的故障。此类故障的发生原因较隐蔽，很难找出其规律性，故常称之为“软故障”。随机性故障的原因分析与故障诊断比较困难，一般而言，故障的发生往往与部件的安装质量、参数的设定、元器件的

品质、软件的设计、工作环境的影响等诸多因素有关。

随机性故障具有可恢复性，在故障发生后，通过重新开机等措施，机床通常可恢复正常，但在运行过程中，又可能发生同样的故障。

加强数控系统的维护检查，确保电气箱的密封，可靠的安装、连接，正确的接地和屏蔽是减少、避免此类故障发生的重要措施。

3. 按故障的指示形式分类

（1）有报警显示的故障。数控机床的故障显示可分为指示灯显示报警与显示器显示报警两种情况。

1）指示灯显示报警是指通过控制系统各单元上的状态指示灯（一般由发光二极管或小型指示灯组成）显示的报警。即使在显示器故障时，根据数控系统的状态指示灯，仍可大致分析并判断故障发生的部位与性质。因此，在维修、排除故障过程中应认真检查这些状态指示灯。

2）显示器显示报警是指可以通过 CNC 显示器显示报警号和报警信息的报警。由于数控系统一般都具有较强的自诊断功能，如果系统的诊断软件以及显示电路工作正常，一旦系统出现故障，可以在显示器上以报警号及文本的形式显示故障信息。数控系统能进行显示的报警少则几十种，多则上千种，它是故障诊断的重要信息。

显示器显示报警又可分为 CNC 的报警和 PMC 的报警两类。前者为数控生产厂家设置的故障显示，它可对照系统的《维修手册》来确定可能产生该故障的原因。后者是由数控机床生产厂家设置的 PMC 报警信息文本，属于机床侧的故障显示，它可对照机床生产厂家所提供的《机床维修手册》中的有关内容确定故障原因。

（2）无报警显示的故障。这类故障发生时，机床与系统均无报警显示，其分析诊断难度通常较大，需要通过仔细、认真的分析和判断才能予以确认。特别是对于一些早期的数控系统，由于系统本身的诊断功能不强或无 PMC 报警信息文本，出现无报警显示的故障情况则更多。

对于无报警显示的故障，通常要具体情况具体分析。根据故障发生前后的变化进行分析判断，原理分析法与 PMC 程序分析法是解决无报警显示的故障的主要方法。

4. 按故障产生的原因分类

（1）数控机床自身故障。这类故障是由数控机床自身的原因引起的，与外部使用环境条件无关。数控机床所发生的绝大多数故障均属此类故障。

（2）数控机床外部故障。这类故障是由外部原因所引起的：供电电压过低、过高或波动过大；电源相序不正确或三相输入电压不平衡；环境温度过高；有害气体、潮气、粉尘侵入；外来振动和干扰等都是引起故障的原因。

此外，人为因素也是造成数控机床故障的外部原因之一。据有关资料统计，首次使用数控机床或由不熟练工人来操作数控机床，在使用的第一年，由于操作不当所造成的外部故障要占机床总故障的三分之一以上。

除上述常见故障分类方法外，还有其他多种不同的分类方法。如：按故障发生时有无破

坏性可分为破坏性故障和非破坏性故障两种；按故障发生与需要维修的具体功能部位可分为数控装置故障、进给伺服系统故障、主轴驱动系统故障、自动换刀系统故障等。这一分类方法在维修时常用。

二、数控机床故障诊断与维修工作任务

现代数控机床是机、电、液、光的高度融合与集成，其机电接口也高度复杂，是典型的技术密集型机电一体化产品，要求数控机床装调维修工在机械和微电子方面具备一定的理论基础与维修、调整技能，这就在以下各方面对他们提出了较高的要求。

1. 机械功能部件与整机装配

（1）能读懂数控机床总装配图或部件装配图。

（2）能绘制连接件装配图。

（3）能根据整机装配调试要求准备工具、工装。

（4）能完成两种以上机械功能部件的装配或一种以上型号的数控机床总装配。

（5）能进行数控机床总装后集合精度、工作精度的检测和调整。

（6）能读懂三坐标测量报告、激光检测报告，并进行一般误差分析和调整。

2. 机械功能部件与整机调整

（1）能读懂数控机床电气原理图、电气接线图。

（2）机床通电试车时，能完成机床数控系统初始化后的资料输入。

（3）能进行系统操作面板、机床操作面板的功能调整。

（4）能进行数控机床试车（如空运转）。

（5）能通过修改常用参数调整机床性能。

（6）能进行两种型号以上数控系统的操作。

（7）能进行两种型号以上数控系统的加工编程。

（8）能根据零件加工工艺要求准备刀具、夹具。

（9）能完成试车工件的加工。

（10）能使用通用量具对所加工的工件进行检测，并进行误差分析和调整。

3. 机械功能部件与整机维修

（1）能读懂机床总装配图或部件装配图。

（2）能读懂数控机床电气原理图、电气接线图。

（3）能读懂数控机床液压与气动原理图。

（4）能拆卸、组装整台数控机床（如数控机床主轴箱与床身的拆装、床鞍与床身的拆装、加工中心主轴箱与床身的拆装、工作台与床身的拆装等）。

（5）能通过数控机床诊断功能判断常见机械、电气、液压（气动）故障。

（6）能排除数控机床的机械故障。

（7）能排除数控机床的强电故障。

4. 整机电气装配

（1）能读懂数控机床电气装配图、电气原理图、电气接线图。

（2）能读懂机床总装配图。

（3）能读懂数控机床液压与气动原理图。

（4）能读懂与电气相关的机械图（数控刀架、刀库与机械手等）。

（5）能按照电气图要求安装两种型号以上数控机床全部电路，包括配电板、电气柜、操作台、主轴变频器、机床各部件之间电缆线的连接等。

5. 整机电气调整

（1）能读懂 PMC 梯形图，并能修改其中的错误。

（2）能使用系统参数、PMC 参数、变频器参数等对数控机床进行调整。

（3）能在数控机床通电试车时，通过机床通信口将机床参数与 PMC 程序（如梯形图）传入 CNC 控制器中。

（4）能进行数控机床各种功能的调试。

（5）能应用数控系统编制加工程序。

（6）能对数控机床进行试车调整。

（7）能选用常用刀具加工试车工件。

（8）能对机床进行水平调整。

（9）能进行数控机床几何精度检测。

（10）能读懂三坐标测量报告，并进行一般分析（如垂直度、平行度、同轴度、位置度等）。

（11）能使用通用量具对轴类、盘类工件进行检测，并进行误差分析。

6. 整机电气维修

（1）能读懂数控机床电气装配图、电气原理图、电气接线图。

（2）能读懂数控机床总装配图。

（3）能读懂液压与气动原理图。

（4）能读懂与电气相关的机械图（数控刀架、刀库与机械手等）。

（5）能通过仪器、仪表检查故障点。

（6）能通过数控系统诊断功能、PMC 梯形图等诊断数控机床常见机械、电气、液压（气动）故障。

（7）能完成两种规格以上数控机床常见强弱电气故障的维修。

【任务实施】

（1）按照随机分组方式，分成各个学习小组，并选出一名组长，将本组成员的个人信息填入表 1-1 中。

（2）根据观摩及讨论的结果，在表 1-1 中记录数控机床故障诊断与维修的故障分类、故障现象及故障诊断方法等。

（3）对观摩过程中的疑难问题进行充分讨论，并将出现问题及解决办法填入表 1-1 中。

表 1–1　数控机床故障诊断与维修任务工单

任务工单				
姓名：		班级：		学号：
工作过程（观摩）				
序号	故障分类	故障现象	故障诊断方法	备注
出现问题			解决办法	

【任务评价】

序号	评价内容	分值	得分
1	能够熟练说出数控机床故障的分类	25	
2	能够正确绘制、熟练说出数控机床装调维修工各方面应具备的技能	25	
3	能够运用数控机床故障诊断与维修工作任务进行数控机床故障分析	25	
4	具有有效沟通、表达及团结协作的职业精神	25	

任务二　数控机床故障诊断与维修的一般步骤

【学习目标】

（1）熟悉数控机床维修前的注意事项。

（2）掌握数控机床故障诊断与维修的一般步骤。

（3）掌握数控机床故障诊断与维修应遵循的原则。

（4）具有自主分析问题和解决问题的能力。

（5）掌握工程工作方法，培养严谨的工作作风，遵守 5S 管理制度。

【任务描述】

以小组为单位学习本任务，并将数控机床故障诊断与维修时的注意事项、一般步骤及原则等填写到任务工单中，要求学习过程中务必注意安全，遵守实训场所相关操作规范。

【知识链接】

一、维修数控机床前的注意事项

故障诊断始于机械设备故障诊断，主要指制造设备和制造过程的状态监测与故障诊断。制造设备主要指加工机床、夹具、量具和刀具；制造过程指工艺过程和工艺参数。机械设备运行时的状态监测与故障诊断包含两方面内容：一是对设备的运行状态进行监测；二是在发现异常情况后对设备的故障进行分析、诊断。在使用数控机床之前，应仔细阅读机床使用说明书以及其他有关资料，以便正确操作使用机床，并注意以下几点。

（1）机床操作、维修人员必须是掌握相应机床专业知识的专业人员或经过技术培训的人员，且必须按安全操作规程及安全操作规定操作机床。

（2）非专业人员不得打开电柜门。打开电柜门前必须确认已经关闭了机床总电源开关。只有专业维修人员才允许打开电柜门，进行通电检修。

（3）除一些供用户使用并可以改动的参数外，其他系统参数、主轴参数、伺服参数等，用户不能私自修改，否则将给操作者带来设备、工件损坏以及人身伤害。

（4）在修改参数后，进行第一次加工时，机床在不装刀具和工件的情况下用机床锁住、单程序段等方式进行试运行，确认机床正常后再使用机床。

（5）机床的 PMC 程序是机床制造商按机床需要设计的，不需要修改。不正确地修改、操作机床可能造成机床的损坏，甚至伤害操作者。

（6）建议机床连续运行最多 24 小时，如果连续运行时间太长会影响电气系统和部分机械部件的寿命，从而会影响机床的精度。

（7）机床的全部连接器、接头等不允许带电拔、插操作，否则将引起严重的后果。

二、数控机床故障诊断步骤

数控机床故障诊断与维修的一般步骤

无论数控机床处于哪一个故障期，数控机床故障诊断的一般步骤都是相同的。当数控机床发生故障时，除非出现危险及数控机床损坏或人身安全的紧急情况，一般不要关断电源，要尽可能地保持机床原来的状态不变，并对出现的一些信号和现象做好记录，主要包括对故障现象的详细记录、故障发生的操作方式及内容、故障号及故障指示灯的显示内容。故障诊断一般按下列步骤进行。

（1）详细了解故障情况。例如，当数控机床发生颤振、振动或超调现象时，要弄清楚是发生在全部轴还是某一轴；如果是某一轴，是全程还是某一位置；是一运动就发生还是仅在快速、进给状态某速度、加速或减速的某个状态下发生。为了进一步了解故障情况，要对数控机床进行初步检查，并着重检查荧光屏上的显示内容，控制柜中的故障指示灯、状态指

示灯等。当故障情况允许时，最好开机试验，详细观察故障情况。

（2）根据故障情况进行分析，缩小范围，确定故障源查找的方向和手段。对故障现象进行全面了解后，下一步可根据故障现象分析故障可能存在的位置。有些故障与其他部分联系较少，容易确定查找的方向；而有些故障原因很多，难以用简单的方法确定故障源的查找方向，这就要仔细查阅数控机床的相关资料，弄清与故障有关的各种因素，确定若干个查找方向，并逐一进行查找。

（3）由表及里进行故障源查找。故障源查找一般是从易到难、从外围到内部逐步进行。所谓难易，包括技术上的复杂程度和拆卸装配方面的难易程度。技术上的复杂程度是指判断其是否有故障存在的难易程度。在故障诊断的过程中，首先应该检查可直接接近或经过简单的拆卸即可进行检查的那些零件，然后检查需要进行大量的拆卸工作之后才能接近和进行检查的那些零件。

三、数控机床故障诊断与维修原则

1. 先方案后操作

故障发生后，维修人员应先向机床操作者了解故障发生的整个过程，查阅机床的技术说明和相关技术图样，考虑好故障的解决方案后再动手维修。

2. 先检查后通电

确定好解决方案后，不要急于通电，要先对机床进行检验、测试和分析，以确定故障的性质是恶性的破坏性故障还是非恶性的破坏性故障。

如果确认是破坏性的故障，必须先将危险排除；如果确认是非破坏性故障，可给机床通电，然后对运转的机床做进一步的动态观察、检验和测试，从而可以找到故障的发生部位。

3. 先软件后硬件

数控系统的软件工作不正常同样可以导致数控机床故障。比如，软件参数丢失，软件的使用方式、操作方法不正确等。

因此，机床通电后，应先确认软件是否正常工作，以免发生更大的故障。

4. 先外部后内部

数控机床发生故障后，维修人员首先应检查机械部件是否发生故障，如行程开关、按钮开关工作是否正常。

确认机械部件没有问题后，再检查液压器件是否发生异常，如液压元器件的连接是否松动。

最后，应检查电气接触部件是否松动，如印制电路板插座、电控柜的插座等。这些部位往往由于机械振动、油污、粉尘、温湿度的变化造成信号接触不良，使信号传递失真，从而导致数控机床故障。

5. 先机械后电气

数控机床是由机械系统、液压系统、电气控制系统组成的高度自动化、复杂的先进机械加工设备。

对数控机床的诊断应该按一定的顺序进行。经验表明，大部分故障是由机械系统动作失灵造成的，如行程开关不能正常工作。此外，机械故障容易察觉，电气故障较难诊断。

因此，在维修时应先逐一检查机械故障，往往能够达到事半功倍的功效。

6. 先公用后专用

公用性的故障影响面大，是主要矛盾；专用性的故障只影响局部，是次要矛盾。

例如，数控机床的所有坐标轴都不能做进给运动、电网或主电源发生故障，这些都是公用性的故障。只有公用性的故障得以排除，专用性的故障才可能得到解决。

7. 先简单后复杂

数控机床可能同时发生多种故障，故障的复杂程度不尽相同，难度小的故障比较容易维修，难度大的故障维修起来比较困难。维修人员应先维修小的容易解决的故障，在维修的过程中可能受到启发，对复杂故障有了清晰的认识，或者复杂的故障变成了小故障，利于复杂故障的解决。

8. 先一般后特殊

导致数控机床发生某一故障的原因可能多种多样，在维修时要优先考虑可能导致故障的常见因素，最后再分析特殊不常见的因素。

【任务实施】

（1）按照随机分组方式，分成各个学习小组，并选出一名组长，将本组成员的个人信息填入表 1-2 中。

（2）根据观摩及讨论的结果，在表 1-2 中记录数控机床故障诊断与维修时的注意事项、一般步骤、原则及适用场合等。

（3）对观摩过程中的疑难问题进行充分讨论，并将问题及解决办法填入表 1-2 中。

表 1-2　数控机床故障诊断与维修的一般步骤学习任务工单

任务工单				
姓名：		班级：	学号：	
工作过程（观摩）				
序号	注意事项	一般步骤	原则	适用场合
出现问题			解决办法	

【任务评价】

序号	评价内容	分值	得分
1	能够熟练说出数控机床故障诊断与维修的注意事项	25	
2	能够熟练说出数控机床故障诊断与维修的一般步骤及原则	25	
3	能够运用数控机床故障诊断与维修的一般步骤并进行数控机床故障诊断与维修	25	
4	具有有效沟通、表达及团结协作的职业精神	25	

任务三　数控机床故障诊断与维修常用工量具介绍

【学习目标】

（1）了解数控机床故障诊断与维修常用工量具。

（2）掌握数控机床故障诊断与维修常用工量具的使用方法。

（3）具有自主分析问题和解决问题的能力。

（4）掌握工程工作方法，培养严谨的工作作风，遵守5S管理制度。

【任务描述】

以小组为单位，学习使用各类数控机床故障诊断与维修常用工量具，并将数控机床故障诊断与维修常用工量具分类、名称、主要功能及应用等填写到任务工单中。要求在学习过程中务必注意安全，遵守实训场所相关操作规范。

【知识链接】

一、拆卸及装配工具

（1）单头钩形扳手（图1-1）：可分为固定式和调节式，可用于扳动在圆周方向上开有直槽或孔的圆螺母。

（2）端面带槽或孔的圆螺母扳手（图1-2）：可分为套筒式扳手和双销叉形扳手。

（3）弹性挡圈装拆用钳子（图1-3）：可分为轴用弹性挡圈装拆用钳子和孔用弹性挡圈装拆用钳子。

（4）弹性手锤：可分为木锤和铜锤（图1-4）。

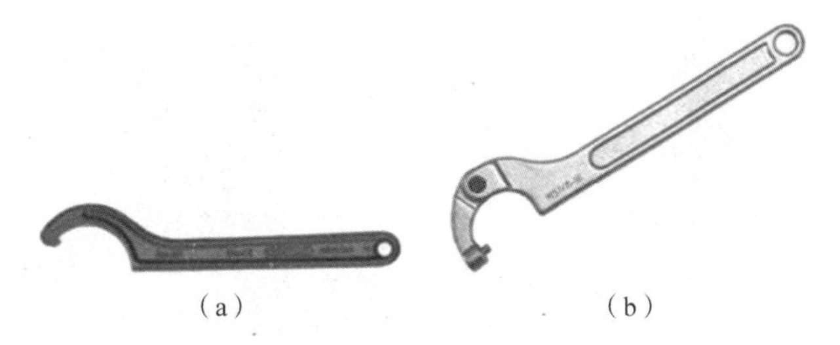

（a）　　（b）

图 1-1　单头钩形扳手

（a）固定式钩形扳手；（b）调节式钩形扳手

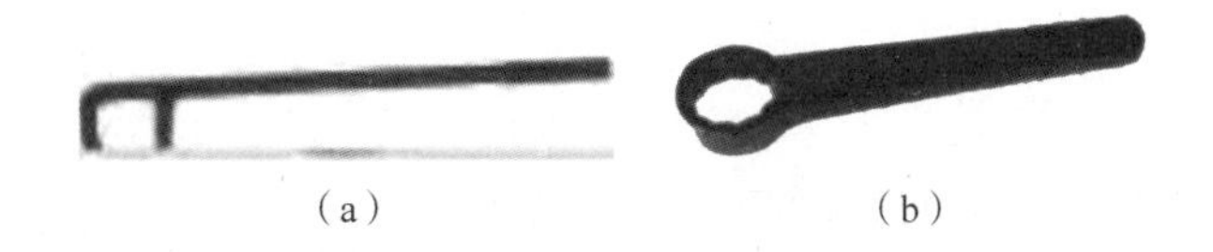

（a）　　（b）

图 1-2　圆螺母扳手

（a）端面带槽的圆螺母扳手；（b）端面带孔的圆螺母扳手

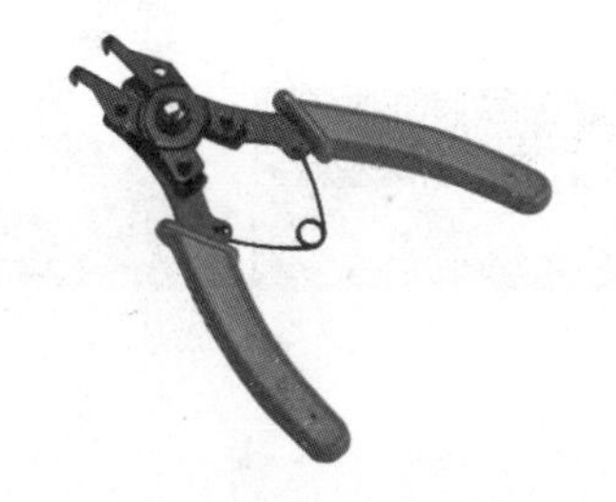

图 1-3　弹性挡圈装拆用钳子

图 1-4　铜锤

（5）拉带锥度平键工具（图 1-5）：可分为冲击式拉锥度平键工具和抵拉式拉锥度平键工具。

（6）拉带内螺纹的小轴、圆锥销工具（俗称拨销器），如图 1-6 所示。

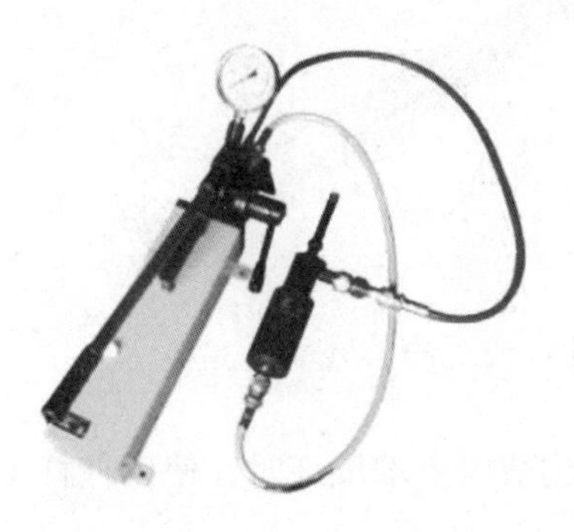

图 1-5　拉带锥度平键工具

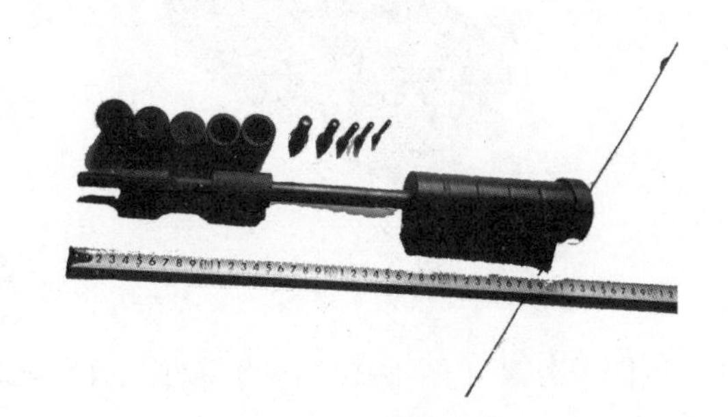

图 1-6　拨销器

（7）拉卸工具：拆装在轴上的滚动轴承、皮带轮式联轴器等零件时，常用拉卸工具，拉卸工具常分为螺杆式及液压式两类，螺杆式拉卸工具分两爪、三爪（图 1-7）和铰链式。

（8）销子冲头，如图 1-8 所示。

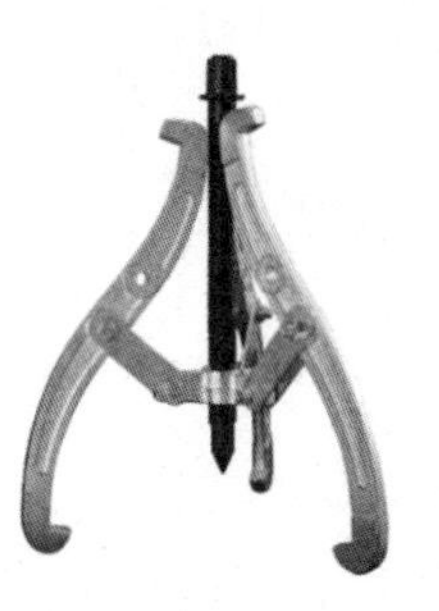

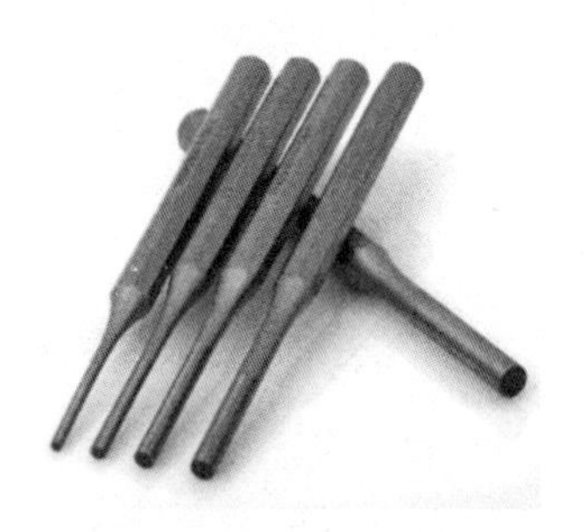

图 1-7　螺杆式三爪拉卸工具　　图 1-8　销子冲头

二、常用的机械维修工具

（1）尺：可分为平尺、刀口尺和 90°角尺，如图 1-9 所示。

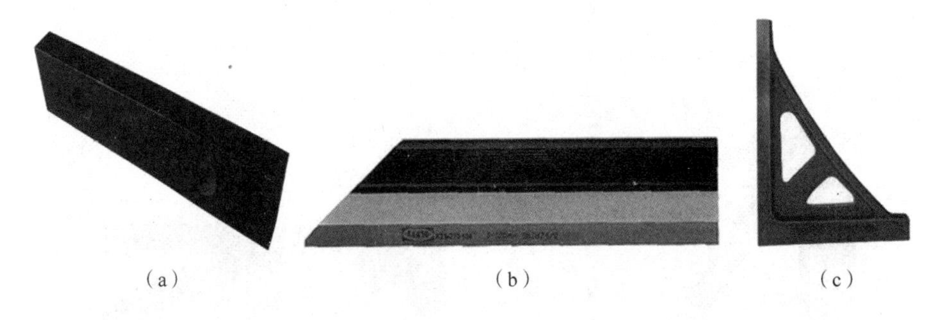

（a）　（b）　（c）

图 1-9　尺

（a）平尺；（b）刀口尺；（3）90°角尺

（2）垫铁（图 1-10）：可分为面为 90°的垫铁、面为 55°的垫铁和水平仪垫铁。

（3）检验棒：可分为带标准锥柄检验棒（图 1-11）、圆柱检验棒和专用检验棒。

图 1-10　垫铁　　图 1-11　带标准锥柄检验棒

（4）杠杆千分尺（图 1-12）：当零件的几何形状精度要求较高时，使用杠杆千分尺可满足其测量要求，其测量精度可达 0.001mm。

（5）万能角度尺（图 1-13）：用来测量工件内外角度的量具，按其游标读数值可分为 2ft① 和 5ft 两种，按其尺身的形状可分为圆形和扇形两种。

---

① 1ft = 0.304 8 m。

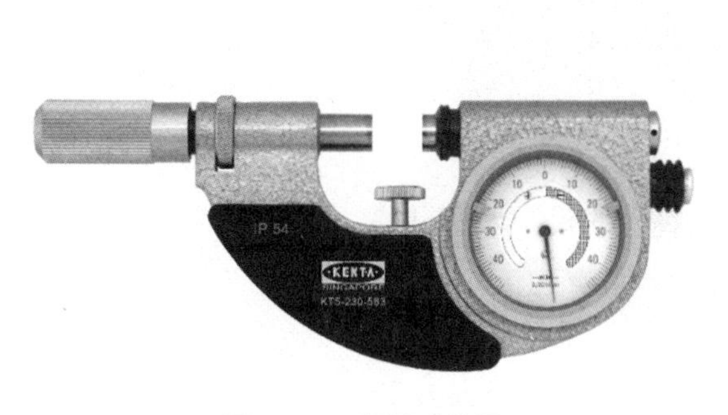

图 1-12　杠杆千分尺

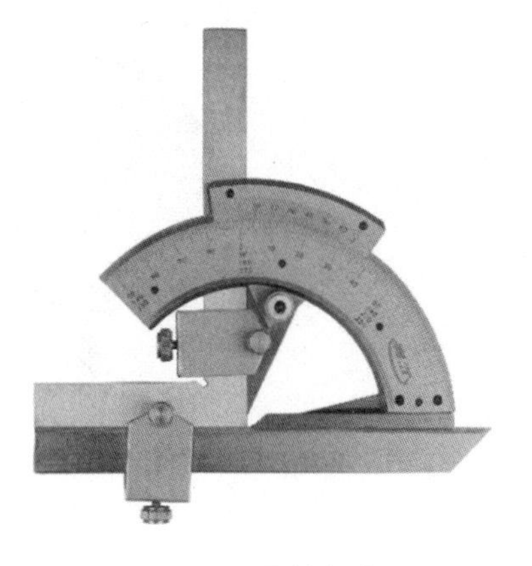

图 1-13　万能角度尺

三、常用的数控机床维修仪表

1. 百分表

百分表（图 1-14）用于测量零件相互之间的平行度、轴线与导轨的平行度、导轨的直线度、工作台台面平面度以及主轴的端面圆跳动、径向圆跳动和轴向窜动。

2. 杠杆百分表

杠杆百分表（图 1-15）用于受空间限制的工件，如内孔跳动、键槽等。使用时应注意使测量运动方向与测头中心垂直，以免产生测量误差。

图 1-14　百分表

图 1-15　杠杆百分表

3. 千分表及杠杆千分表

千分表及杠杆千分表（图 1-16）的工作原理与百分表和杠杆百分表一样，只是分度值不同，常用于精密机床的修理。

4. 比较仪

比较仪（图 1-17）可分为扭簧比较仪与杠杆齿轮比较仪。扭簧比较仪特别适用于精度

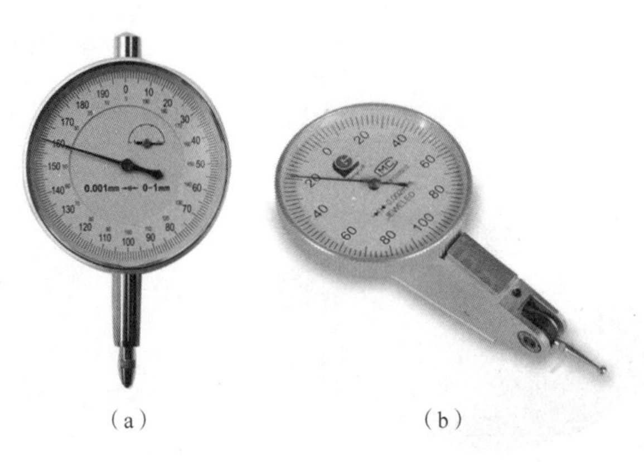

（a）　　　　　　（b）

图 1-16　千分表及杠杆千分表

（a）千分表；（b）杠杆千分表

要求较高的跳动量的测量。

5. 水平仪

水平仪是机床制造和修理中最常用的测量仪器之一，用来测量导轨在垂直面内的直线度、工作台台面的平面度以及零件相互之间的垂直度、平行度等。水平仪按其工作原理可分为水准式水平仪和电子水平仪。水准式水平仪有条式水平仪（图 1-18）、框式水平仪和合像水平仪 3 种结构形式。

图 1-17　比较仪

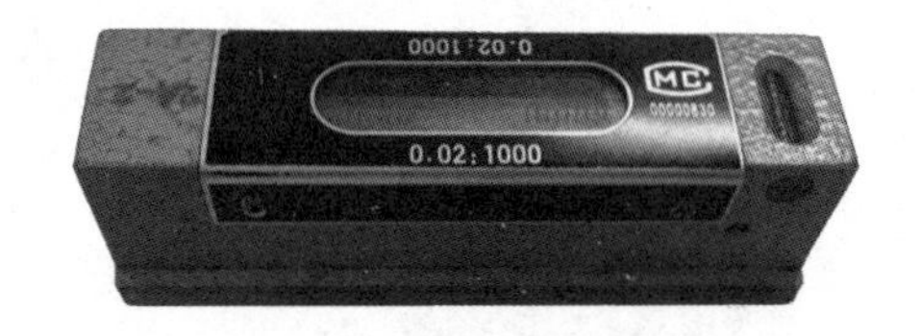

图 1-18　条式水平仪

图 1-19　光学平直仪

6. 光学平直仪

光学平直仪在机械维修中常用来检查床身导轨在水平面内和垂直面内的直线度、检验用平板的平面度，是当前导轨直线度测量方法中较先进的仪器之一，如图 1-19 所示。

7. 经纬仪

经纬仪（图 1-20）是机床精度检查和维修中常用的高精度仪器之一，常用于数控铣床和加工中心的水平转台和万能转台的分度精度的精确测量，通常与平行光管组成光学系统来使用。

8. 转速表

转速表常用于测量伺服电机的转速，是检查伺服调速系统的重要依据之一，常用的转速表有离心式转速表（图 1-21）和数字式转速表等。

图 1-20 经纬仪

图 1-21 离心式转速表

四、常用的数控机床维修仪器

在数控机床的故障检测过程中，借助一些仪器是必要的，这些仪器能从定量分析角度直接反映故障点状况，起到决定性的作用。

1. 测振仪

测振仪用来测量数控机床主轴的运行情况、电机的运行情况，甚至整机的运行情况，可根据所需测定的参数、振动频率和动态范围、传感器的安装条件、机床的轴承型式（滚动轴承或滑动轴承）等因素，分别选用不同类型的传感器。常用的传感器有涡流式位移传感器、磁电式速度传感器和压电加速度传感器。

测振判断的标准，一般情况下在现场最便于使用的是绝对判断标准，它是针对各种典型对象制定的，如国际通用的机械振动标准 ISO20816 系列。

2. 红外测温仪

红外测温是利用红外辐射原理，将对物体表面温度的测量转换成对其辐射功率的测量，采用红外探测器和相应的光学系统接收被测物不可见的红外辐射能量，并将其变成便于检测的其他能量形式予以显示和记录。

按红外辐射的不同响应形式，红外测温仪可分为光电探测器和热敏探测器两类。红外测温仪用于检测数控机床容易发热的部件，如功率模块、导线接点、主轴轴承等。主要制造厂商的产品有中国昆明物理研究所的 HCW 系列，中国西北光学仪器厂的 HCW-1、HCW-2，美国 LAND 公司的 CYCLOPS、SOLD 等。

利用红外测温的仪器还有红外热电视、光机扫描热像仪以及焦平面热像仪等。红外诊断的判定主要有温度判断法、同类比较法、档案分析法、相对温差法以及热像异常法。

3. 激光干涉仪

激光干涉仪可对机床、三测机及各种定位装置进行高精度的（位置和几何）精度校正，可完成各项参数的测量，如线形位置精度、重复定位精度、角度、直线度、垂直度、平行度及平面度等。激光干涉仪还具有一些选择功能，如自动螺距误差补偿（适用大部分数控系统）、机床动态特性测量与评估、回转坐标分度精度标定、触发脉冲输入输出功能等。

激光干涉仪用于机床精度的检测及长度、角度、直线度、直角等的测量，其测量精度高、效率高、使用方便，测量长度可达十几米甚至几十米，测量精度达微米级。

【任务实施】

（1）按照随机分组方式，分成各个学习小组，并选出一名组长，将本组成员的个人信息填入表 1-3 中。

（2）根据观摩及讨论的结果，在表 1-3 中记录数控机床故障诊断与维修常用工量具的分类、名称、主要功能及应用等。

（3）对观摩过程中的疑难问题进行充分讨论，并将问题及解决办法填入表 1-3 中。

表 1-3　数控机床故障诊断与维修常用工量具学习任务工单

<table>
<tr><td colspan="5">任务工单</td></tr>
<tr><td colspan="2">姓名：</td><td colspan="2">班级：</td><td>学号：</td></tr>
<tr><td colspan="5">工作过程（观摩）</td></tr>
<tr><td>序号</td><td>分类</td><td>名称</td><td>主要功能</td><td>应用</td></tr>
<tr><td></td><td></td><td></td><td></td><td></td></tr>
<tr><td></td><td></td><td></td><td></td><td></td></tr>
<tr><td></td><td></td><td></td><td></td><td></td></tr>
<tr><td></td><td></td><td></td><td></td><td></td></tr>
<tr><td colspan="3">出现问题</td><td colspan="2">解决办法</td></tr>
<tr><td colspan="3"></td><td colspan="2"></td></tr>
</table>

【任务评价】

序号	评价内容	分值	得分
1	能够熟练说出数控机床故障诊断与维修常用工量具的名称	25	
2	能够正确说出数控机床故障诊断与维修常用工量具的用途	25	

续表

序号	评价内容	分值	得分
3	能够熟练使用数控机床故障诊断与维修常用工量具	25	
4	具有有效沟通、表达及团结协作的职业精神	25	

(1) 数控机床故障的分类是什么?

(2) 数控机床故障诊断与维修的主要工作任务是什么?

(3) 数控机床故障诊断与维修应遵循的原则是什么?

(4) 数控机床故障诊断与维修的一般步骤是什么?

(5) 数控机床故障诊断与维修常用的工量具及仪器是什么?

(6) 数控机床故障诊断与维修的基本方法是什么?

(7) 数控机床故障诊断与维修技术的发展方向是什么?

模块二 亚龙YL-569型0i MF数控设备整体认知

实训设备是开展实践教学环节必不可少的载体，在设备使用过程中教师和学生均应充分了解设备的技术参数，严格遵守设备的使用说明。本模块主要介绍亚龙 YL-569 型 0i MF 数控铣床的基本组成、技术参数、可开展的实训项目、使用须知及通电流程等具体内容。

任务一 亚龙YL-569型0i MF数控设备介绍

亚龙 YL-569 型 0I MF 数控设备介绍

【学习目标】

（1）了解亚龙 YL-569 型 0i MF（以下简称 YL-569 型 0i MF）数控铣床基本组成及技术参数。

（2）掌握亚龙 YL-569 型 0i MF 数控铣床具体功能单元。

（3）掌握亚龙 YL-569 型 0i MF 数控铣床设备可开展的实训项目。

（4）具有自主分析问题和解决问题的能力。

（5）掌握工程工作方法，培养严谨的工作作风，遵守 5S 管理制度。

【任务描述】

以小组为单位，观摩亚龙 YL-569 型 0i MF 数控设备，并将设备的配置名称、主要部件、规格及数量等填写到任务工单中。要求在观摩过程中务必注意安全，遵守实训场所相关操作规范。

【知识链接】

一、设备概述

亚龙 YL-569 型系列数控设备适合于职业院校的数控装调维修专业、数控加工专业、机电一体化专业的教学与实训，该设备是根据职业院校专业的教学特点，结合企业的实际需求以及相关岗位技能要求而开发的具有生产型功能与学习型功能的设备。该设备采用模块化结构，通过不同的组合，可以进行数控机床的电气装调、系统调试、数控机床功能部件机械几何精度检测及维修等实训项目，满足企业对这类人才的需求。该设备还

适用于数控机床装调与维修技能大赛、数控装调工的职业技能鉴定。YL-569 型数控铣床外形如图 2-1 所示。

图 2-1　YL-569 型数控铣床外形

二、设备功能说明

亚龙 YL-569 型系列数控机床由电气控制单元、电气安装实训单元、机械单元、刀库单元、工具检具等组成。

电气控制单元主要包含数控系统、进给驱动、主轴单元、PMC 单元，冷却控制电路、接口单元、保护电路、电源电路等组成。该单元包含完整的数控机床电气控制部分，内嵌有数控机床智能化考核系统，可以直接与数控机床功能部件进行连接用于真实的电气调试与维修实训；立式结构符合真实的数控电气安装环境，器件布局与实际机床一致，符合《机械电气安全 机械电气设备 第 1 部分：通用技术条件》（GB/T 5226.1—2019）标准，可以更贴合实际岗位要求进行技能训练。

配置的智能化故障维修系统可以通过产生故障、故障分析、故障诊断、线路检查、故障点确定等过程训练学生数控机床维修能力，可以配合计算机软件进行学生登录、自动评分、成绩统计、实训结果评价等，还可以通过网络连接进行数控技术的应知考核，大幅减轻教师的故障设定、评分、统计等工作量，是一套集实施、检查、评估于一体的数控技术教育装备。

电气安装实训单元主要是针对数控机床电气安装技能反复训练的需要而设计的，包含 PMC 接口单元、冷却控制电路、接口单元、保护电路、电源电路等组成。电气安装板采用可以反复使用的网孔板结构，通过接口转换单元和电气控制单元进行连接，完成对数控机床功能部件的调试。电源部分设置漏电开关和缺相保护电路，在发生漏电、短路、缺相时，设备保护电路自动动作。

机械单元是一台 *X*、*Y*、*Z* 三轴伺服控制立式机床，结构及外形尺寸紧凑合理。机床主轴为伺服电机驱动。底座、滑座、工作台、立柱、主轴箱等主要基础件采用高刚性的铸铁结构。底座的内部分布着加强筋。滑座为箱体式结构，保证了基础件的高刚性和抗弯减振性能；基础件采用树脂砂造型并经过时效处理，确保机床长期使用精度的稳定性，为机床性能的稳定性提供了保障。

X、*Y*、*Z* 向导轨均为直线导轨，配合自动润滑，实现机床的高速运动；*X*、*Y*、*Z* 三个方向进给采用高精度、高强度的滚珠丝杠，进给速度高；驱动电机通过弹性联轴器与丝杠直联，进给伺服电机直接将动力传递给高精度滚珠丝杠，无背隙，保证机床的定位精度；采用知名品牌高速、高精度、高刚性主轴单元，其轴向和径向承载能力强，转速达 10 000 r/min；主轴采用中心吹气结构，在主轴松刀后迅速用中心高压气体对主轴内锥进行清洁，保证刀具夹持的精度。

X、*Y*、*Z* 三个方向导轨、丝杠都采用密闭防护，保证丝杠及导轨的清洁，确保机床的传动及运动精度；采用先进的集中自动润滑装置，定时、定量地自动间歇式润滑，工作稳定可靠。

斗笠式刀库应用较多，此换刀装置的优点是结构简单、成本较低、换刀可靠性较好。

机床工量具主要用于完成国标中规定的检测实训项目。

三、设备实训项目

（1）数控机床电气组成的学习。

（2）数控机床电气控制及 PMC 的学习。

（3）主轴与进给轴控制的学习。

（4）数控机床控制电路安装实训。

（5）电气原理图及装配图的识图与绘制。

（6）主轴、进给轴、系统、伺服驱动等参数设置。

（7）数据备份。

（8）主轴、进给轴、冷却、润滑等模块的基本功能调试。

（9）数控机床故障诊断与维修。

（10）数控机床几何精度检测。

（11）行程限位参数设定实训。

（12）回零参数设置实训。

（13）螺距补偿设置实训。

（14）刀库的安装与调试实训。

（15）数控机床升级、改造与维修实训。

（16）数控零件的铣削加工。

（17）数控机床编程与操作。

四、设备配置

设备标准配置如表 2-1 所示，工量具如表 2-2 所示，材料如表 2-3 所示。

表 2-1　设备标准配置

序号	名称	主要部件、器件及规格	数量
1	电气平台	569 型数控机床实训设备	1 台
2	数控系统	发那科 0i MF	1 台
3	驱动单元	发那科交流伺服系统	1 套

续表

序号	名称	主要部件、器件及规格	数量
4	手轮单元	手摇脉冲发生器	1只
5	刀库单元	12把刀斗笠式刀库	1台
6	润滑单元	电动润滑泵	1台
7	数控机床机械单元	YL-557A	1台

表2-2　工量具

序号	名称	型号及规格	数量
1	剥线钳	HS-700D	1只
2	斜口钳	DL2206	1只
3	压线钳	HS-30J	1只
4	压线钳	HS-06WF	1只
5	尖嘴钳	DL2106	1只
6	剪刀	民用型	1把
7	万用表	MY60	1块
8	十字螺丝刀	3 mm×75 mm	1把
9	十字螺丝刀	5 mm×150 mm	1把
10	一字螺丝刀	3 mm×75 mm	1把
11	一字螺丝刀	5 mm×150 mm	1把
12	试电笔	氖管式	1支
13	内六角扳手	7件套	1套
14	方尺	（0级大理石）	1块
15	棉布	—	1条
16	润滑脂	—	1份
17	杠杆百分表	0~0.8 mm/0.01 mm	1块
18	磁性表座	CZ-6A	1只
19	橡皮锤	圆头	1只
20	条式水平仪	200 mm	2个
21	紫铜棒	ϕ25 mm×240 mm	1条
22	工具箱	430 mm×230 mm×200 mm	1只
23	记号笔	0.8~3 mm	1只
24	百分表	0~10 mm/0.01 mm	1块
25	主轴检棒	与主轴锥度配套	1只

表2-3 材料

序号	名称	型号	数量
1	多芯软线	RV1.5 mm 黑	1卷
2	多芯软线	RV0.75 mm 黑	1卷
3	多芯软线	RV0.75 mm 红	1卷
4	多芯软线	RV0.75 mm 蓝	1卷
5	多芯软线	RV0.75 mm 白	1卷
6	接地线	RV1.5 mm 黄绿线	10 m
7	绝缘端子	QE1008 压 0.75 线	1包
8	冷压端子	SV2-4 压 2.5 线	1包
9	冷压端子	SV1.25-4 压 0.75 线	1包
10	扎带	150 黑色	100条
11	号码管	ϕ3.5（空白）	3 m
12	号码管	ϕ5.5（空白）	3 m

五、技术参数

1. 电气控制单元参数

（1）电源：三相五线、AC 380 V×(1±10%)、50 Hz；
（2）数控控制台尺寸：长×宽×高＝800 mm×600 mm×1 800 mm；
（3）漏电保护：漏电动作电流≤30 mA；
（4）缺相自动保护、过载保护。

2. 电气安装实训单元参数

（1）电源：三相五线 AC 380 V×(1±10%)、50 Hz；
（2）数控控制台尺寸：长×宽×高＝800 mm×600 mm×1 600 mm。

3. 机床参数

工作台规格：700 mm×420 mm；
X 坐标行程：550 mm；
Y 坐标行程：400 mm；
Z 坐标行程：450 mm；
主轴中心至立柱导轨滑块安装面距离：~453 mm；
主轴端面至工作台面距离：110~560 mm；
X、*Y*、*Z* 快速移动速度：0~48 000 mm/min；

主轴最高转速：10 000 r/min；

主轴锥孔：BT40；

主轴功率：5.5 kW；

刀库形式：斗笠式；

刀库容量：12 把；

工作台 T 形槽：3 mm×14 mm×110 mm；

工作台承重：300 kg；

X、*Y*、*Z* 坐标定位精度（国标）：0.008 mm；

X、*Y*、*Z* 坐标重复定位精度（国标）：0.006 mm；

气源压力：0.6～1 MPa；

光机外形尺寸：约 1 600 mm×1 600 mm×2 100 mm；

防护：简易半防护。

机床精度完全符合国家标准，电气控制和操作完全符合标准机床的要求。

数控铣床维修实训系统需严格按照《加工中心检验条件 第 1 部分：卧式和带附加主轴头机床几何精度检验（水平 *Z* 轴）》（GB/T 18400.1—2010）、《加工中心检验条件 第 4 部分：线性和回转轴线的定位精度和重复定位精度检验》（GB/T 18400.4—2010）执行，同时严格按照《加工中心检验条件 第 6 部分：进给率速度和插补精度检验》（GB/T 18400.6—2001）检验，适用于精度要求高、形状复杂、工序多、循环周期长、品种多变的零件加工。

正常使用条件如下。

环境温度：0～40 ℃；湿度：≤85%。

【任务实施】

（1）按照随机分组方式，分成各个观摩小组，并选出一名组长，将本组成员的个人信息填入表 2-4 中。

（2）根据观摩及讨论的结果，在表 2-4 中记录亚龙 YL-569 型 0i MF 数控设备的配置名称、数量、型号或规格等。

（3）对观摩过程中的疑难问题进行充分讨论，并将问题及解决办法填入表 2-4 中。

表 2-4　亚龙 YL-569 型 0i MF 数控设备观摩任务工单

<table>
<tr><td colspan="5">任务工单</td></tr>
<tr><td colspan="2">姓名：</td><td colspan="2">班级：</td><td>学号：</td></tr>
<tr><td colspan="5">所需材料、工量具、设备列表</td></tr>
<tr><td>序号</td><td>名称</td><td>数量</td><td>型号或规格</td><td>备注</td></tr>
<tr><td></td><td></td><td></td><td></td><td></td></tr>
<tr><td></td><td></td><td></td><td></td><td></td></tr>
<tr><td></td><td></td><td></td><td></td><td></td></tr>
</table>

续表

工作过程（观摩）				
序号	配置名称	主要部件	规格	数量
出现问题			解决办法	

【任务评价】

序号	评价内容	分值	得分
1	能够熟练说出亚龙 YL-569 型 0i MF 数控设备的标准配置	25	
2	能够正确说出亚龙 YL-569 型 0i MF 数控设备的技术参数	25	
3	能够概述亚龙 YL-569 型 0i MF 数控设备	25	
4	具有有效沟通、表达及团结协作的职业精神	25	

任务二　设备使用须知

数控设备使用须知

【学习目标】

（1）了解亚龙 YL-569 型数控铣床的组成。

（2）掌握亚龙 YL-569 型数控铣床具体使用过程中的注意事项。

（3）养成良好的安全文明生产习惯。

（4）掌握工程工作方法，培养严谨的工作作风，遵守 5S 管理制度。

【任务描述】

以小组为单位，观摩亚龙 YL-569 型数控铣床，并将亚龙 YL-569 型数控铣床设备具体使用过程中的注意事项、应会技能、主要功能及适用场合等填写到任务工单中。要求在观摩过程中务必注意安全，遵守实训场所相关操作规范。

【知识链接】

设备使用过程注意事项

实训设备是教育改革进程的重要环节，在教学实验与实习、技能培训和考核、理论与实践相结合、教学与生产相联系及培养学生动手能力、思维能力、创新能力等方面有着不可替代的作用。正确使用及保养至关重要，不仅能方便工作和学习，而且能延长使用寿命和应用周期，更能发挥有形资产的功能，培育无形资产的人才。设备在使用过程中应注意以下事项。

（1）使用设备前必须熟悉产品技术说明书、使用说明书和实验指导书，按厂方提出的技术规范和程序进行操作和实验，特别对文字和图形符号应引起关注，设备使用后按顺序关断电源、水源和气源。

（2）注重设备的环境保护，减少暴晒、水浸及腐蚀物的侵袭，确保设备的绝缘电阻、耐压系数、接地装置及室内的温度、湿度和净化度符合相关要求，学会在安全用电状态下工作。

（3）提倡设备在常规技术参数要求范围下工作，谨防在极限技术参数要求范围下操作，禁止设备在超越技术要求范围外工作，即做常规性实验，限做极限性实验，禁做破坏性实验。

（4）在实验、培训时，应对搭建的各种电路进行检查，无误后才能通电。

（5）严防重物、重力、机械物撞击和超越设备的承载能力和受冲击能力，防止设备变形，防止损坏。

（6）对于各种单元板、单元模块和仪表要轻拿、稳放，切勿产生拖、摔、砸等现象，以免损坏。

（7）如设备出现漏电、缺相、短路，各种仪表、灯光显示异常及电火花、机械噪音或异味、冒烟等现象，应使用急停开关并立即断电、待查，进行设备检查与维护，切勿带故障操作和使用。

（8）减少电灾害、磁干扰及振动对设备允许范围外的伤害。

（9）机械运动必须注意摩擦、撞击等异常阻力，必要时要实施润滑处理。

（10）长期不使用的设备，要做定期检查、维护，才能进行工作。

【任务实施】

（1）按照随机分组方式，分成各个学习小组，并选出一名组长，将本组成员的个人信息填入表2-5中。

（2）根据观摩及讨论的结果，在表2-5中记录亚龙YL-569型数控铣床具体使用过程中的注意事项、应会技能、主要功能及适用场合等。

（3）对观摩过程中的疑难问题进行充分讨论，并将问题及解决办法填入表2-5中。

表 2-5　设备使用须知任务工单

<table>
<tr><td colspan="6">任务工单</td></tr>
<tr><td colspan="2">姓名：</td><td colspan="2">班级：</td><td colspan="2">学号：</td></tr>
<tr><td colspan="6">所需材料、工量具、设备列表</td></tr>
<tr><td>序号</td><td colspan="2">名称</td><td>数量</td><td>型号或规格</td><td>备注</td></tr>
<tr><td></td><td colspan="2"></td><td></td><td></td><td></td></tr>
<tr><td></td><td colspan="2"></td><td></td><td></td><td></td></tr>
<tr><td></td><td colspan="2"></td><td></td><td></td><td></td></tr>
<tr><td colspan="6">工作过程（观摩）</td></tr>
<tr><td>序号</td><td>注意事项</td><td>应会技能</td><td colspan="2">主要功能</td><td>适用场合</td></tr>
<tr><td></td><td></td><td></td><td colspan="2"></td><td></td></tr>
<tr><td></td><td></td><td></td><td colspan="2"></td><td></td></tr>
<tr><td></td><td></td><td></td><td colspan="2"></td><td></td></tr>
<tr><td></td><td></td><td></td><td colspan="2"></td><td></td></tr>
<tr><td colspan="3">出现问题</td><td colspan="3">解决办法</td></tr>
<tr><td colspan="3"></td><td colspan="3"></td></tr>
</table>

【任务评价】

序号	评价内容	分值	得分
1	能够熟练说出亚龙 YL-569 型数控铣床的组成	25	
2	能够掌握亚龙 YL-569 型数控铣床具体使用过程中的注意事项	25	
3	能够养成良好的安全文明生产习惯	25	
4	具有有效沟通、表达及团结协作的职业精神	25	

任务三　开机通电操作流程

【学习目标】

（1）了解亚龙 YL-569 型数控铣床安装要求。

（2）了解亚龙 YL-569 型数控铣床主电源要求。

（3）了解亚龙 YL-569 型数控铣床电气柜放置要求。

（4）养成良好的安全文明生产习惯。

（5）掌握工程工作方法，培养严谨的工作作风，遵守5S管理制度。

【任务描述】

以小组为单位，观摩亚龙YL-569型数控铣床，并将其安装要求、主电源要求、电气柜放置要求等填写到任务工单中。要求在观摩过程中务必注意安全，遵守实训场所相关操作规范。

【知识链接】

一、机床安装要求

为保证机床精度，机床在安装时应满足下列条件：

（1）应在海平面1 000 m以下；

（2）环境温度为5~40 ℃；

（3）环境温度为40 ℃时，湿度不能大于50%；

（4）工作环境的光线不能低于500 lx；

（5）安装环境应尽可能处于无酸、无腐蚀、无油雾、无降尘的环境；

（6）安装位置应避免阳光直射或过度振动；

（7）未调节好机床水平前，请勿加工。

二、主电源要求

（1）电压：三相五线、AC 380 V；

（2）电压波动：最大±10%；

（3）电源频率：（50±1）Hz；

（4）严禁从有干扰源的配电盘引出主电源（例如，该配电盘为电焊机、电火花机床供电），这将导致电源无法正常工作；

（5）机床应单独接保护地（禁止使用零线替代保护地接入机床），如果使用公共地，其他设备不能产生大的干扰值（如电焊机、电火花机床等设备）。

三、电气柜放置要求

（1）确保铁屑、冷却液、油不会溅到电气箱；

（2）电柜应放在机床的右侧位置；

（3）连接航空插头连接器时，注意确保连接到位。

【任务实施】

（1）按照随机分组方式，分成各个观摩小组，并选出一名组长，将本组成员的个人信息填入表2-6中。

（2）根据观摩及讨论的结果，在表2-6中记录亚龙YL-569型数控铣床安装要求、主电源要求、电气柜放置要求及开机操作流程等。

（3）对观摩过程中的疑难问题进行充分讨论，并将问题及解决办法填入表2-6中。

表 2-6 开机通电操作流程任务工单

<table>
<tr><td colspan="5">任务工单</td></tr>
<tr><td colspan="2">姓名：</td><td colspan="2">班级：</td><td>学号：</td></tr>
<tr><td colspan="5">所需材料、工量具、设备列表</td></tr>
<tr><td>序号</td><td>名称</td><td>数量</td><td>型号或规格</td><td>备注</td></tr>
<tr><td></td><td></td><td></td><td></td><td></td></tr>
<tr><td></td><td></td><td></td><td></td><td></td></tr>
<tr><td></td><td></td><td></td><td></td><td></td></tr>
<tr><td colspan="5">工作过程（观摩）</td></tr>
<tr><td>序号</td><td>安装要求</td><td>主电源要求</td><td>电气柜放置要求</td><td>开机操作流程</td></tr>
<tr><td></td><td></td><td></td><td></td><td></td></tr>
<tr><td></td><td></td><td></td><td></td><td></td></tr>
<tr><td></td><td></td><td></td><td></td><td></td></tr>
<tr><td></td><td></td><td></td><td></td><td></td></tr>
<tr><td colspan="3">出现问题</td><td colspan="2">解决办法</td></tr>
<tr><td colspan="3"></td><td colspan="2"></td></tr>
</table>

【任务评价】

序号	评价内容	分值	得分
1	能够了解亚龙 YL-569 型数控铣床安装要求	25	
2	能够了解亚龙 YL-569 型数控铣床主电源要求	25	
3	能够了解亚龙 YL-569 型数控铣床电气柜放置要求	25	
4	具有有效沟通、表达及团结协作的职业精神	25	

复习思考题

（1）亚龙 YL-569 型数控铣床基本组成有哪些？

（2）亚龙 YL-569 型数控铣床主要技术参数有哪些？

（3）亚龙 YL-569 型数控铣床可开展的实训项目有哪些？

（4）亚龙 YL-569 型数控铣床具体使用过程中的注意事项有哪些？

（5）亚龙 YL-569 型数控铣床安装要求有哪些？

模块三　数控机床电气装调

数控机床是高度自动化的设备，综合了机械、自动化、计算机、测量、微电子等最新技术。数控机床电气系统是实现机床自动化的基础。本模块主要介绍数控机床常用的电器、电气原理图的绘制及读图、典型机床电路介绍，进行冷却泵启/停电路设计与实施、刀库换刀电路设计与实施相关实训内容。

任务一　典型电器认知

【学习目标】

（1）了解数控机床常用电器的分类。

（2）掌握数控机床常用电器的一般结构和工作原理。

（3）掌握数控机床常用电器的选用方法。

（4）熟悉常用电器的规格、型号及其意义。

（5）具有自主分析问题和解决问题的能力。

（6）掌握工程工作方法，培养严谨的工作作风，遵守5S管理制度。

【任务描述】

以小组为单位，观摩亚龙YL-569型数控铣床的电气控制单元，并将机床各电器的名称、结构特征、主要功能及适用场合等填写到任务工单中。要求在观摩过程中务必注意安全，遵守实训场所相关操作规范。

【知识链接】

一、电器的基本知识

1. 电器的定义

能根据操作信号或外界信号（机械力、电动力和其他物理量），自动或手动接通和断开电路，从而连续或断续地改变电路参数或状态，实现对电路或用电设备的切换、控制、保

护、检测、变换和调节用的电气设备或元器件。

2. 电器的分类

（1）按电压高低分类。

高压电器：工作电压高于交流电压 1 200 V 或高于直流电压 1 500 V 的各种电器，如高压断路器、隔离开关、电抗器、电压互感器、电流互感器、避雷针等。

低压电器：工作电压在交流电压 1 200 V 或直流电压 1 500 V 以下的各种电器，如接触器、启动器、自动开关、熔断器、继电器、主令电器等。

（2）按控制作用分类。

主令电器：用来发出信号指令的电器，如按钮、主令控制器、转换开关等。

配电电器：用于电能的输送和分配的电器，如开关、低压断路器、隔离器等。

执行电器：用来完成某种动作或传递功率，如接触器、电磁阀、电磁铁。

控制电器：用来控制电路的通断，如继电器。

保护电器：用来保护电源、电路及用电设备的安全，使它们不致在短路、过载状态下运行，免遭损坏，如熔断器、热继电器、过（欠）电流（压）继电器、漏电保护器等。

（3）按操作方式分类。

手动电器：通过人力操作而动作的电器，如刀开关、隔离开关、按钮开关等。

自动电器：按照信号或某个物理量的变化而自动动作的电器，如高、低压断路器、接触器、继电器等。

（4）按动作原理分类。

电磁式电器：根据电磁铁的原理来工作，如接触器、继电器等。

非电磁式电器：依靠外力（人力或机械力）或某种非电量的变化而动作的电器，如按钮、行程开关、速度继电器、热继电器等。

3. 低压电器的发展方向

目前，低压电器正沿着体积小、质量轻、安全可靠、使用方便的方向发展，应大力发展电子化的新型控制电器，如接近开关、光电开关、电子式时间继电器、固态继电器与接触器等以适应控制系统迅速电子化的需要。

二、主令电器

主令电器是在自动控制系统中发出指令或信号的电器，主要用来控制接触器、继电器或其他电器线圈，使电路接通或分断，从而达到控制生产机械的目的，如控制按钮、行程开关、接近开关、万能转换开关、主令控制器及其他主令电器（如脚踏开关、旋钮开关、紧急开关）等。

典型电器元件介绍
——主令电器

1. 按钮开关（按钮）

按钮开关通常用作短路时接通或断开小电流控制电路的开关。按钮开关是由按钮帽、复位弹簧、桥式触点和外壳等组成的，通常制成具有动合触点和复合式结构，如图 3-1 所示。指示灯式按钮内可装入信号灯以显示信号；紧急式按钮装有蘑菇形按钮帽，以便于紧急操作；按钮式按钮是用手扭动旋转来进行操作的。按钮开关外形如图 3-2 所示。

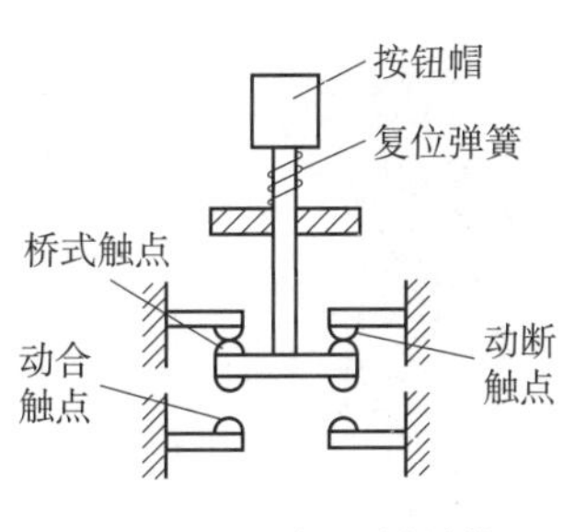

图 3-1　按钮开关结构

图 3-2　按钮开关外形

按钮开关的额定电压为交流 380 V、直流 220 V，额定电流为 5 A。按钮帽有多种颜色，一般红色的用作停止按钮，绿色的用作启动按钮。按钮的选择主要根据所需要的触点数、使用场合及颜色来进行。按钮开关的图形和文字符号如图 3-3 所示。

SB　　SB　　SB

（a）　　（b）　　（c）

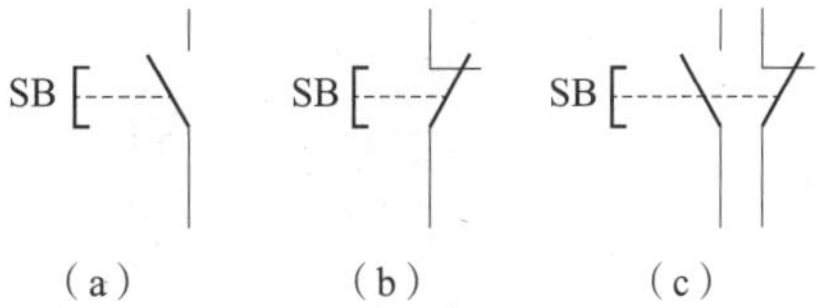

图 3-3　按钮开关的图形及文字符号

（a）启动按钮；（b）停止按钮；（c）复合按钮

2. 行程开关

行程开关又称限位开关或位置开关，是一种利用生产机械某些运动部件对开关操作机构的碰触而使触点动作，发出控制信号的主令电器，其外形如图 3-4 所示。行程开关主要用来控制生产机械的运动方向、行程及位置保护。

行程开关的图形和文字符号如图 3-5 所示。

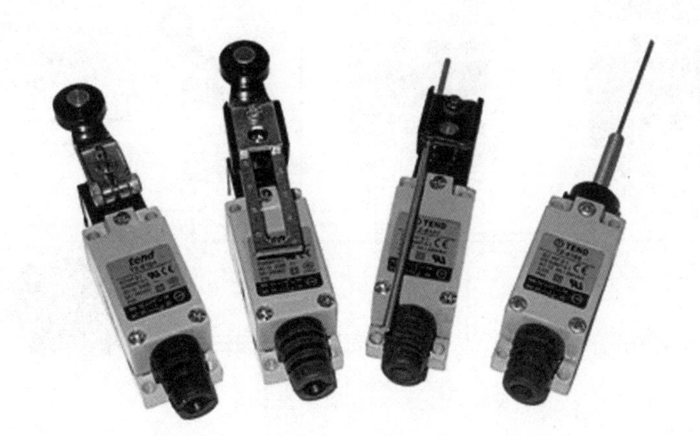

图 3-4　行程开关外形

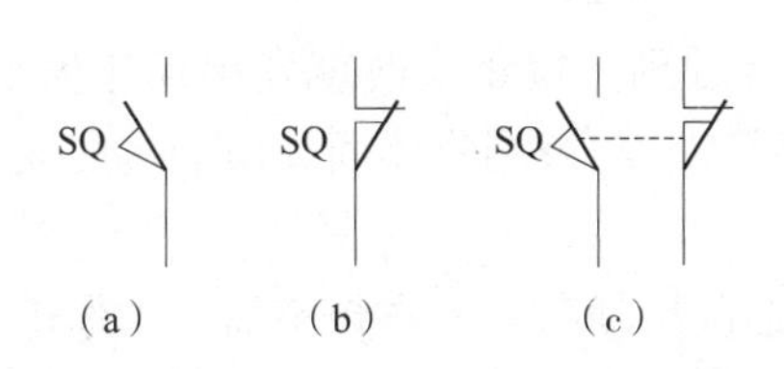

图 3-5　行程开关的图形和文字符号

（a）常开触点；（b）常闭触点；（c）复合触点

当运动部件机械的挡铁碰触行程开关的滚轮时，传动杠杆连同转轴一起转动，使凸轮推动撞块。当撞块被压到一定位置时，推动微动开关快速动作，使其常闭触点分断、常开触点闭合。当滚轮上的挡铁移开后，复位弹簧就使行程开关各部分恢复原始位置。行程开关的结构及工作原理如图 3-6 所示。行程开关的应用如图 3-7 所示。

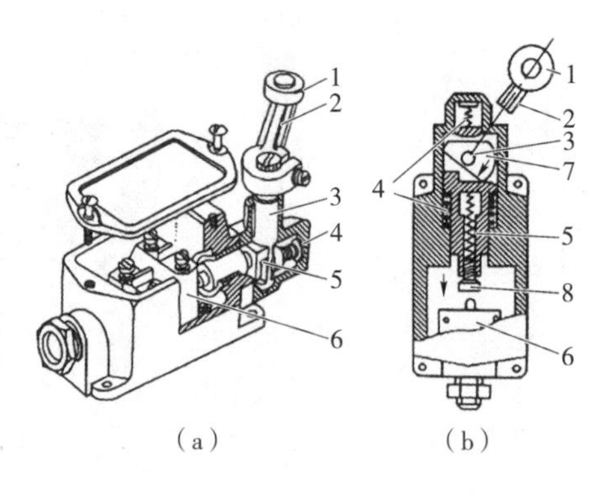

图 3-6　行程开关的结构及工作原理

（a）结构；（b）工作原理

1—滚轮；2—杠杆；3—转轴；4—复位弹簧；5—撞块；6—微动开关；7—凸轮；8—调节螺钉

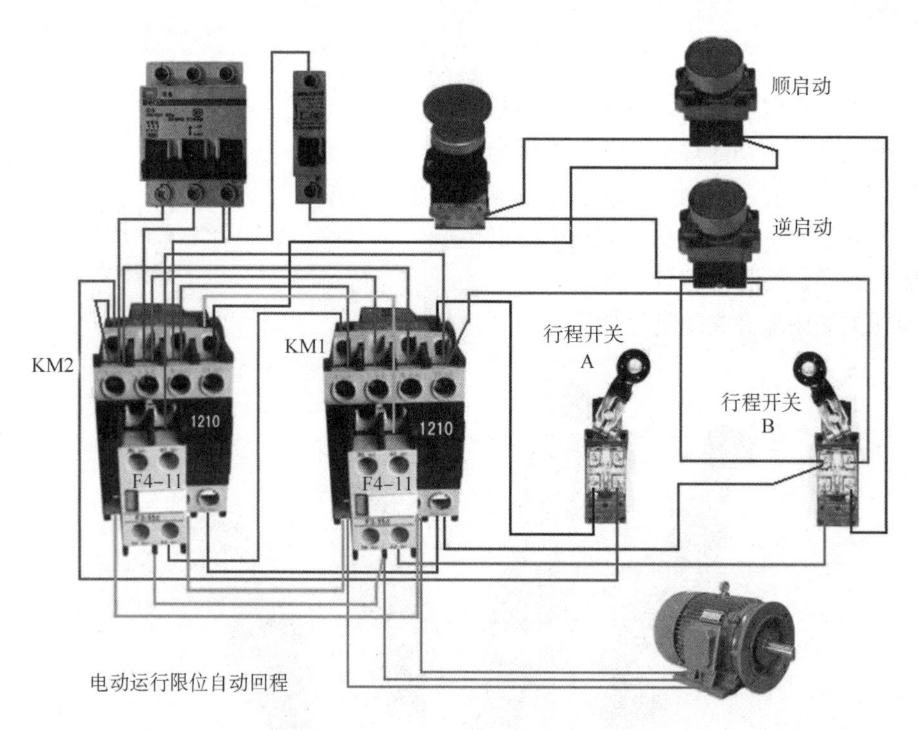

图 3-7　行程开关的应用

3. 接近开关

接近开关又称作无触点行程开关。当某种物体与之接近到一定距离时就发出动作信号，它不像机械行程开关那样需要施加机械力，而是通过其感应头与被测物体间介质能量的变化来获取信号。图 3-8 所示为接近开关。

接近开关因具有工作稳定可靠、使用寿命长、重复定位精度高、操作频率高、动作迅速等优点，故应用越来越广泛。接近开关图形符号如图 3-9 所示。

接近开关按其工作原理分为高频振荡型、电容型、感应电桥型、永久磁铁型、霍尔效应型等，其中高频振荡型最为常用。高频振荡型接近开关的电路主要由振荡电路、放大电路和输出电路三部分组成，其工作原理如图 3-10 所示。

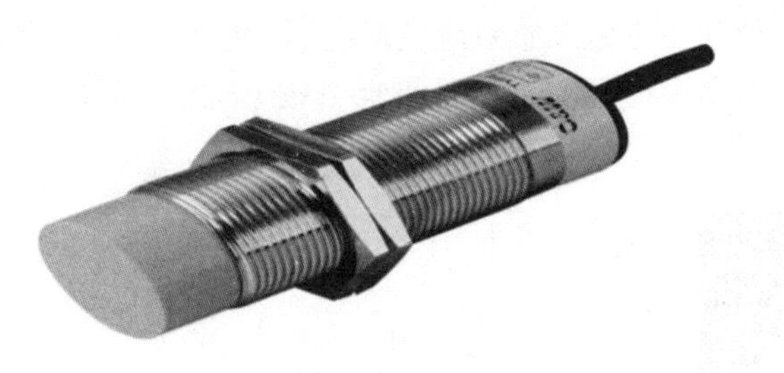

图 3-8　接近开关

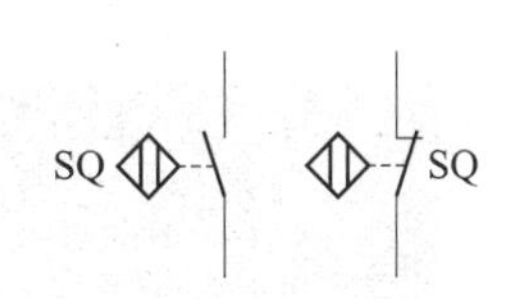

图 3-9　接近开关图形符号

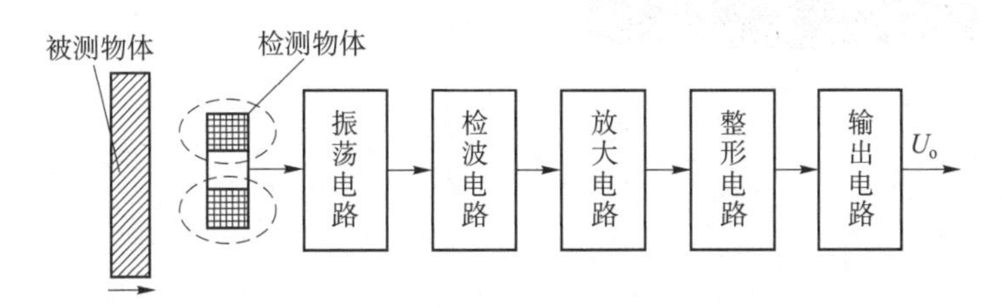

图 3-10　高频振荡型接近开关工作原理

三、配电电器

配电电器用于电能的输送和分配的电器，如开关电源、变压器、低压断路器和隔离器等。

1. 开关电源

开关电源（switching mode power supply）的功能是将一种形式的电能转换为另一种形式的电能。在 THWSKW-2A 型数控机床维修技术综合实训装置中使用的开关电源规格为输入 AC220 V，输出 DC24 V、5A，其外形和图形符号如图 3-11 所示。

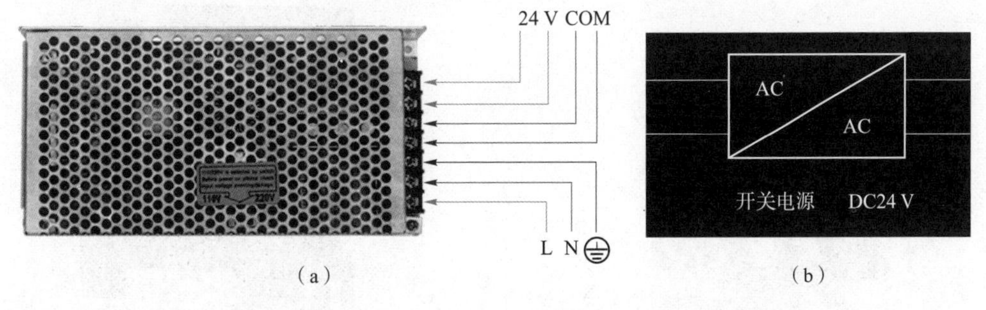

（a）　（b）

图 3-11　开关电源

（a）开关电源外形；（b）开关电源图形符号

2. 变压器

变压器是变换交流电压、电流和阻抗的器件。当初级线圈中通交流电流时，铁芯（或磁芯）中便产生交流磁通，使次级线圈中感应出电压（或电流）。变压器由铁芯（或磁芯）和线圈组成，线圈有两个或两个以上的绕组，其中接电源的绕组叫作初级线圈，其余的绕组叫作次级线圈。在发电机中，不管是线圈运动通过磁场或磁场运动通过固定线圈，均能在线圈中感应出电动势。在这两种情况下，磁通的值均不变，但与线圈相交链的磁通数量却有变

动，这就是互感原理。变压器就是一种利用电磁互感效应变换电压、电流和阻抗的器件。

变压器（transformer）的作用是隔离、变压、变流、变阻。变压器外形及图形符号如图 3-12 所示。

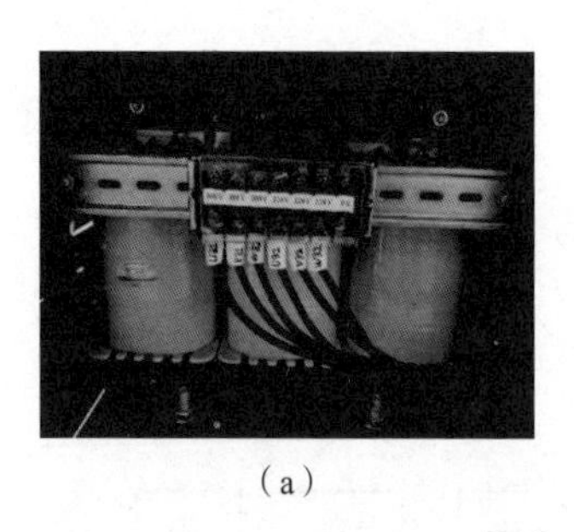
(a)

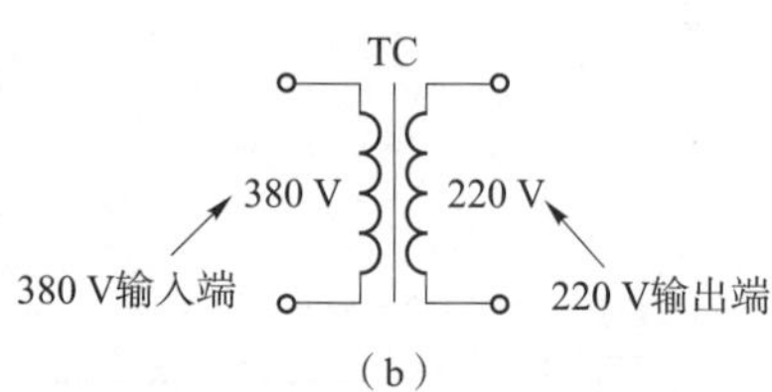

(b)

图 3-12　变压器

(a) 变压器外形；(b) 变压器图形符号

3. 低压断路器

低压断路器又称自动空气开关，用于分配电能、不频繁地启动异步电机以及对电源线路及电机等的保护。当发生严重的过载、短路或欠电压等故障时，低压断路器能自动切断电路，是低压配电线路应用非常广泛的一种保护电器，其外形及图形符号如图 3-13 所示。

四、执行电器

执行电器用来完成某种动作或传递功率，如接触器、电磁阀、电磁铁。

接触器是一种用于频繁地接通或断开主电路或大容量控制电路的自动切换电器。它主要应用于电力、配电与用电。接触器通常分为交流接触器和直流接触器。图 3-14 所示为接触器外形。

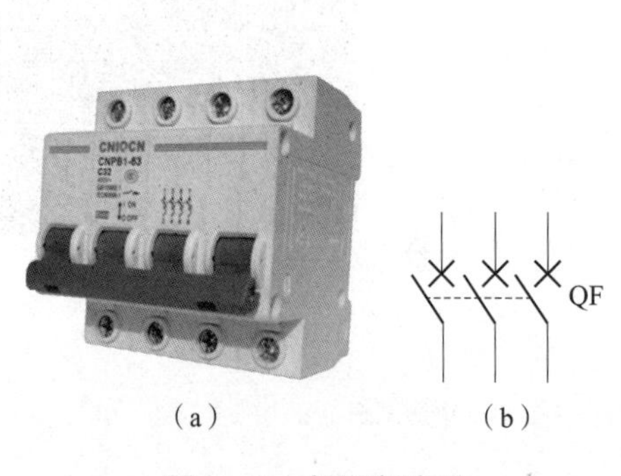

(a)　(b)

图 3-13　低压断路器

(a) 低压断路器外形；(b) 低压断路器图形符号

图 3-14　接触器外形

当电磁线圈通电后，铁芯被磁化产生磁通，从而在衔铁气隙处产生电磁力将衔铁吸合，主触点在衔铁的带动下闭合，接通主电路。同时衔铁还带动辅助触点动作，动断辅助触点首先断开，接着动合辅助触点闭合。当电磁线圈失电或电压显著降低后，吸力消失或减弱（小于反力），在反力弹簧的作用下衔铁释放，主触点、辅助触点又恢复到原来的状态。

接触器图形符号如图 3-15 所示。

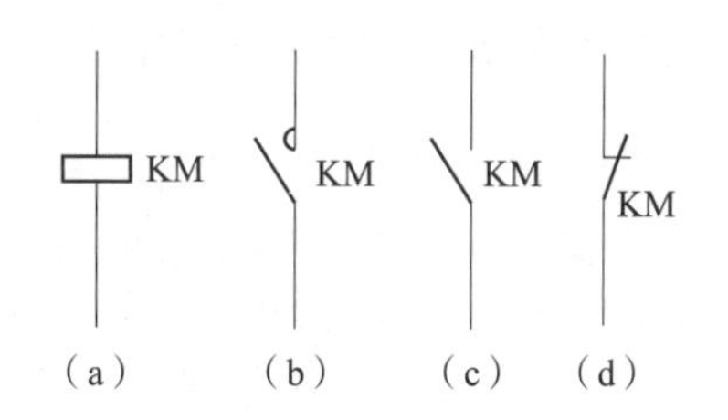

图 3-15　接触器图形符号

(a) 线圈；(b) 主触点；(c) 辅助常开触点；(d) 辅助常闭触点

五、控制电器

控制电器是用来控制电路通断的，包括开关电器和起控制作用的继电器。开关电器广泛用于配电系统和电力拖动控制系统，用作电源的隔离、电气设备的保护和控制。控制电器可分为手动开关和自动开关两大类，包括刀开关、组合开关及自动开关等。

1. 电磁式继电器

电磁式继电器是指主要借助电磁力或某个物理量的继电器，它利用各种物理量的变化将电量或非电量信号转化为电磁力或使输出状态发生阶跃变化，从而通过其触点或突变量促使在同一电路或另一电路中的其他器件或装置动作的一种控制元件。

电磁式继电器与接触器的主要区别如下。

(1) 电磁式继电器可对多种输入量的变化做出反应，而接触器只有在一定的电压信号作用下动作；

(2) 电磁式继电器用于切换小电流的控制电路和保护电路，而接触器用来控制大电流电路；

(3) 电磁式继电器没有灭弧装置，也无主辅触点之分等。

2. 中间继电器

中间继电器是一种电压继电器，通常用于传递信号和同时控制多个电路，也可直接用它来控制小容量电机或其他电气执行元件。其主要用途是当其他继电器的触点数量或触点容量不够时，可借助中间继电器来扩大触点容量（触点并联）或触点数量，起到中间转换的作用。

中间继电器图形符号如图 3-16 所示。

3. 电流继电器

电流继电器根据输入（线圈）电流大小而动作，其线圈串接于电路中，导线粗、匝数少、阻抗小。

电流继电器图形符号如图 3-17 所示。

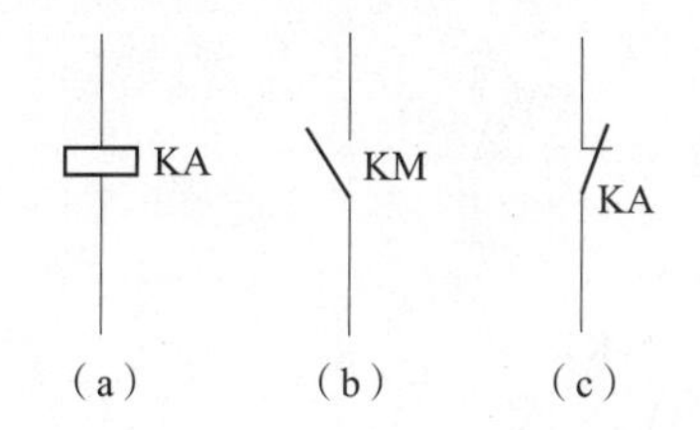

图 3-16　中间继电器图形符号

(a) 线圈；(b) 常开触点；(c) 常闭触点

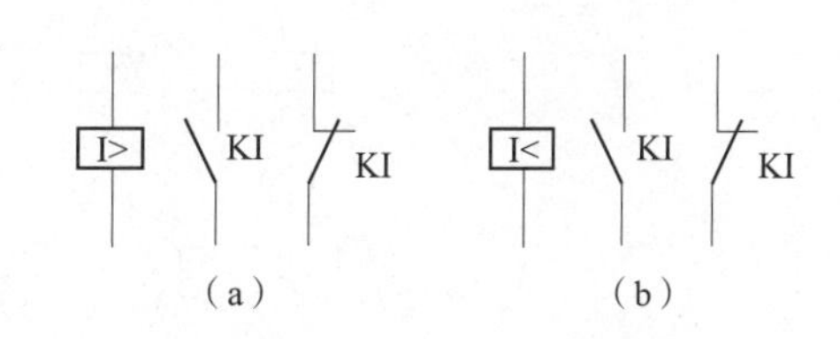

图 3-17　电流继电器图形符号

(a) 过电流继电器；(b) 欠电流继电器

电流继电器主要技术指标如下。

动作电流 I_q：使电流继电器开始动作所需的电流值；

返回电流 I_f：电流继电器动作后返回原状态时的电流值；

返回系数 K_f：返回值与动作值之比，$K_f=I_f/I_q$。

4. 电压继电器

根据输入（线圈）电压大小而动作的继电器称为电压继电器。电压继电器线圈并联在电路中，匝数多、导线细。

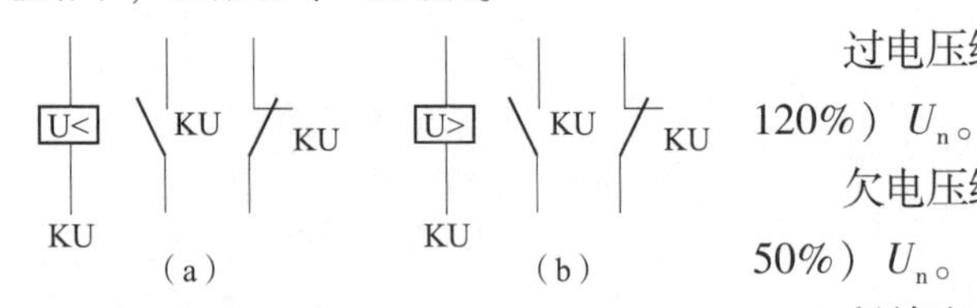

图 3-18 电压继电器图形符号

(a) 欠电压继电器；(b) 过电压继电器

过电压继电器：动作电压的调整范围为（105%～120%）U_n。

欠电压继电器：吸合电压的调整范围为（30%～50%）U_n。

释放电压的调整范围为（7%～20%）U_n。

电压继电器图形符号如图 3-18 所示。

5. 时间继电器

时间继电器是一种在接收信号后经过一定的延时才能输出信号，实现触点延时接通或断开的控制电器。时间继电器外形如图 3-19 所示。

图 3-19 时间继电器外形

(a) 空气阻尼式时间继电器（JS7 系列）；(b) 晶体管式时间继电器（JS14 系列）

时间继电器图形符号如图 3-20 所示。

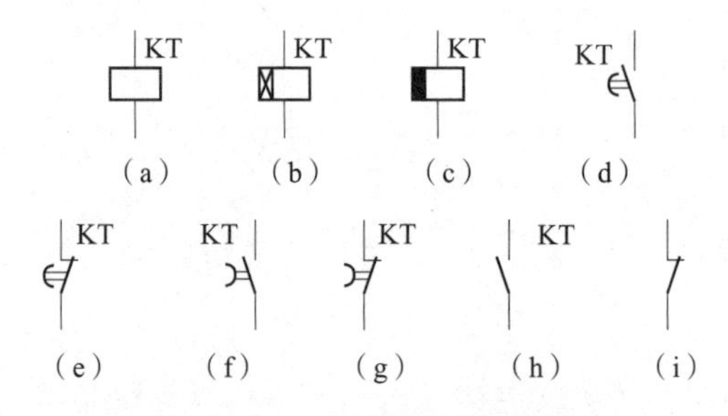

图 3-20 时间继电器图形符号

(a) 线圈一般符号；(b) 通电延时线圈；(c) 断电延时线圈；(d) 延时闭合动合触点；(e) 延时断开动断触点；(f) 延时断开动合触点；(g) 延时闭合动断触点；(h) 瞬时动合触点；(i) 瞬时动断触点

6. 热继电器

热继电器是一种利用电流的热效应原理进行工作的保护电器。电机在实际运行中，短时过载是允许的，但如果长期过载，欠电压运行或断相运行等都可能使电机的电流超过其额定值，这将导致电机发热。绕组温升超过额定温升，将损坏绕组的绝缘，缩短电机的使用寿命，严重时甚至会烧毁电机绕组。因此必须采取过载保护，最常用的是利用热继电器进行过载保护。热继电器外形如图 3-21 所示。

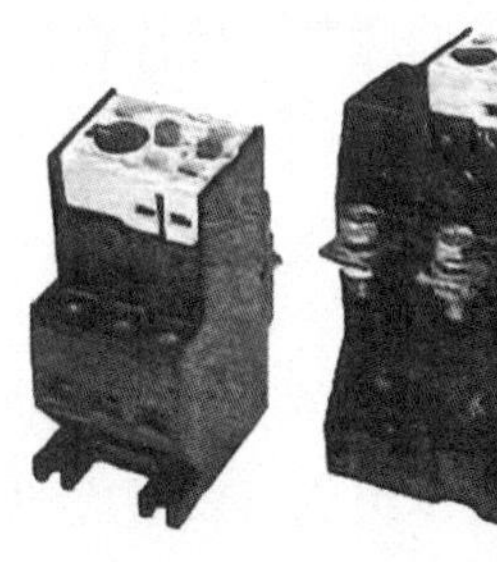

图 3-21　热继电器外形

热继电器主要由热元件、双金属片、触点和动作机构等组成。图 3-22 所示为双金属片式热继电器结构。

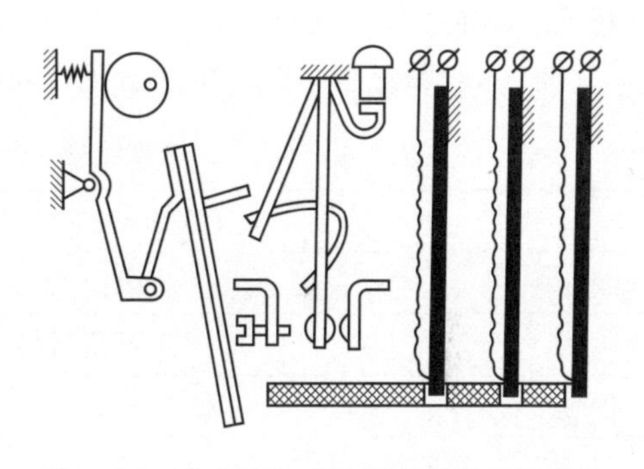

图 3-22　双金属片式热继电器结构示意图

在图 3-22 中，片簧及弓簧构成一组跳跃机构；凸轮可用来调节动作电流；补偿双金属片可补偿周围环境温度变化的影响。当周围环境温度变化时，主双金属片和与之采用相同材料制成的补偿双金属片会产生同一方向的弯曲，可使导板与补偿双金属片之间的推动距离保持不变。此外，热继电器可通过调节螺钉选择自动复位或手动复位。

7. 速度继电器

速度继电器用于转速的检测，常用在三相交流异步电机反接制动。当转速接近 0 时，自动切除反相序电源。

六、保护电器

熔断器是保护电器的一种，俗称保险丝，在低压配电线路中主要作为短路和过载时的保护用。它具有结构简单、体积小、质量轻、工作可靠、价格低廉等优点，所以，在强电、弱电系统都得到广泛的应用。熔断器主要由熔体和放置熔体的绝缘管或绝缘底座（亦称熔壳）组成。当熔断器串入电路时，负载电流流过熔体，熔体电阻上的耗损使其发热，温度上升。

当电路正常工作时，其发热温度低于熔化温度，故长期不熔断。

当电路发生过载或短路时，电流大于熔体允许的正常发热电流，使熔体温度急剧上升，

超过其熔点而熔断，从而分断电路，保护了电路和设备。

熔断器灭弧的方法大致有两种：一种是将熔体装在一个密封绝缘管内，绝缘管由高强度材料制成，并且，这种材料在电弧的高温下，能分解出大量的气体，使管内产生很高的压力，用以压缩电弧和增加电弧的电位梯度，以达到灭弧的目的；另一种是将熔体装在有绝缘砂粒填料（如石英砂）的熔管内，在熔体断开电路产生电弧时，石英砂可以吸收电弧能量，金属蒸气可以散发到砂粒的缝隙中，熔体很快冷却下来，从而达到灭弧的目的。图 3-23 所示为熔断器结构。

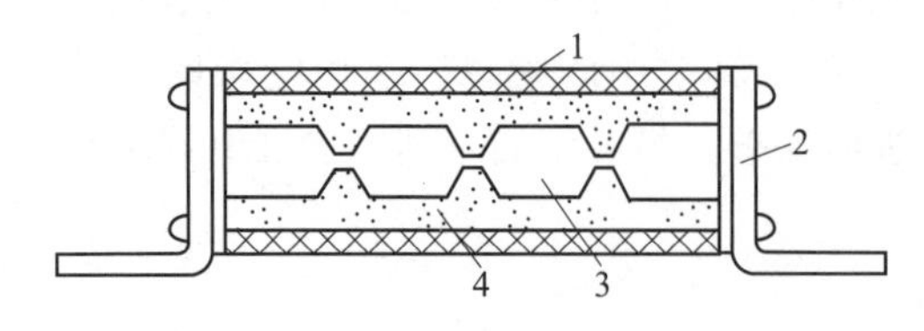

图 3-23　熔断器结构

1—熔管；2—端盖及接线板；3—熔片；4—石英砂

【任务实施】

（1）按照随机分组方式，分成各个观摩小组，并选出一名组长，将本组成员及对应的个人信息填入表 3-1 中。

（2）根据观摩及讨论的结果，在表 3-1 中记录亚龙 YL-569 型数控机床电器的名称、结构特征、功能及适用场合等。

（3）对观摩过程中的疑难问题进行充分讨论，并将问题及解决办法填入表 3-1 中。

表 3-1　机床电器观摩任务工单

<table>
<tr><td colspan="6">任务工单</td></tr>
<tr><td colspan="2">姓名：</td><td colspan="2">班级：</td><td colspan="2">学号：</td></tr>
<tr><td colspan="6">所需材料、工量具、设备列表</td></tr>
<tr><td>序号</td><td colspan="2">名称</td><td>数量</td><td>型号或规格</td><td>备注</td></tr>
<tr><td></td><td colspan="2"></td><td></td><td></td><td></td></tr>
<tr><td></td><td colspan="2"></td><td></td><td></td><td></td></tr>
<tr><td></td><td colspan="2"></td><td></td><td></td><td></td></tr>
<tr><td colspan="6">工作过程（观摩）</td></tr>
<tr><td>序号</td><td>电器名称</td><td>结构特征</td><td colspan="2">主要功能</td><td>适用场合</td></tr>
<tr><td></td><td></td><td></td><td colspan="2"></td><td></td></tr>
<tr><td></td><td></td><td></td><td colspan="2"></td><td></td></tr>
<tr><td></td><td></td><td></td><td colspan="2"></td><td></td></tr>
<tr><td></td><td></td><td></td><td colspan="2"></td><td></td></tr>
<tr><td colspan="3">出现问题</td><td colspan="3">解决办法</td></tr>
<tr><td colspan="3"></td><td colspan="3"></td></tr>
</table>

【任务评价】

序号	评价内容	分值	得分
1	能够熟练说出机床电器的名称	25	
2	能够正确绘制常用电器的图形符号	25	
3	能够掌握数控机床常用电器的一般结构和工作原理	25	
4	具有有效沟通、表达及团结协作的职业精神	25	

任务二　电气原理图识读

【学习目标】

（1）了解电气控制系统图的分类。
（2）了解电气系统图绘制原则。
（3）掌握电气原理图的制图与识图。
（4）掌握工程工作方法，培养严谨的工作作风，遵守5S管理制度。

【任务描述】

以小组为单位，观摩数控机床的电气原理图，并将各典型电路名称、工作过程、主要功能及适用场合等填写到任务工单中。要求观摩过程中务必注意安全，遵守实训场所相关操作规范。

【知识链接】

一、电气控制系统图

根据机械运动形式对电气控制系统的要求，采用国家统一规定的电气图形符号和文字符号，按照电气设备和电器的工作顺序，详细表示电路、设备或成套装置的全部基本组成和连接关系的图形就是电气控制系统图。

常见的电气控制系统图有电气原理图、接插件定义图与接插件连接图。在机床电气控制原理分析中最常用的是电气原理图。

电气原理图、接插件定义图、接插件连接图，如图3-24所示。

1. 电气原理图识图步骤

（1）准备：了解生产过程和工艺对电路提出的要求；了解各种用电设备和控制电器的位置及用途；了解图中的图形符号及文字符号的意义。

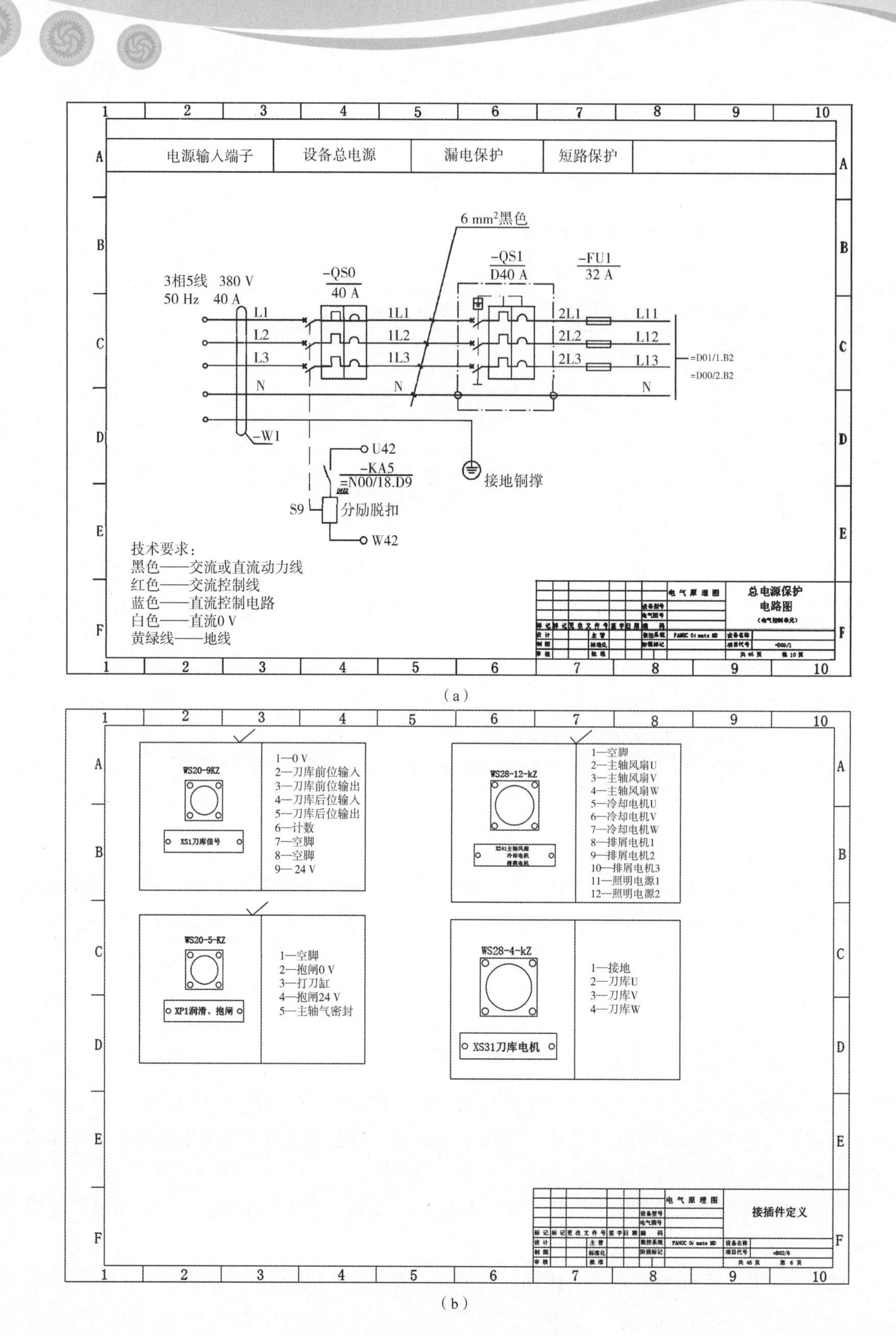

(a)

(b)

图 3-24　电气控制系统图

(a) 总电源保护电器原理图；(b) 部分电气接插件定义图

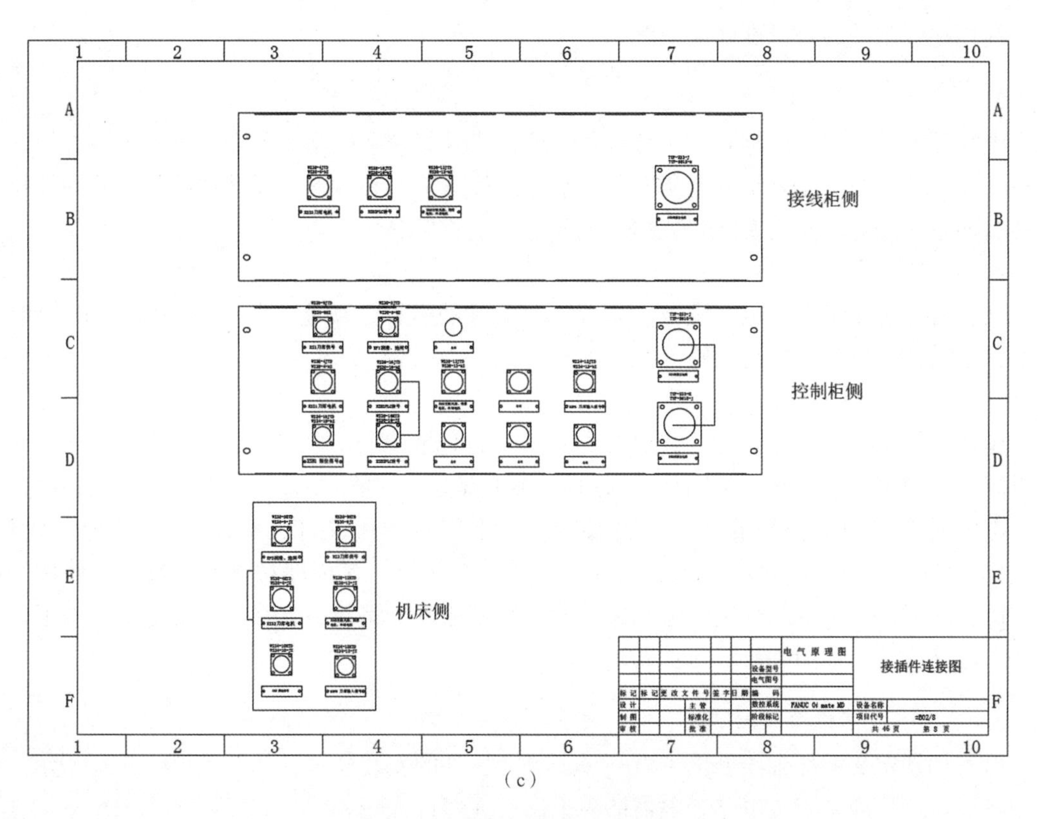

（c）

图 3-24　电气控制系统图（续）

（c）部分电气接插件接线图

（2）主电路：首先要仔细看一遍电气图，弄清电路的性质，是交流电路还是直流电路。然后从主电路入手，根据各用电设备和控制电器的组合判断电机的工作状况，如电机的起停、正反转等。

（3）控制电路：分析完主电路后，再分析控制电路。要按动作顺序对每条小回路逐一分析研究，然后再全面分析各条回路间的联系和制约关系，要特别注意回路和机械与液压部件的动作关系。

（4）最后阅读保护、照明、信号指示、检测等部分。

2. 电气原理图识图方法

（1）结合电工基础知识识图。

在掌握电工基础知识的基础上，准确、迅速地识别电气图。若改变电机电源相序，即可改变其旋转方向的控制。

（2）结合典型电路识图。

典型电路就是常见的基本电路，如电机的启动、制动、顺序控制等。不管多复杂的电路，几乎都是由若干基本电路组成的。因此，熟悉各种典型电路，是看懂较复杂电气图的基础。

（3）结合制图要求识图。

在绘制电气图时，为了加强图纸的规范性、通用性和示意性，必须遵循一些规则和要

求，利用这些制图的知识准确地识图。

3. 电气原理图

电气原理图是用图形符号、文字符号、项目代号等表示电路的各个电气元器件之间的关系和工作原理的图。其表示电流从电源到负载的传送情况和各电气元器件的动作原理及相互关系，而不考虑各元器件实际安装的位置和实际连线情况。

（1）文字符号。

文字符号是用来表示电气设备、装置、元器件的名称、功能、状态和特征的字符代码。例如，FR 表示热继电器。

（2）图形符号。

图形符号用来表示一台设备或概念的图形、标记或字符。例如，“~”表示交流，*R* 表示电阻等。国家标准《电气简图用图形符号 第 1 部分：一般要求》（GB/T 4728—2018）规定了电气简图中图形符号的画法。国家标准《电气技术用文件的编制 第 1 部分：规则》（GB/T 6988. 1—2024）于 2025 年 4 月 1 日正式开始执行，该标准对电气图技术中使用的简图、图和图表提出了专门规则。电气原理图一般分电源电路、主电路和辅助电路，本书以电源保护电路为例，如图 3-24（a）所示。

4. 电气接插件定义图

电气插件定义图详细绘制出了电气设备、元器件的安装位置，如图 3-24（b）所示。该图中各电器代号应与有关电路和电器清单上所有元器件代号相同。

5. 电气接插件连接图

电气接插件连接图用来表明电气设备各单元之间的接线关系，如图 3-24（c）所示。该图表明了电气设备外部元器件的相对位置及它们之间的电气连接，是实际安装接线的依据。

二、电气控制系统图的图幅分区

为了查找和确定图纸上某元器件或设备的位置，方便阅读，往往需要将图幅分区。图幅分区的方法：在图的边框处，从标题栏相对的左上角开始，竖边方向用大写拉丁字母、横边方向用阿拉伯数字，依次编号，这样就将图幅分成了若干个图区。图幅分区式样如图 3-25 所示。

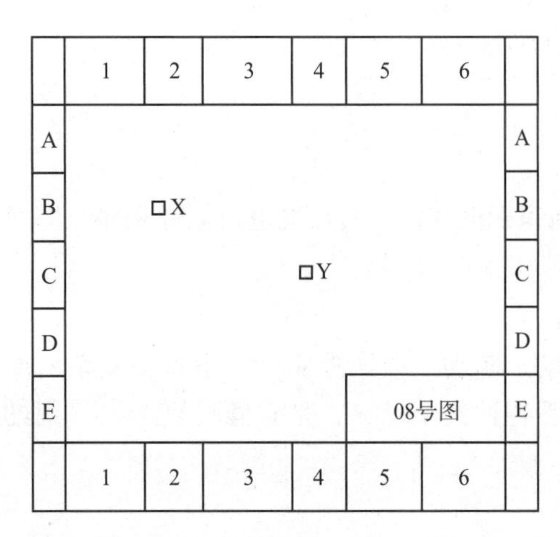

图 3-25　图幅分区式样

图幅分区后，相当于在图样上建立了一个坐标。电气图上项目和连接线的位置则由此“坐标”唯一地确定，用“图号/行、列或区号”标注。

图 3-25 中 X 元器件位于 B2 区，可标注为 08/B2；Y 元器件位于 C4 区，标注为 08/C4。

在较简单的机床电气原理图中，图幅竖边

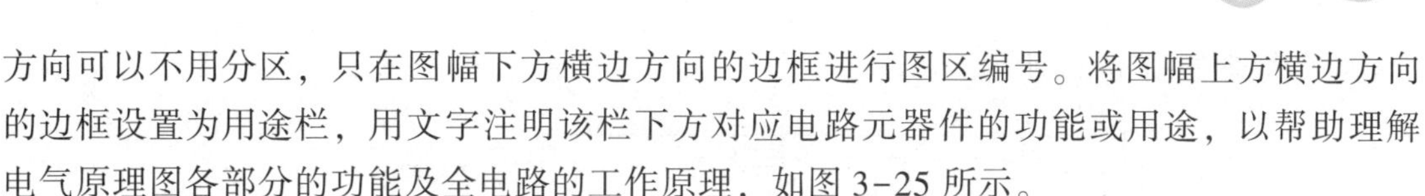

方向可以不用分区，只在图幅下方横边方向的边框进行图区编号。将图幅上方横边方向的边框设置为用途栏，用文字注明该栏下方对应电路元器件的功能或用途，以帮助理解电气原理图各部分的功能及全电路的工作原理，如图 3-25 所示。

三、电气原理图绘制规则

（1）电源电路画成水平线，三相交流电源相序 L1、L2、L3 自上而下依次画出，中线和保护地线依次画在相线之下。直流电源“+”端在上，“-”端在下。

（2）主电路在电路图的左侧并垂直电源电路。它由主熔断器、接触器的主触点、热继器的热元件及电机等组成。它可通过较大的电机工作电流。

（3）辅助电路一般包括控制主电路工作状态的控制电路、显示主电路工作状态的指示电路、提供机床局部照明的照明电路等。它由主令电器的触点、接触器线圈及辅助触点、继电器线圈及触点、指示灯和照明灯等组成。一般按照从左至右、从上至下的排列来表示操作顺序。辅助电路通过的电流较小，一般不超过 5 A。

（4）电路图中，各电器的触点位置都是按电路未通电或电器未受外力作用时的常态位置画出的。

（5）电路图中，同一电器的各元器件不按实际位置画在一起，而是按其在电路中所起的作用分别画在不同电路中，但它们的动作却是相互关联的，用相同的文字符号标注。

（6）主电路编号时，在电源开关的出线端按相序依次编号 U11、V11、W11。然后按从上至下、从左至右的顺序，每经过一个电气元器件后编号递增；辅助电路编号时，按“等电位”原则从上至下、从左至右的顺序用数字依次编号，每经过一个电气元器件后，编号依次递增。

（7）电路图按功能划分成若干个图区。通常是一条回路或一条支路划为一个图区，并从左向右依次用阿拉伯数字编号，标注在电路图下部的图区栏中。

（8）电路图中的每个电路在机床电气操作中的用途，用文字在电路图上部的用途栏内标明。

（9）在电路图中每个接触器线圈的文字符号 KM 的下面画两条竖线，分成左、中、右三栏，对应主触点、动合触点、动断触点。把受其控制而动作的触点所处的图区号按规定填入相应栏内。对备而未用的触点，在相应的栏中用“×”标出或不标出任何符号。

（10）在电路图中每个继电器线圈的文字符号下面画一条竖线，分成左、右两栏对应动合触点、动断触点，把受其控制而动作的触点所处的图区号按规定填入相应栏内。对备而未用的触点，在相应的栏中用“×”标出或不标出任何符号。

（11）电路图中接触器或继电器触点文字符号下面的数字，表示该接触器或继电器线圈所处的图区号。

四、典型电路分析

1. 点动控制

（1）点动控制电路：常用于机床主轴或工作台的调整，如机床的试车、检修等。点动

控制电气原理如图 3-26 所示。

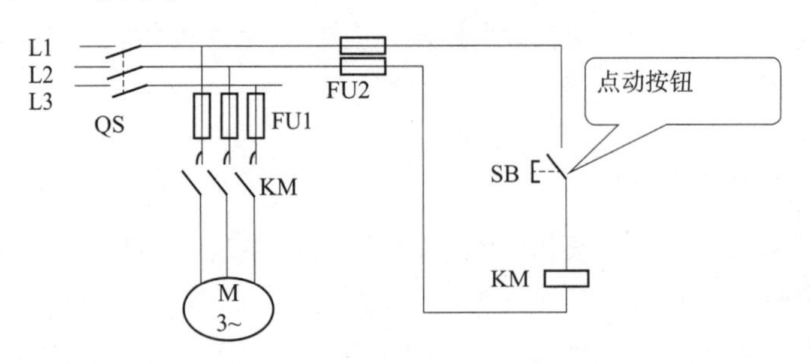

图 3-26　点动控制电气原理图

（2）工作过程。

先接通电源开关 QS。

按下 SB ⟶KM 线圈得电⟶KM 主触头闭合⟶电机 M 通电启动。

松开 SB ⟶KM 线圈失电⟶KM 主触头复位⟶电机 M 断电停转

2. 连续控制

（1）连续控制电气原理如图 3-27 所示。

保护环节：

1）短路保护：FU1、FU2。

2）过载保护：FR。

3）欠压失压保护：KM 自锁环节。

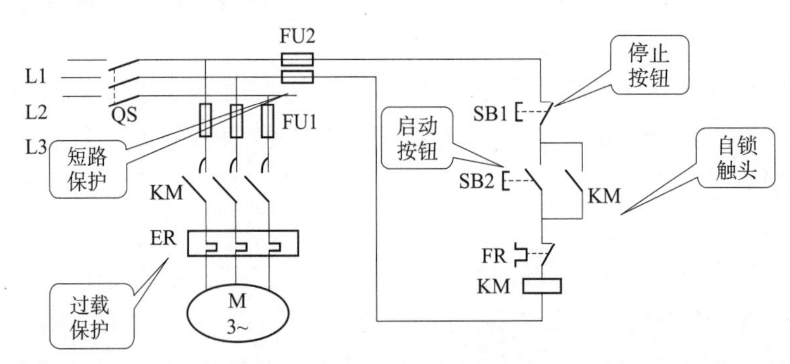

图 3-27　连续控制电气原理图

（2）工作过程。

先接通电源开关 QS。

按下 SB2 ⟶KM 线圈得电⟶ {KM 主触头闭合⟶电机 M 通电启动。
KM 辅助触头闭合自锁。}

按下 SB1 ⟶KM 线圈失电⟶电机 M 断电停转。

3. 电机正反转控制

（1）电机正反转控制介绍。

用于机床工作台的前进与后退或主轴的正反转等。由电机工作原理可知，只要把电机的

三相电源进线中的任意两相对调，就可改变电机的转动方向。需要用两个接触器来实现这一要求。当正转接触器工作时，电机正转；当反转接触器工作时，将电机接到电源的任意两根连线对调一下，电机反转。图 3-28 所示为电机正反转控制原理。

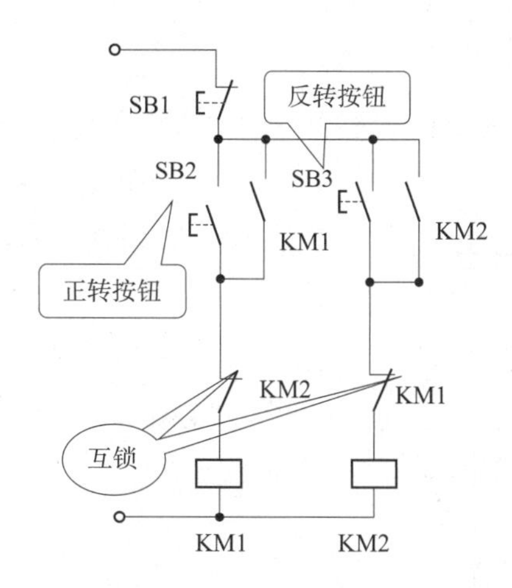

图 3-28　电机正反转控制电气原理图

（2）工作过程。

1）正转控制。

按下 SB2 ⟶KM1 线圈得电⟶ {KM1 主触头闭合⟶电机 M 正转。
KM1 自锁触头闭合。
KM1 互锁触头断开。}

2）反转控制。

按下 SB3→KM2 线圈得电→电机 M 反转。

4. 电机 Y-△降压启动

（1）电机 Y-△降压启动电气原理如图 3-29 所示。

（2）工作过程。

先接通电源开关 QS。

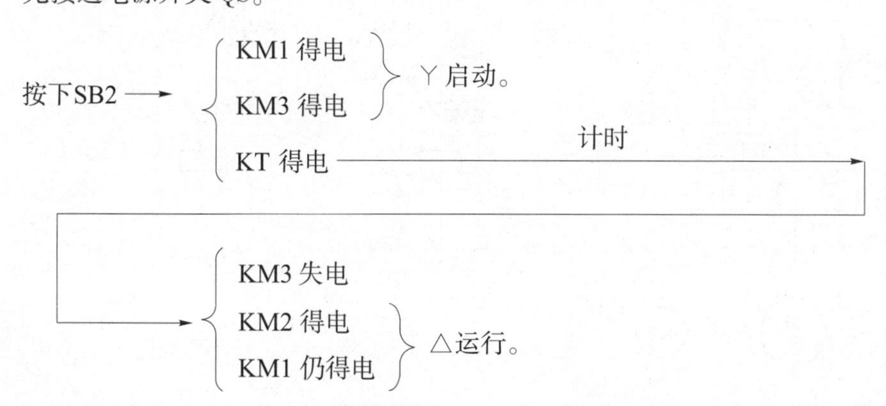

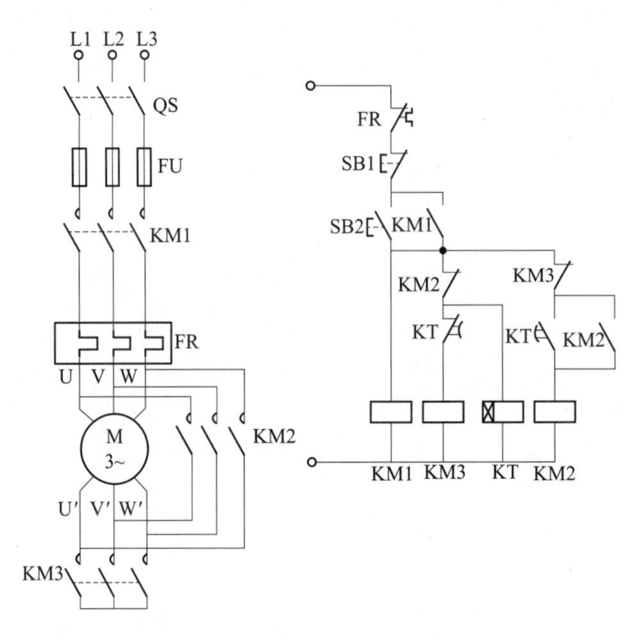

图 3-29　电机 Y-△降压启动电气原理图

5. 电机反接制动

(1) 电机反接制动电气原理图。

反接制动是电机的一种制动方式，它通过反接相序，使电机产生起阻滞作用的反转矩以便制动电机。反接制动制动力强、制动迅速、控制电路简单、设备投资少，但制动准确性差，制动过程中冲击力强烈，易损坏传动部件。图 3-30 所示为电机反接制动电气原理图。

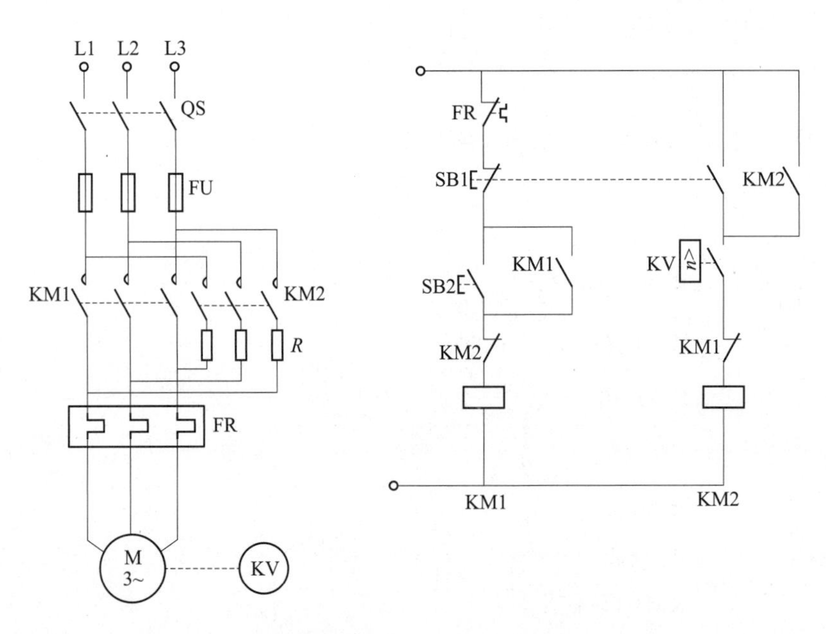

图 3-30　电机反接制动电气原理图

(2) 工作过程。

按下 SB2 ⟶KM1 线圈得电⟶电机 M 正转运行⟶KV 动合触点闭合。

按下 SB1 ⟶ {KM1 断电
KM2 得电（开始制动）⟶$n \approx 0$，KV 复位。
KM2 断电（制动结束）。}

【任务实施】

(1) 按照随机分组方式，分成各个观摩小组，并选出一名组长，将本组成员的个人信息填入表 3-2 中。

(2) 根据观摩及讨论的结果，在表 3-2 中记录典型电路名称、工作过程、主要功能及适用场合等。

(3) 对观摩过程中的疑难问题进行充分讨论，并将问题及解决办法填入表 3-2 中。

表 3-2　典型电路分析任务工单

<table>
<tr><td colspan="6">任务工单</td></tr>
<tr><td colspan="6">姓名：　　　　班级：　　　　学号：</td></tr>
<tr><td colspan="6">所需材料、工量具、设备列表</td></tr>
<tr><td>序号</td><td colspan="2">名称</td><td>数量</td><td>型号或规格</td><td>备注</td></tr>
<tr><td></td><td colspan="2"></td><td></td><td></td><td></td></tr>
<tr><td></td><td colspan="2"></td><td></td><td></td><td></td></tr>
<tr><td></td><td colspan="2"></td><td></td><td></td><td></td></tr>
<tr><td colspan="6">工作过程（观摩）</td></tr>
<tr><td>序号</td><td>电路名称</td><td>工作过程</td><td colspan="2">主要功能</td><td>适用场合</td></tr>
<tr><td></td><td></td><td></td><td colspan="2"></td><td></td></tr>
<tr><td></td><td></td><td></td><td colspan="2"></td><td></td></tr>
<tr><td></td><td></td><td></td><td colspan="2"></td><td></td></tr>
<tr><td></td><td></td><td></td><td colspan="2"></td><td></td></tr>
<tr><td colspan="3">出现问题</td><td colspan="3">解决办法</td></tr>
<tr><td colspan="3"></td><td colspan="3"></td></tr>
</table>

【任务评价】

序号	评价内容	分值	得分
1	能够熟练说出电气控制系统图的分类	25	
2	能够正确绘制电气系统图	25	
3	能够运用电气原理图对典型电路进行分析	25	
4	具有有效沟通、表达及团结协作的职业精神	25	

任务三　刀库换刀电路设计

【学习目标】

（1）了解斗笠式刀库的换刀过程及工作原理。
（2）掌握斗笠式刀库电路结构。
（3）掌握斗笠式刀库电路的装调方法。
（4）具有自主分析斗笠式刀库电路故障诊断的能力。
（5）掌握工程工作方法，培养严谨的工作作风，遵守5S管理制度。

【任务描述】

以小组为单位，观摩斗笠式刀库换刀电路单元，并将刀库换刀电路的换刀过程、工作原理、电路结构及适用场合等填写到任务工单中。要求观摩过程中务必注意安全，遵守实训场所相关操作规范。

【知识链接】

一、斗笠式刀库换刀过程

1. 斗笠式刀库的换刀动作

（1）刀库转到换刀坐标处，如图3-31（a）所示。
（2）主轴准停。
（3）刀库前进（抓旧刀），如图3-31（b）所示。
（4）主轴松刀。
（5）Z 轴向上移动（让出刀库旋转尺寸空间），如图3-31（c）所示。

(6) 刀库旋转（选刀），如图 3-31（d）所示。

(7) Z 轴向下移动（移动至换刀位置），如图 3-31（e）所示。

(8) 主轴紧刀（抓新刀）。

(9) 刀库后退（换刀结束），如图 3-31（f）所示。

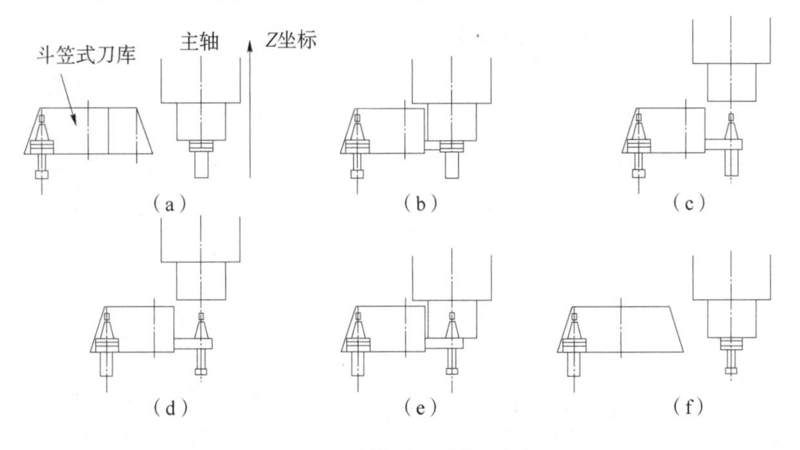

图 3-31 斗笠式刀库换刀过程

2. 主轴的定位精度

斗笠式刀库在换刀过程中对主轴的定位精度要求是非常高的，这是由标准刀具 BT600 和主轴的结构决定的。标准刀具 BT600 有两个对称的定位键槽，对应主轴上的两个配合键。这就要求主轴在换刀过程中始终定位到同一位置。只用 PMC 控制是不能实现上述要求的，还需要主轴和 CNC 程序的配合才能实现。当系统检测到该程序执行 M6TXX 信号时，CNC 程序发出换刀准备和主轴定位信号。PMC 控制主轴变频器使主轴定位，由主轴编码器反馈主轴位置给 CNC 程序，由 CNC 程序检测主轴是否定位到要求范围内，定位完成后 CNC 程序发出换刀开始信号并传输给 PMC。图 3-32 所示为控制流程。

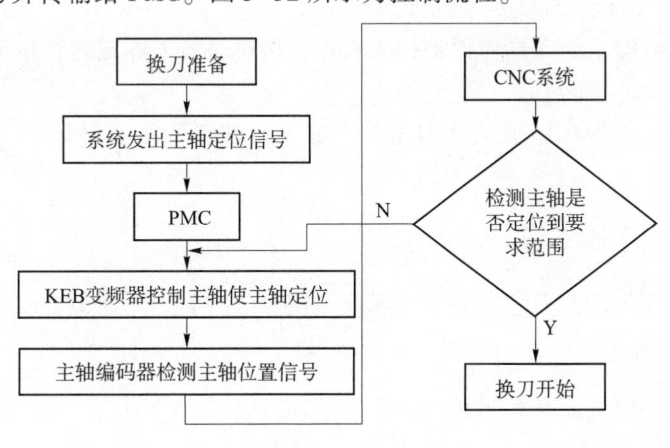

图 3-32 控制流程图

3. PMC 与 CNC 程序的配合

PMC 在整个换刀过程中主要控制刀库的正反转、刀库的前进/后退、松刀/紧刀阀的动

作以及换刀动作顺序。主轴的上升、下降和定位都是由 CNC 程序控制的，所以在整个换刀过程中 PMC 与 CNC 程序的配合是重点也是难点。

在整个换刀过程中动作顺序是由 PMC 控制的，其应用 PMC 中的 D 存储器来放置的不同数字代表换刀过程的不同动作，如表 3-3 所示。

表 3-3　换刀动作与数字

D800	换刀动作顺序	D800	换刀动作顺序
1	换刀开始	5	刀库回转到位
2	气缸前进到位	6	Z 轴换刀点到位
3	送刀完成	7	紧刀完成
4	刀位信号读取完毕	8	换刀完成

从表 3-3 和图 3-31 中可以知道，CNC 程序和 PMC 的配合应该在 D800 = 3 和 D800 = 5 的动作。当 D800=3 时，动作如图 3-31（b）所示，下一个动作如图 3-31（c）所示，Z 轴上升，这就要求 CNC 程序和 PMC 的配合。所以当松刀完成后，PMC 会发出一个信号给 CNC 程序，CNC 程序将控制 Z 轴上升。当 D800=5，动作如图 3-31（d）所示，下一个动作如图 3-31（e）所示。当轴下降到换刀点，这就要求 CNC 程序和 PMC 的配合。所以当刀盘回转到位后，PMC 会发出一个信号给 CNC 程序，CNC 程序将控制 Z 轴下降到换刀点。当 D800 = 8 时，换刀完成，PMC 会发出一个信号给 CNC 程序，CNC 程序将执行 M6TXX 后的命令。

PMC 和 CNC 程序的配合体现在当换刀过程中需要 Z 轴动作时，PMC 就会提前告诉 CNC 程序，CNC 程序会根据 PMC 所给的条件判断该执行什么命令；CNC 程序执行完毕后会通知 PMC 命令执行完毕，PMC 将顺序执行下一个动作。PMC 和 CNC 程序的配合在整个换刀过程中至关重要，只有配合得恰到好处才能顺利完成整个换刀过程。

电路设计和 PC 编程时要满足下列要求。

手动操作方式：按下机床操作面板的刀库正转按键时，刀库正转；按下机床操作面板的刀库反转按键时，刀库反转。

自动操作方式：在程序中输入换刀指令，如“M6T1”，按下循环启动按键，完成换刀过程。

二、刀库换刀电路设计

刀库换刀电路由刀库换刀主电路、刀库换刀控制电路和刀库换刀 PMC 的 I/O 电路组成。

1. 刀库换刀主电路

刀库换刀主电路电气原理如图 3-33 所示。

2. 刀库换刀控制电路

刀库换刀控制电路电气原理如图 3-34 所示。

3. 刀库换刀 PMC 的 I/O 电路

刀库换刀 PMC 的 I/O 电路电气原理如图 3-35 所示。

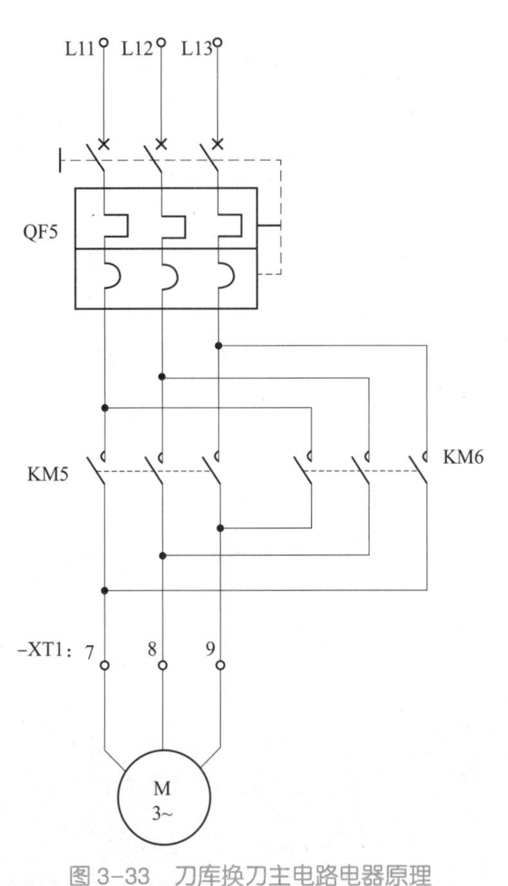

图 3-33　刀库换刀主电路电器原理

KM5—刀库电机正转的交流接触器；KM6—刀库电机反转的交流接触器

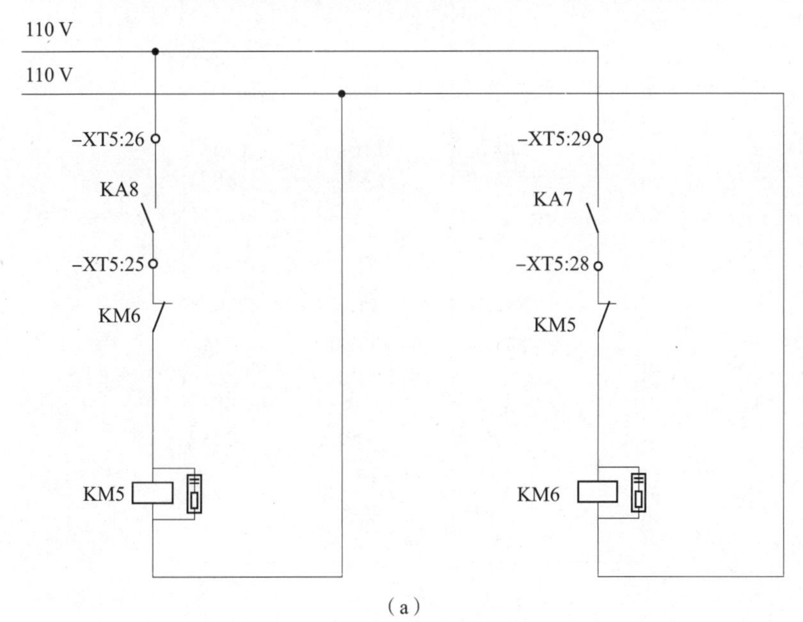

（a）

图 3-34　刀库换刀控制电路电气原理

（a）刀库电机正转、反转互锁电路；

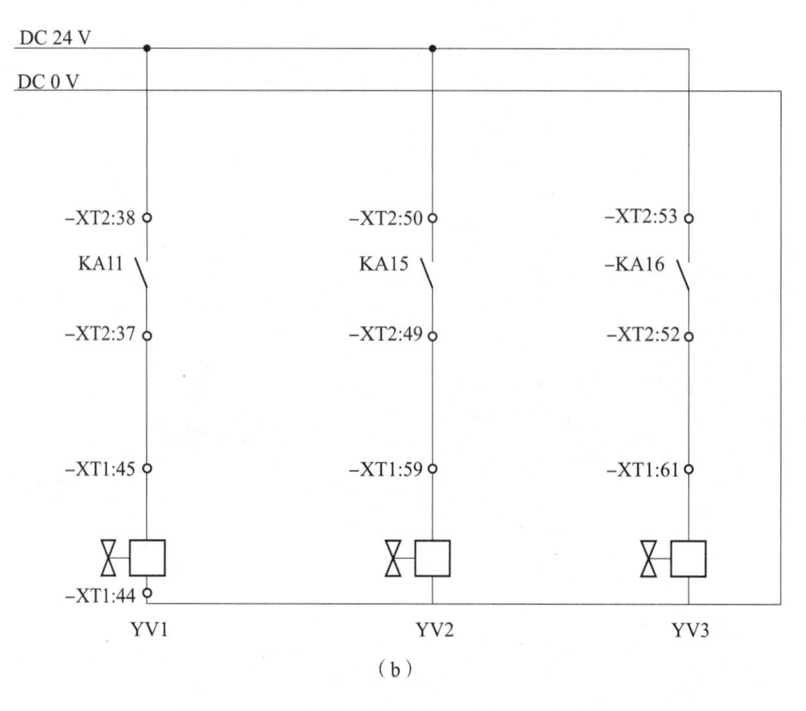

（b）

图 3-34　刀库换刀控制电路电气原理（续）

（b）主轴松刀、刀库接近主轴与刀库远离主轴电路

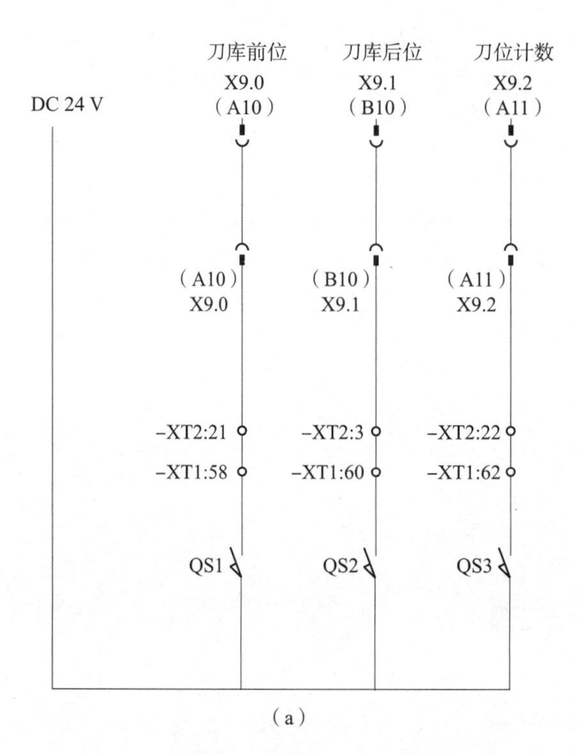

（a）

图 3-35　刀库换刀 PMC 的 I/O 电路电气原理

（a）PMC 输入信号；

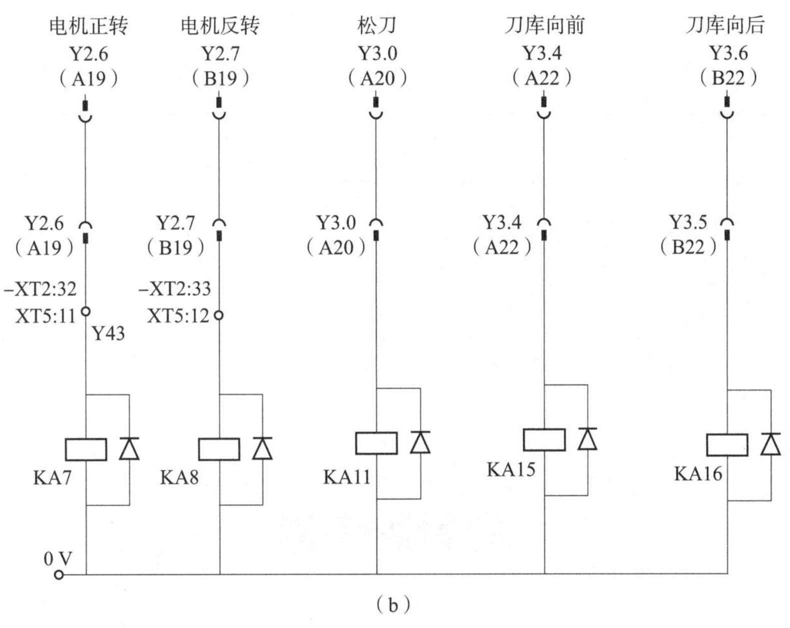

（b）

图 3-35　刀库换刀 PMC 的 I/O 电路电气原理（续）

（b）PMC 输出信号

【任务实施】

（1）按照随机分组方式，分成各个观摩小组，并选出一名组长，将本组成员的个人信息填入表 3-4 中。

（2）根据观摩及讨论的结果，在表 3-4 中记录刀库换刀电路的换刀过程、工作原理、电路结构及适用场合等。

（3）对观摩过程中的疑难问题进行充分讨论，并将问题及解决办法填入表 3-4 中。

表 3-4　刀库换刀电路设计任务工单

<table>
<tr><td colspan="6">任务工单</td></tr>
<tr><td colspan="2">姓名：</td><td colspan="2">班级：</td><td colspan="2">学号：</td></tr>
<tr><td colspan="6">所需材料、工量具、设备列表</td></tr>
<tr><td>序号</td><td colspan="2">名称</td><td>数量</td><td>型号或规格</td><td>备注</td></tr>
<tr><td></td><td colspan="2"></td><td></td><td></td><td></td></tr>
<tr><td></td><td colspan="2"></td><td></td><td></td><td></td></tr>
<tr><td></td><td colspan="2"></td><td></td><td></td><td></td></tr>
<tr><td colspan="6">工作过程（观摩）</td></tr>
<tr><td>序号</td><td>换刀过程</td><td>工作原理</td><td colspan="2">电路结构</td><td>适用场合</td></tr>
<tr><td></td><td></td><td></td><td colspan="2"></td><td></td></tr>
<tr><td></td><td></td><td></td><td colspan="2"></td><td></td></tr>
<tr><td></td><td></td><td></td><td colspan="2"></td><td></td></tr>
<tr><td colspan="3">出现问题</td><td colspan="3">解决办法</td></tr>
<tr><td colspan="3"></td><td colspan="3"></td></tr>
</table>

【任务评价】

序号	评价内容	分值	得分
1	熟练掌握斗笠式刀库的换刀过程及工作原理	25	
2	熟练掌握斗笠式刀库电路结构	25	
3	能够运用刀库换刀电气原理图自主进行斗笠式刀库电路的故障诊断	25	
4	具有有效沟通、表达及团结协作的职业精神	25	

复习思考题

（1）简述数控机床常用电器分类。
（2）简述数控机床常用电器的选用方法。
（3）简述电气控制系统图的分类。
（4）简述电气系统图绘制原则。
（5）简述斗笠式刀库的换刀过程及工作原理。
（6）简述斗笠式刀库电路结构。

模块四 FANUC 0i-F数控系统介绍

FANUC 数控系统简介

FANUC 0i-F（包含 FANUC Series 0i-MF 和 Series 0i-TF）数控系统是FANUC 公司 2017 年推出的新产品，目前已全面推广应用。该系统是在高端的 31iB 系统平台下构建的，在这个平台强大的智能化功能的支持下完美呈现机床运行所需的高效率及便利性，其所连接的 αi、βi 伺服放大器也全面升级为 B 系列高速化、智能化伺服放大器，用来驱动升级后的 B 系列的主轴电机及伺服电机，满足高速、高精加工。FANUC 0i-F 系统具有高度集成化的内部结构，通过最新的 FSSB 高速接口，与 αi-B 系列、βi-B 系列的主轴及伺服单元进行连接，同时使用升级后的 I/O Link 接口与 I/O 模块进行连接，实现智能化、高速化的通信。其主板上集成轴卡功能，并安装有 FROM/SRAM 板以及可检测转速的系统冷却风扇，和用在断电时保持系统 SRAM 资料的电池。此外该系统支持最大两槽扩展，充分满足当前机床行业自动化、网络化的功能需求。本模块针对 FANUC 0i-F 数控系统硬件连接、基本参数、PMC、数据传输与备份等进行全面的剖析，同时针对 FANUC 0i-F 数控系统伺服硬件连接、参数设定、PMC 程序编程的具体内容进行详细的讲解。

任务一　系统硬件连接

【学习目标】

（1）掌握 FANUC 0i-F 数控系统的硬件的结构及接口。

（2）了解 FANUC 数控系统的硬件综合连接。

（3）了解 FANUC 数控系统的基本构成。

（4）具有辨识 FANUC 0i-F 数控系统硬件接口定义的能力。

（5）掌握 FANUC 0i-F 数控系统的主板结构与接口。

（6）掌握工程工作方法，培养严谨的工作作风，遵守 5S 管理制度。

【任务描述】

以小组为单位，观摩 FANUC 0i-F 数控系统的硬件连接，并将 FANUC 0i-F 数控系统主板硬件连接的接口名称、定义说明及主要功能等填写到任务工单中。要求观摩过程中务必注

意安全，遵守实训场所相关操作规范。

【知识链接】

一、FANUC 数控系统典型硬件的结构及接口

1. 数控系统的主板结构与接口

FANUC 0i-F 数控系统的主板结构与接口如图 4-1 所示。主板上方有两个风扇，便于主板散热。主板右下方 CP1 连接的是 DC 3 V 的锂电池，为存储器的后备电池。用户所编制的零件加工程序、刀具偏置量，以及系统参数等存储在控制单元的 CMOS 存储器中，当系统主电源切断时，依靠锂电池记忆这些数据。因此当电池电压下降到一定程度，显示器上出现“BAT”报警时，应及时更换电池，以防数据丢失。

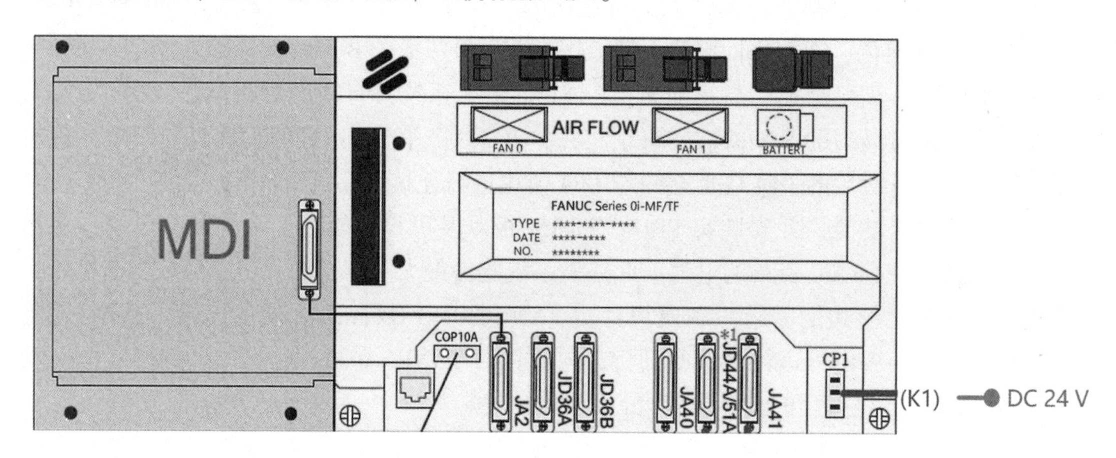

图 4-1　FANUC 0i-F 数控系统的主板结构与接口

FANUC 0i-F 数控系统的主板上有以下接口。

（1）电源接口 CP1。

数控系统控制单元主板正常工作时需要外部提供 DC 24 V 电源。外部 AC 200 V 电源经过开关电源整流后变为 DC 24 V，通过 CP1 接口输入，供主板工作。

（2）串行主轴/编码器 JA41。

机床上使用测量信号，如机床上的数字测量仪或者刀具测量信号，追求高速精确测量时，最多可以连接八个高速输入点。

（3）模拟主轴/跳转信号 JA40。

当机床使用模拟主轴时，主轴的速度命令转化为±10 V 的模拟量，由 JA40 输出连接至变频器的速度命令端口。但为了降低外部干扰引起的回路的检测报警，指令需要选择屏蔽电缆、屏蔽层接地处理，同时如果模拟主轴采用主轴编码器反馈时，不同于串行主轴连接时接至主轴驱动器，主轴编码器需要反馈至主板接口的 JA41 上。

（4）I/O 模块通信 JD44A/JD51A。

对于数控机床各坐标轴的运动控制，即在用户加工程序中 G、F 指令部分，由数控系统

控制实现；而对于数控机床的顺序逻辑动作，即在用户加工程序中 M、S、T 指令部分，由 PMC 控制实现，其中包括主轴速度控制、刀具选择、工作台更换、转台分度、工件夹紧与松开等。这些来自机床侧的输入、输出信号与 CNC 之间是通过 I/O Link 建立通信联系的。根据 PMC 控制点数不同，需要通过 I/O Link 连接电缆连接多个 I/O 模块。I/O Link 的两个接口分别为 JD51A（JD1A）、JD1B，电缆总是从一个单元 JD51A（JD1A）连接下一个单元 JD1B。CNC、I/O 模块和机床控制信号的连接，如图 4-2 所示。

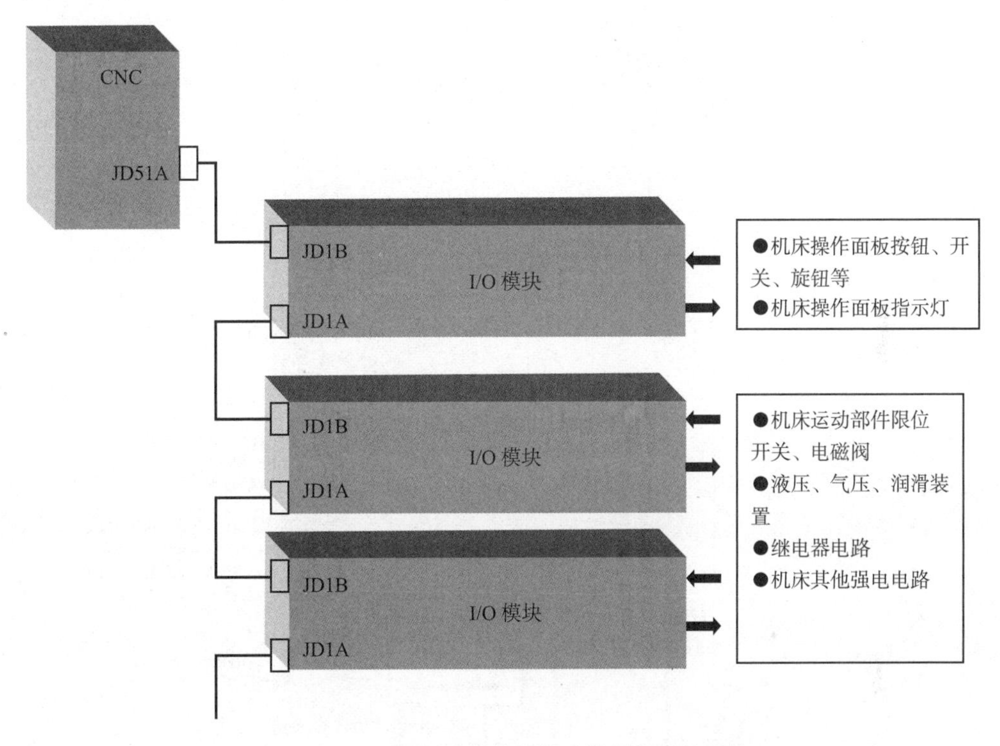

图 4-2 CNC、I/O 模块和机床控制信号的连接

（5）伺服放大器接口 COP10A。

系统 FSSB 光纤接口安装在主板上的 COP10A 接口，通过光纤连接至主轴或伺服驱动的 COP10B，然后依次由主轴或伺服驱动器上的 COP10A 连接至下一级伺服。

（6）MDI 接口 JA2。

它是 MDI 键盘与数控系统连接接口，JA2 通过电缆连接至键盘的 CK27 端口。

（7）内嵌以太网 CD38A。

CD38A 连接至计算机的以太网端口完成机床数据及信息的传输和采集。另外的通信接口为 RS232 串行通信接口，该接口有两个，分别是 JD36A（1 口）和 JD36B（2 口），可通过参数设定 1 口和 2 口，作为经常使用的端口为 JD36A（1 口）。

（8）触摸屏 JD36B。

当使用触摸屏时，该接口将连接至触摸屏，不作为 RS232 通信口使用。

2. 伺服放大器

（1）伺服放大器的作用。

要加工出各种形状的工件，达到零件图样要求的形状、位置、表面质量精度要求，刀具

和工件之间必须按照给定的进给速度、给定的进给方向、一定的切削深度作相对运动。这个相对运动是由一台或几台伺服电机驱动的。伺服放大器接受从控制单元 CNC 发出伺服轴的进给运动指令，经过转换和放大后驱动伺服电机，实现所要求的进给运动。

（2）βi-B 伺服接口定义及连接。

电源模块、主轴模块、伺服模块共同构成一体化装置，其相互之间的连接关系如图 4-3 所示。

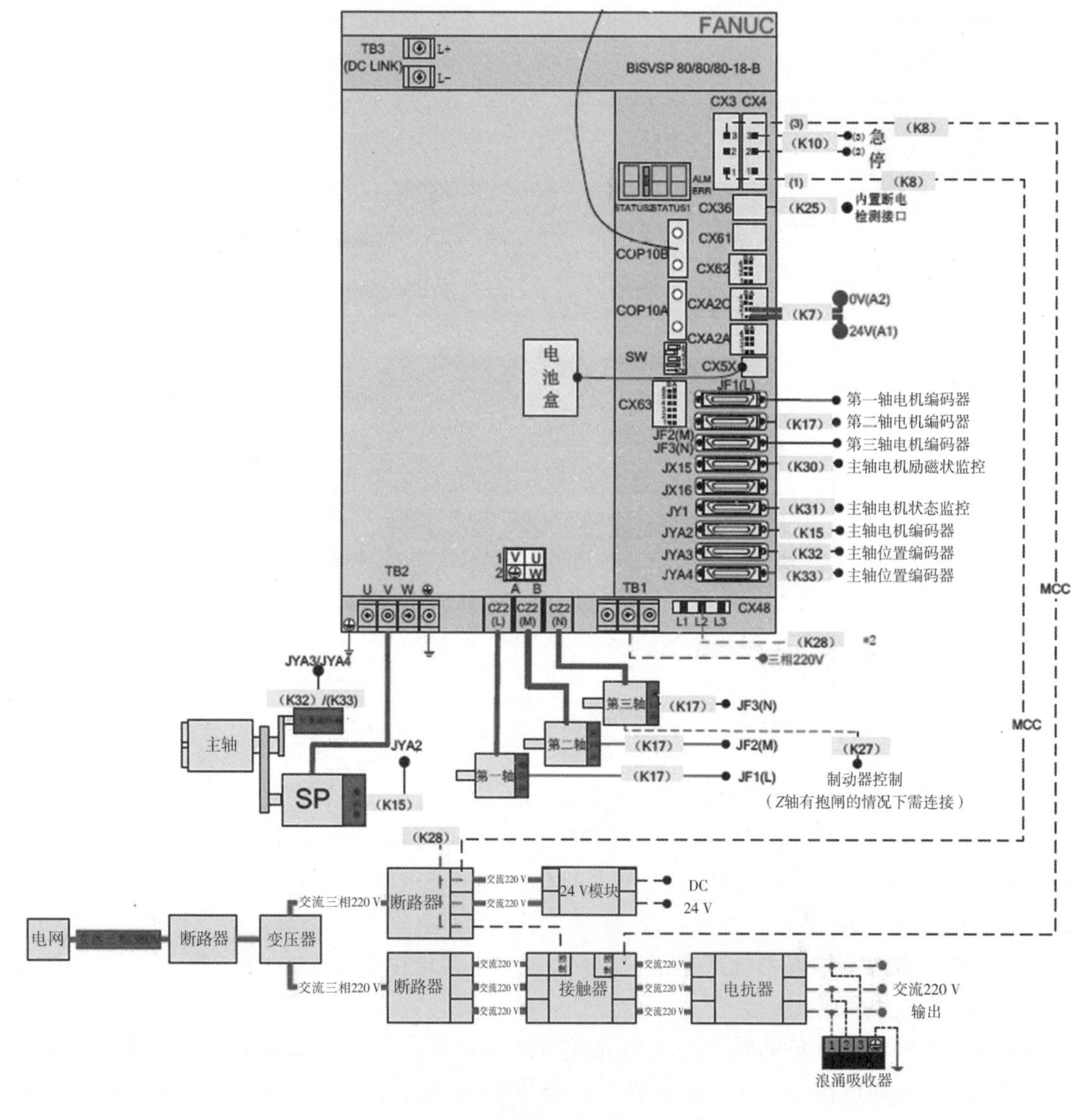

图 4-3 βi-B 伺服接口连接关系图

3. 数控系统的 I/O 装置接口与连接

（1）FANUC 系统常用的 I/O 装置。

FANUC 系统常用的 I/O 装置一般有内置 I/O 模块、外置 I/O 模块、分线盘 I/O 模块、机床操作面板 I/O 卡、系统 I/O 模块，如图 4-4 所示。

（2）I/O 装置接口定义。

I/O 板接口如图 4-5 所示。

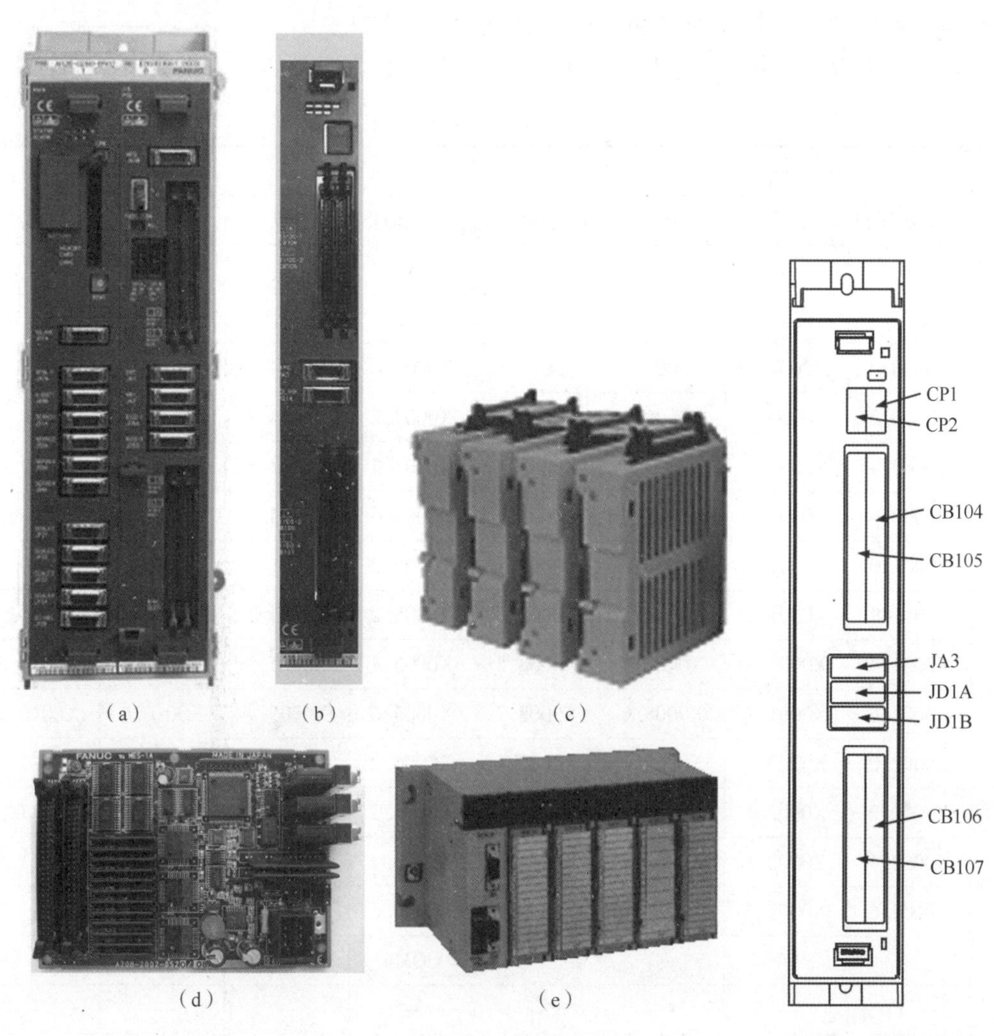

图 4-4　FANUC 系统常用的 I/O 设备

（a）内置 I/O 模块；（b）外置 I/O 模块；（c）分线盘 I/O 模块；（d）机床操作面板 I/O 卡；（e）系统 I/O 模块

图 4-5　I/O 板接口

POWER：电源指示灯；

CP1：DC24 V 电源输入；

CB104：DI/DO-1；

CB105：DI/DO-2；

JA3：MPG 电子手轮；

JD1A：I/O Link；

JD1B：I/O Link（连接系统板 JD51A/JD44A）；

CB106：DI/DO-3；

CB107：DI/DO-4。

（3）I/O 板接口定义说明。

1）CB104~CB107：每个数字量输入/输出接口为 24 点输入，16 点输出。

2）CB104~CB107：共有 96 点输入，64 点输出。

若 X、Y 地址从 0 开始分配，则 CB104~CB107 接口针脚定义如表 4-1 所示。

表 4-1　接口针脚定义

接口	CB104		CB105		CB106		CB107	
针脚	HIROSE50 针脚		HIROSE50 针脚		HIROSE50 针脚		HIROSE50 针脚	
电压端	A	B	A	B	A	B	A	B
01	0 V	+24 V	0 V	+24 V	0 V	+24 V	0 V	+24 V
02	X0000. 0	X0000. 1	X0003. 0	X0003. 1	X0004. 0	X0004. 1	X0007. 0	X0007. 1
03	X0000. 2	X0000. 3	X0003. 2	X0003. 3	X0004. 2	X0004. 3	X0007. 2	X0007. 3
04	X0000. 4	X0000. 5	X0003. 4	X0003. 5	X0004. 4	X0004. 5	X0007. 4	X0007. 5
05	X0000. 6	X0000. 7	X0003. 6	X0003. 7	X0004. 6	X0004. 7	X0007. 6	X0007. 7
06	X0001. 0	X0001. 1	X0008. 0	X0008. 1	X0005. 0	X0005. 1	X0010. 0	X0010. 1
07	X0001. 2	X0001. 3	X0008. 2	X0008. 3	X0005. 2	X0005. 3	X0010. 2	X0010. 3
08	X0001. 4	X0001. 5	X0008. 4	X0008. 5	X0005. 4	X0005. 5	X0010. 4	X0010. 5
09	X0001. 6	X0001. 7	X0008. 6	X0008. 7	X0005. 6	X0005. 7	X0010. 6	X0010. 7
10	X0002. 0	X0002. 1	X0009. 0	X0009. 1	X0006. 0	X0006. 1	X0011. 0	X0011. 1
11	X0002. 2	X0002. 3	X0009. 2	X0009. 3	X0006. 2	X0006. 3	X0011. 2	X0011. 3
12	X0002. 4	X0002. 5	X0009. 4	X0009. 5	X0006. 4	X0006. 5	X0011. 4	X0011. 5
13	X0002. 6	X0002. 7	X0009. 6	X0009. 7	X0006. 6	X0006. 7	X0011. 6	X0011. 7
14	—	—	—	—	COM4	—	—	—
15	—	—	—	—	—	—	—	—
16	Y1000. 0	Y1000. 1	Y1002. 0	Y1002. 1	Y1004. 0	Y1004. 1	Y1006. 0	Y1006. 1
17	Y1000. 2	Y1000. 3	Y1002. 2	Y1002. 3	Y1004. 2	Y1004. 3	Y1006. 2	Y1006. 3
18	Y1000. 4	Y1000. 5	Y1002. 4	Y1002. 5	Y1004. 4	Y1004. 5	Y1006. 4	Y1006. 5
19	Y1000. 6	Y1000. 7	Y1002. 6	Y1002. 7	Y1004. 6	Y1004. 7	Y1006. 6	Y1006. 7
20	Y1001. 0	Y1001. 1	Y1003. 0	Y1003. 1	Y1005. 0	Y1005. 1	Y1007. 0	Y1007. 1
21	Y1001. 2	Y1001. 3	Y1003. 2	Y1003. 3	Y1005. 2	Y1005. 3	Y1007. 2	Y1007. 3
22	Y1001. 4	Y1001. 5	Y1003. 4	Y1003. 5	Y1005. 4	Y1005. 5	Y1007. 4	Y1007. 5
23	Y1001. 6	Y1001. 7	Y1003. 6	Y1003. 7	Y1005. 6	Y1005. 7	Y1007. 6	Y1007. 7
24	DOCOM	DOCOM	DOCOM	DOCOM	DOCOM	DOCOM	DOCOM	DOCOM
25	DOCOM	DOCOM	DOCOM	DOCOM	DOCOM	DOCOM	DOCOM	DOCOM

【任务实施】

（1）按照随机分组方式，分成各个观摩小组，并选出一名组长，将本组成员的个人信息填入表 4-2 中。

（2）根据观摩及讨论的结果，在表 4-2 中记录 FANUC 0i-F 数控系统主板硬件连接的接口名称、定义说明、主要功能及适用场合等。

（3）对观摩过程中的疑难问题进行充分讨论，并将问题及解决办法填入表 4-2 中。

表 4-2 FANUC 0i-F 数控系统主板接口任务工单

<table>
<tr><td colspan="6">任务工单</td></tr>
<tr><td colspan="2">姓名：</td><td colspan="2">班级：</td><td colspan="2">学号：</td></tr>
<tr><td colspan="6">所需材料、工量具、设备列表</td></tr>
<tr><td>序号</td><td colspan="2">名称</td><td>数量</td><td>型号或规格</td><td>备注</td></tr>
<tr><td></td><td colspan="2"></td><td></td><td></td><td></td></tr>
<tr><td></td><td colspan="2"></td><td></td><td></td><td></td></tr>
<tr><td></td><td colspan="2"></td><td></td><td></td><td></td></tr>
<tr><td colspan="6">工作过程（观摩）</td></tr>
<tr><td>序号</td><td>接口名称</td><td>定义说明</td><td colspan="2">主要功能</td><td>适用场合</td></tr>
<tr><td></td><td></td><td></td><td colspan="2"></td><td></td></tr>
<tr><td></td><td></td><td></td><td colspan="2"></td><td></td></tr>
<tr><td></td><td></td><td></td><td colspan="2"></td><td></td></tr>
<tr><td></td><td></td><td></td><td colspan="2"></td><td></td></tr>
<tr><td colspan="3">出现问题</td><td colspan="3">解决办法</td></tr>
<tr><td colspan="3"></td><td colspan="3"></td></tr>
</table>

【任务评价】

序号	评价内容	分值	得分
1	能够熟练说出 FANUC 0i-F 数控系统的硬件的结构及接口	25	
2	能够正确辨识 FANUC 0i-F 数控系统硬件接口定义	25	
3	能够掌握 FANUC 0i-F 数控系统的主板结构与接口	25	
4	具有有效沟通、表达及团结协作的职业精神	25	

任务二　系统基本参数设定

系统参数编辑

【学习目标】

（1）了解系统参数的表达方式。

（2）掌握系统参数页面的显示及参数查询功能。

（3）掌握基本参数的设定。

（4）理解在调试中出现的报警含义，并能排除报警。

（5）通过对数控系统参数调试，具备故障排除的思维及方法能力。

（6）掌握工程工作方法，培养严谨的工作作风，遵守 5S 管理制度。

【任务描述】

以小组为单位，观摩 FANUC 0i-F 数控系统的系统基本参数，并将参数设定页面中各项系统参数的参数号、简要说明、设定说明及机床设定值等填写到任务工单中。要求观摩过程中务必注意安全，遵守实训场所相关操作规范。

【知识链接】

一、系统参数概述

1. 数控系统参数的定义及作用

数控系统的参数是数控系统用来匹配机床及数控功能的一系列数据。在 FANUC 0i-F 数控系统中参数可分为系统参数、PMC 参数。系统参数功能由 FANUC 公司定义；PMC 参数是数控机床的 PMC 程序中使用的数据，如计时器、计数器、保持型继电器的数据，这些参数由机床厂家来定义。这两类参数是数控机床正常工作的前提条件。

2. 典型参数表达方式

（1）位型以及位（机械组/路径/轴/主轴）型参数。

位型参数格式是用 8 位的二进制数表示参数的位为 0 或为 1 的状态，位 1 与位 0 对应，第 8 位与位 7 对应。

0000				EIA	NCR	ISP	CTV	TVC

数据号　　数据#0～#7表示位(bit)

（2）其他参数。

除位型参数外，其他参数的表达方式如下。

1023	各轴的伺服轴号
数据号	

二、参数页面的显示及参数编辑

1. 系统参数的调用和显示

（1）数控装置的 MDI 键盘，按 MDI 键盘上的功能键【SYSTEM】一次后，再单击软键【参数】，进入参数显示界面。

（2）用 MDI 键盘上的翻页键或光标移动键，逐页找到期望的参数。

（3）还可通过 MDI 键盘输入参数号，再单击软键【搜索号】，这样可以显示指定参数所在的界面，光标同时处于指定参数位置。

2. 系统参数编辑

（1）参数改写状态。

解除数控系统参数设定完成后，处于写保护状态，在该状态下不允许更改参数。要想修改或调整参数，应使参数置于可写状态，即需要解除写保护，操作步骤如下。

1）将数控系统置于 MDI 模式或急停模式。

2）按 MDI 键盘上的功能键 OFFSET SETTING 数次后，或者按功能键 OFFSET SETTING 一次后再单击软键【设定】，可显示“设定”界面主页，如图 4-6 所示。

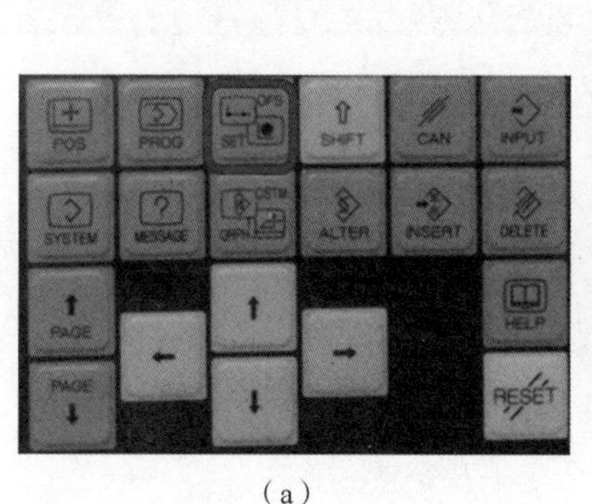

（a）

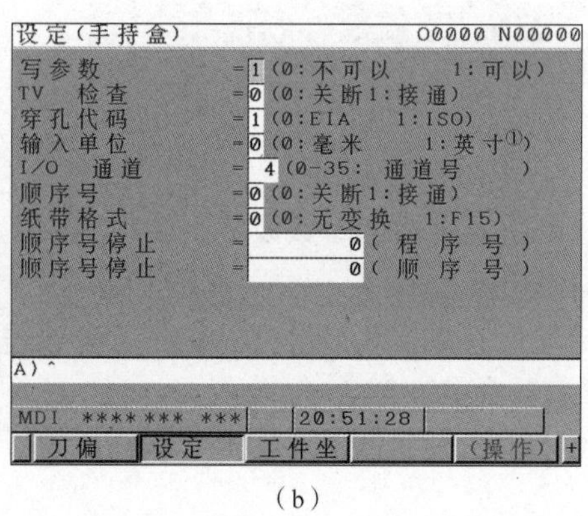

（b）

图 4-6　写参数允许界面

（a）功能键；（b）“设定”界面

3）将光标移至“写参数”行。

4）单击软键【操作】。

5）输入“1”，再单击软键【输入】，使“写参数=1”，这样参数处于可写状态。

（2）参数常规设定方式。

参数常规设定步骤如下。

---

① 英寸为英制长度单位，符号为 in，1 in=2.54 cm。

1）进入参数设定界面，将光标置于需要设定的参数位置上，输入数据，然后按软键【输入】，输入的数据将被设定到光标指定的参数中。

2）对于位型参数：单击软键【ON：1】，则将光标位置置1；单击软键【OFF：0】，则将光标位置置0。

3）输入参数值后，单击软键【+输入】，则把输入值加到原来的值上。

4）输入参数值后，单击软键【输入】，则可输入新的参数值，如图4-7所示。

5）输入参数后，也可以用MDI键盘上的功能键 INPUT 完成写参数操作。

三、系统参数设定

系统基本参数设定

系统参数设定需调用参数设定界面。首先在紧急停止状态下按功能键【SYSTEM】，然后单击软键【+】几次，直到软键【参数调】出现，选中软键【参数调】出现参数设定支援界面，如图4-8所示。图4-8中的项目就是参数的设定调试步骤。

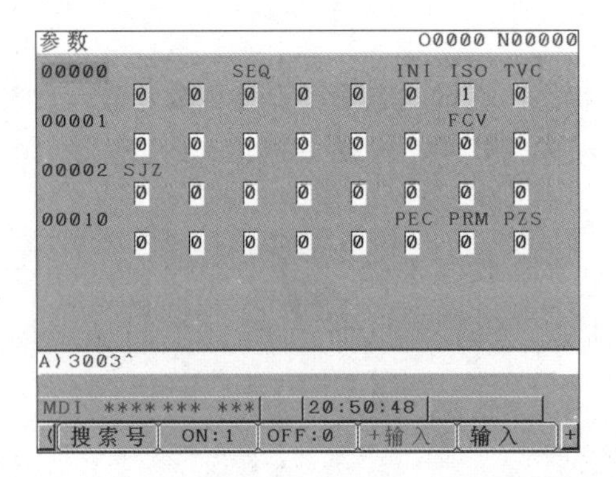

图4-7 参数值输入界面

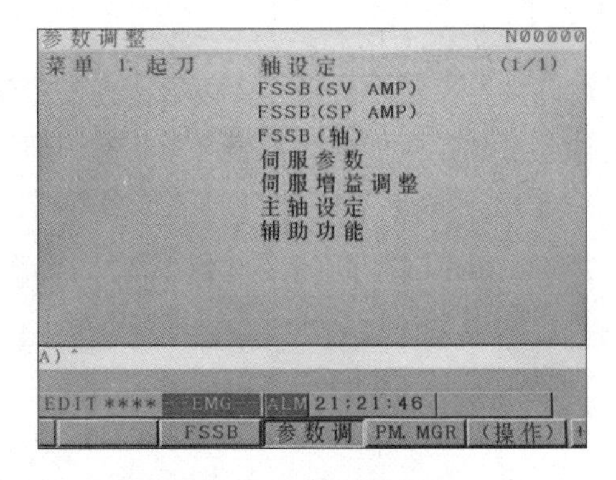

图4-8 参数设定界面

按照参数设定界面的参数设定，依次完成下列各项参数的输入和调试。

（1）轴设定。

轴设定里面有以下几个组，对每一组参数进行设定。

1）基本（basic）组。

有关基本组参数的设定，如表4-3所示。

表4-3 基本组参数

组	参数号	简述	设定说明	设定值		
				X	*Y*	*Z*
基本	1001#0	直线轴的最小移动单位。 0：公制（公制机床）； 1：英制（英制机床）	一般为公制机床	0	0	0
	1013#1	设定最小输入增量和最小指令增量。 0：IS-B（0.001 mm，0.001°，0.000 1 in）； 1：IS-C（0.000 1 mm，0.000 1°，0.000 01 in）	一般设定为0	0	0	0

续表

组	参数号	简述	设定说明	设定值		
				X	*Y*	*Z*
基本	1005#0	参考点没有建立时，在自动运行中指定除了G28以外的移动指令是否发生P/S 224报警。 0：出现报警（No. 224）； 1：不出现报警	为了机床安全，一般设定为0	1	1	1
	1005#1	无挡块回参考点设定功能是否有效。 0：无效（各轴）； 1：有效（各轴）	0为有挡块设定；1为无挡块设定	1	1	1
	1006#0	设定是直线轴还是旋转轴。 0：直线轴； 1：旋转轴	—	0	0	0
	1006#3	设定各轴的移动量类型是半径指定还是直径指定。 0：半径指定； 1：直径指定	车床的*X*轴设定为1	0	0	0
	1006#5	设定各轴返回参考点方向。 0：按正方向； 1：按负方向	脱离挡块后轴的移动方向	1	0	0
	1008#0	设定旋转轴的循环功能是否有效。 0：无效； 1：有效	设定坐标是否循环轮回	0	0	0
	1008#2	相对坐标值。 0：不按每一转的移动量循环显示； 1：按每一转的移动量循环显示	—	0	0	0
	1020	各轴的编程名称	*X*轴：88； *Y*轴：89； *Z*轴：90	88	89	90
	1022	各轴的属性	*X*轴：1； *Y*轴：2； *Z*轴：3	1	2	3
	1023	各轴的伺服轴号	确定CNC轴和伺服电机的关系	1	2	3

续表

组	参数号	简述	设定说明	设定值		
				X	Y	Z
基本	1815#1	分离型位置编码器。 0：不使用； 1：使用	接光栅尺或分离型旋转编码器时设定为1	0	0	0
	1815#4	使用绝对位置检测器时，机械位置与绝对位置检测器的位置。 0：不一致； 1：一致	使用绝对位置检测器时，初次调试时设定为0，通过移动机床使机械位置与绝对位置检测器的位置一致时设定为1	1	1	1
	1815#5	绝对位置检测器。 0：不使用； 1：使用	使用绝对位置检测功能时设定为1，需要硬件支持（绝对编码器）	1	1	1
	1825	各轴的伺服环增益	3 000～8 000，互相插补的轴必须设定一致	5 000		
	1826	各轴的到位宽度	20～50	50		
	1828	各轴移动时的最大允许位置偏差量	500～10 000 设定值=快移速度/(60×回路增益)	10 000		
	1829	各轴停止中的最大允许位置偏差量	50～2 000	2 000		

2）主轴（spindle）组。

有关主轴组参数的设定，如表4-4所示。

表4-4　主轴组参数

组	参数号	简述	设定说明	设定值
主轴	3716#0	指定主轴电机类型。 0：模拟； 1：串行	—	1
	3717	为各个主轴电机设定编号	—	1

3）坐标系（Coordinate）组。

有关坐标系组参数的设定，如表4-5所示。

表 4-5　坐标系组参数

<table>
<tr><th>组</th><th>参数号</th><th>简述</th><th>设定说明</th><th colspan="3">设定值</th></tr>
<tr><td rowspan="5">坐标系</td><td>1240</td><td>在机械坐标系上的各轴第一参考点的坐标值</td><td>确立参考点在机械坐标系中的坐标</td><td colspan="3">0</td></tr>
<tr><td>1241</td><td>在机械坐标系上的各轴第二参考点的坐标值</td><td>确立参考点在机械坐标系中的坐标，在本机床中，Z 值为刀库换刀点</td><td>0</td><td>0</td><td>-101</td></tr>
<tr><td>1260</td><td>旋转每一周的移动量</td><td>一般设为 360 000，说明旋转轴转一圈坐标旋转 360°</td><td colspan="3">0</td></tr>
<tr><td>1320</td><td>各轴存储式行程检测 1 的正方向边界的坐标值</td><td>返回参考点后设定，基准是机床坐标系</td><td colspan="3">999 999. 000</td></tr>
<tr><td>1321</td><td>各轴存储式行程检测 1 的负方向边界的坐标值</td><td>返回参考点后设定，基准是机床坐标系</td><td colspan="3">-999 999. 000</td></tr>
</table>

4）进给速度（feed rate）组。

有关进给速度组参数的设定，如表 4-6 所示。

表 4-6　进给速度组参数

<table>
<tr><th>组</th><th>参数号</th><th>简述</th><th>设定说明</th><th>设定值</th></tr>
<tr><td rowspan="9">坐标系</td><td>1401#6</td><td>快速空运行是否有效。
0：无效；
1：有效</td><td>—</td><td>0</td></tr>
<tr><td>1410</td><td>空运行速度及手动直线，圆弧插补的进给速度</td><td rowspan="8">一般情况下速度的设定单位是 mm/min</td><td>3 000</td></tr>
<tr><td>1420</td><td>各轴快速运行速度</td><td>3 000</td></tr>
<tr><td>1421</td><td>各轴快速运行倍率的 F0 速度</td><td>500</td></tr>
<tr><td>1423</td><td>各轴手动连续进给（JOG 进给）时的进给速度</td><td>3 000</td></tr>
<tr><td>1424</td><td>各轴的手动快速运行速度</td><td>5 000</td></tr>
<tr><td>1425</td><td>各轴返回参考点的 F1 速度</td><td>500</td></tr>
<tr><td>1428</td><td>回参考点速度</td><td>3 000</td></tr>
<tr><td>1430</td><td>最大切削速度</td><td>6 000</td></tr>
</table>

（2）FSSB 设定项。

数控系统通过高速串行伺服总线 FSSB 用一根光纤与多个伺服放大器进行连接，光纤连接实现了光电隔离，提高了可靠性和抗干扰性，也减少了连接电缆的数量。数控及伺服硬件

连接完成后，第一次调试时需要通过FSSB设定参数激活这种连接。

1）FSSB（SV AMP）设定。

进入参数调整界面，单击软键【操作】，将光标移动到“FSSB（SV AMP）”项，单击软键【选择】，出现参数设定界面，设定完成相关项目后，单击软键【操作】，再单击软键【选择】。数控系统如果不能通过FSSB检测到伺服模块，参数界面就不会出现伺服相关信息，需检查硬件问题，如图4-9所示。

2）FSSB（轴）设定。

进入参数调整界面，按软键【操作】，将光标移动到“FSSB（轴）”项，单击软键【选择】，出现参数设定界面（数控机床半闭环连接的情况下，不用修改数据）。设定完成后按软键【设定】。FSSB（轴）设定界面如图4-9、图4-10所示。

伺服放大器设定　　O0000 N00000
HRV 3

号	放大	系列	电流	轴	名称
1-02	A1-L	αi SV	40A	01	X
1-03	A1-M	αi SV	40A	02	Y
1-04	A2-L	αi SV	40A	03	Z

A)^
MDI ****　21:26:42
连接　SV AMP　SP AMP　轴　（操作）

图4-9　FSSB（SV AMP）设定

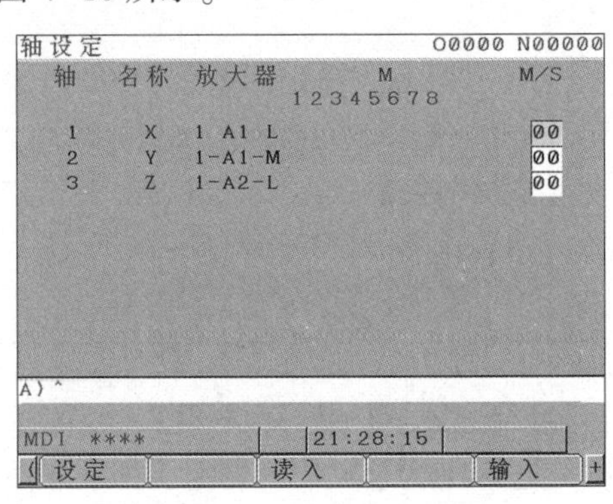

图4-10　FSSB（轴）设定

（3）伺服的初始化设定。

在急停状态下，进入参数调整，单击软键【操作】，将光标移动到“伺服设定”处，单击软键【选择】，出现伺服参数设定界面。单击软键【切换】后，进入初始化设定界面，在此界面进行设定。

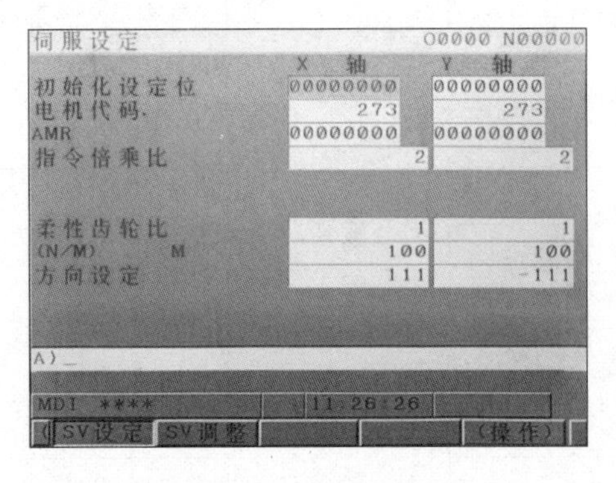

图4-11　参数设定界面

1）轴初始化设定值。

各轴的初始化设定位参数系统自动默认值均为00000010，若需要对伺服参数初始化，则初始化设定位各轴对应为00000000，如图4-11所示。

2）电机代码。

按照伺服电机铭牌电机型号和伺服驱动放大器铭牌型号，查阅《伺服电机参数说明书》，找到相应电机代码，输入该代码数值。

3）AMR。

该参数相当于伺服电机极数参数，设定为00000000。

4）指令倍率。

设定从CNC到伺服系统的移动量的指令倍率。设定值=（指令单位/检测单位）×2，通常指令单位等于检测单位，因此将该参数设为2。

5）柔性齿轮比。

半闭环时柔性齿轮比=电机每旋转一周所需的位置脉冲数/1 000 000。式中，分母是电机每旋转一周轴的移动量。例如，当直线轴电机与滚珠丝杠 1∶1 连接时的柔性齿轮比数据，如表 4-7 所示。

表 4-7 半闭环时柔性齿轮比示例

检测单位/μm	滚珠丝杠的导程/mm					
	6	8	10	12	16	20
1	6/1 000	8/1 000	10/1 000	12/1 000	16/1 000	20/1 000
0. 5	12/1 000	16/1 000	20/1 000	24/1 000	32/1 000	40/1 000
0. 1	60/1 000	80/1 000	100/1 000	120/1 000	160/1 000	200/1 000

6）方向设定。

111：从脉冲编码器看电机轴沿顺时针方向旋转。

-111：从脉冲编码器看电机轴沿逆时针方向旋转。

（4）主轴设定。

在紧急停止状态下，首先按下机床紧急停止按钮，然后按功能键【SYSTEM】，再单击软键【+】几次，直到软键【主轴设定】出现主轴设定界面，单击软键【切换】后，如图 4-12 所示。

单击软键【代码】时显示电机代码一览界面（图 4-13），软键【代码】在光标位于电动型号项目时显示。此外要从电机代码一览页面返回到上一界面，单击软键【返回】。切换到电机代码一览界面时，显示电机型号代码所对应的电机名称和放大器名称。将光标移动到希望设定的代码编号，单击软键【选择】，输入完成。

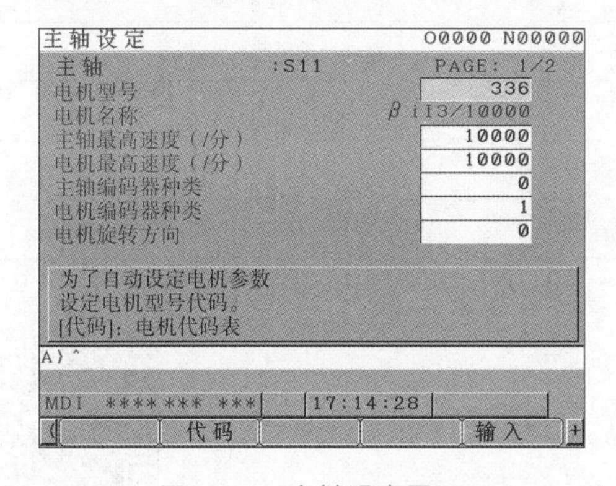

图 4-12 主轴设定界面

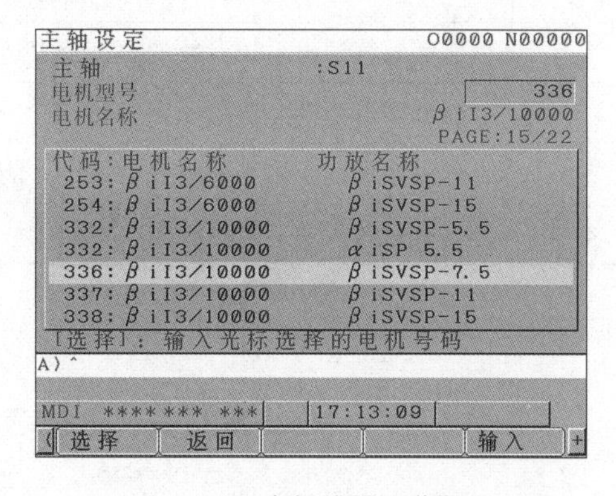

图 4-13 电机代码一览界面

【任务实施】

（1）按照随机分组方式，分成各个观摩小组，并选出一名组长，将本组成员的个人信息填入表 4-8 中。

（2）根据观摩及讨论的结果，在表 4-8 中记录各项系统参数的参数号、简要描述、设定说明及机床设定值等。

（3）对观摩过程中的疑难问题进行充分讨论，并将问题及解决办法填入表 4-8 中。

表 4-8　系统基本参数设定任务工单

<table>
<tr><td colspan="6">任务工单</td></tr>
<tr><td colspan="2">姓名：</td><td colspan="2">班级：</td><td colspan="2">学号：</td></tr>
<tr><td colspan="6">所需材料、工量具、设备列表</td></tr>
<tr><td>序号</td><td colspan="2">名称</td><td>数量</td><td>型号或规格</td><td>备注</td></tr>
<tr><td></td><td colspan="2"></td><td></td><td></td><td></td></tr>
<tr><td></td><td colspan="2"></td><td></td><td></td><td></td></tr>
<tr><td></td><td colspan="2"></td><td></td><td></td><td></td></tr>
<tr><td colspan="6">工作过程（观摩）</td></tr>
<tr><td>序号</td><td>参数号</td><td>简要描述</td><td colspan="2">设定说明</td><td>机床设定值</td></tr>
<tr><td></td><td></td><td></td><td colspan="2"></td><td></td></tr>
<tr><td></td><td></td><td></td><td colspan="2"></td><td></td></tr>
<tr><td></td><td></td><td></td><td colspan="2"></td><td></td></tr>
<tr><td></td><td></td><td></td><td colspan="2"></td><td></td></tr>
<tr><td colspan="3">出现问题</td><td colspan="3">解决办法</td></tr>
<tr><td colspan="3"></td><td colspan="3"></td></tr>
</table>

【任务评价】

序号	评价内容	分值	得分
1	能够熟练说出系统参数的类型及表达方式	25	
2	能够掌握基本参数的设定	25	
3	理解在调试中出现的报警含义，并能排除报警	25	
4	具有有效沟通、表达及团结协作的职业精神	25	

任务三　系统 PMC 介绍

PMC 程序介绍

【学习目标】

（1）了解 PMC 的原理及 I/O 地址分配。
（2）掌握 PMC 程序结构。
（3）掌握斗笠式刀库的换刀流程。
（4）掌握斗笠式刀库换刀的 PMC 编程。
（5）具备 PMC 程序设计和调试能力。

（6）掌握工程工作方法，培养严谨的工作作风，遵守 5S 管理制度。

【任务描述】

以小组为单位，观摩 FANUC 0i-F 数控系统的 PMC 程序，并将各部分 PMC 程序的信号、地址、说明及资源类型等填写到任务工单中。要求观摩过程中务必注意安全，遵守实训场所相关操作规范。

【知识链接】

PMC 原理及 IO 地址分配介绍

一、PMC 原理及 I/O 地址分配介绍

1. PMC 的定义

可编程控制器 PLC 在 FANUC 数控系统中称为可编程机床控制器（programmable machine controller，PMC）。PLC 和 PMC 只是名称上不同，其本质一致，除非特殊需要，在本书中将统一使用 PMC 这一名称。目前 FANUC 的数控产品将 PMC 内置，不需要独立的 PMC 设备，PMC 已成为数控系统的重要组成部分。

2. PMC 程序结构介绍

FANUC 0i-F 数控系统的 PMC 程序通常由第一级程序、第二级程序、第三级程序及子程序组成，如图 4-14 所示。

（1）第一级程序。

PMC 第一级程序从程序开始到 END1 命令，在系统每个梯形图执行周期中执行一次，主要特点是信号采样实时以及输出信号响应快。它主要处理短脉冲信号，如转矩限制 HIGH 指令信号、机床准备就绪信号、急停信号、主轴停信号、各轴互锁信号、抱闸控制信号。在第一级程序中，程序尽可能短，这样可以缩短 PMC 程序执行时间。如果没有输入信号，只需要编写 END1 功能指令。

（2）第二级程序

PMC 第二级程序是 END1 命令之后、END2 命令之前的程序。第二级程序通常包括操作方式选择、主轴功能、进给倍率等程序，为了简洁易读、各功能分工明确，也可以将功能程序编写进子程序，在第二级程序中进行调用。

（3）第三级程序。

PMC 第三级程序是 END2 命令之后、END3 命令之前的程序。第三级程序主要处理低速响应信号，通常用于 PMC 程序报警信号的处理。在编写顺序程序时，可选择是否使用第三级程序，但本书中未使用。

（4）子程序。

子程序是 END3 命令之后、END 命令之前的程序。通常将具有特定功能并且多次使用的程序段作为子程序。主程序中用指令决定具体子程序的执行状况。当主程序调用子程序并

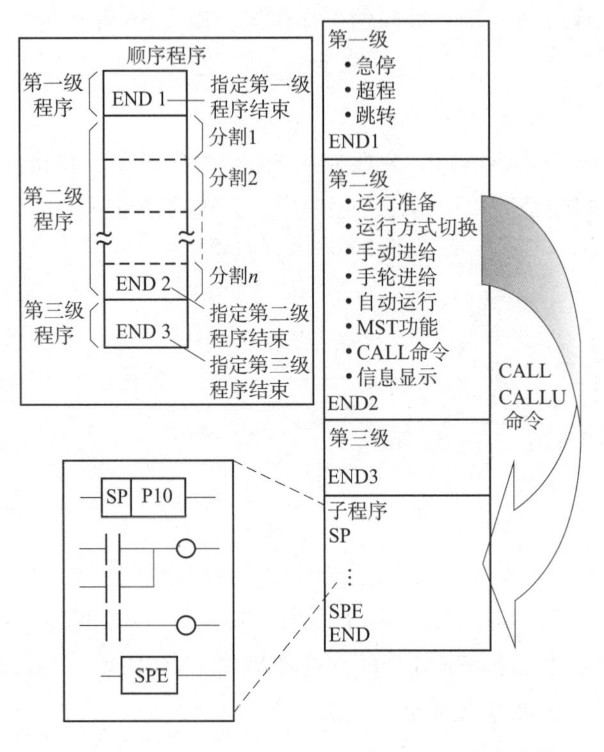

图 4-14 FANUC 0i-F 数控系统的 PMC 程序结构

执行时，子程序执行全部指令直至结束，然后系统将返回调用子程序的主程序。子程序用于为程序分段和分块，使其成为较小的、更易管理的块。在程序调试和维护时，通过使用较小的程序块，对这些区域和整个程序进行简单的调试并排除故障。只有在需要时才调用子程序块，可以更有效地使用 PMC。因为所有的子程序块可能无须执行每次扫描，所以能够缩短 PMC 程序处理时间。

3. FANUC 数控系统顺序程序

FANUC 数控系统顺序程序由第一级程序和第二级程序两部分组成，如图 4-15 所示。

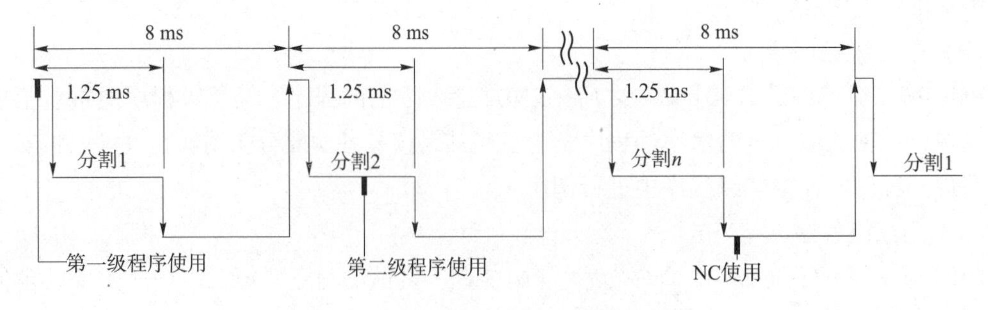

图 4-15 FANUC 数控系统顺序程序

第一级程序仅处理短脉冲信号，如急停、各进给坐标轴超程、机床互锁信号、返回参考点减速、跳步、进给暂停信号、加工中心大型刀库的计数等。

第二级程序包含了数控机床功能的主要内容，如操作方式、辅助功能、换刀等的处理。

4. I/O 地址分配

地址用来区分信号。不同的地址分别对应机床侧的输入输出信号、CNC 侧的输入输出信号、内部继电器、计数器、保持型继电器和数据表等，如图 4-16 所示。

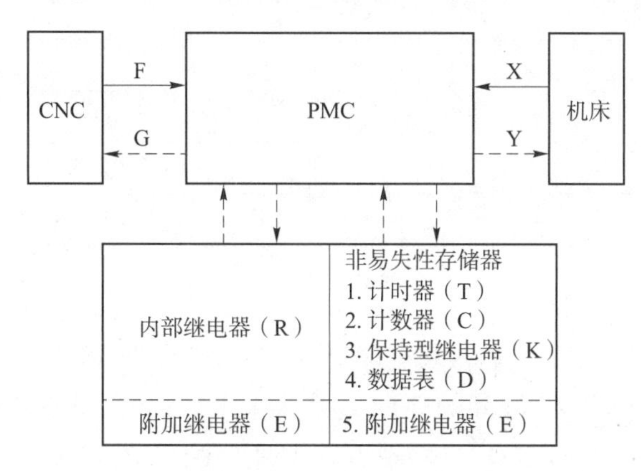

图 4-16　I/O 地址分配

在图 4-16 中，由实线表示的与 PMC 相关的输入输出信号经 I/O 板的接收电路和驱动电路传送；由虚线表示的与 PMC 相关的输入输出信号仅在存储器中传送，如在 RAM 中传送；这些信号的状态都可以在 CRT 上显示。

（1）地址格式和信号类型。

地址用地址号和位号表示，格式如下：

X127 . 7
　　位号0到7
　　地址号（字母后四位数以内）

在地址号的开头必须指定一个字母，用来表示所列的信号类型，如表 4-9 所示。在功能指令中指定字节单位的地址时，位号可以省略，如 X127。

表 4-9　PMC 地址和信号类型

字母	信号类型	备注	
		PMC-SA1	PMC-SA3
X	来自机床侧的信号（MT→PMC）	X0-X127（外部 I/O 模块） X1000-X1003（内部 I/O 模块）	
Y	由 PMC 输出到机床侧的信号（PMC→MT）	Y0-Y127（外部 I/O 模块） Y1000-Y1003（内部 I/O 模块）	
F	来自 CNC 侧的输入信号（CNC→PMC）	F0-F255	
G	由 PMC 输出到 NC 的信号（PMC→CNC）	G0-G255	
R	内部继电器	R0-R999 R9000-R9099	R0-R1499 R9000-R9117
A	信息显示请求信号	A0-A24	
C	计数器	C0-C79	

续表

字母	信号类型	备注	
		PMC-SA1	PMC-SA3
K	保持型继电器	K0-K19	
T	可变定时器	T0-T79	
D	数据表	—	D0-D1859
L	标记号	—	L1-L9999
P	子程序	—	P1-P512

（2）I/O 地址分配的方法。

I/O 地址分配的方法在数控铣床维修实训设备上的操作步骤如下：按下操作面板上的按键【SYSTEM】—两次【>】—【PMCMCNF】—【模块】，在光标在 X0 处输入“0.0.1.OC02I”，光标在 Y0 处输入“0.0.1.OC02O”，实现 I/O 地址分配，X0～X13，Y0～Y13，如图 4-17 所示。

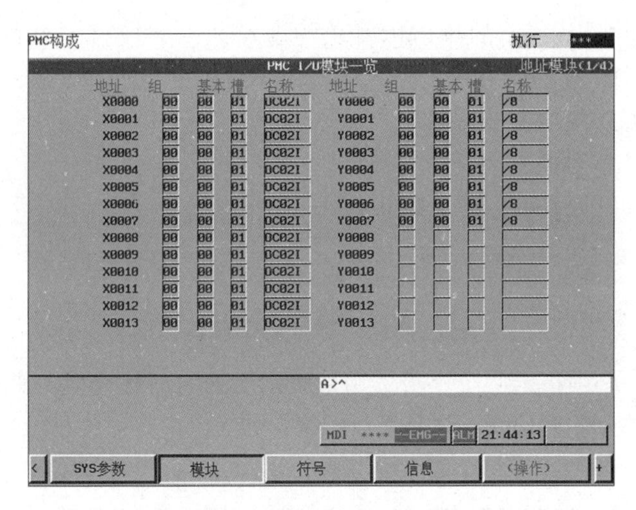

图 4-17　I/O 地址分配方法

二、斗笠式刀库换刀的 PMC 编程

1. 斗笠式刀库工作原理

（1）刀库控制要求。

FANUC 0i Mate MF 数控铣床维修实训设备的刀库使用的是 12 把刀具的斗笠式刀库，因此，本部分内容按立式加工中心斗笠式刀库的控制作介绍，其他类型的加工中心可参考该方法实施。

刀库的控制分为手动控制和程序自动控制两种方式。手动控制主要用于刀库的安装、调试或维护等，主要有手动选刀及主轴刀具夹紧、松开操作等；程序自动控制主要用于生产中的自动换刀控制。刀库自动换刀动作过程如图 4-18 所示。

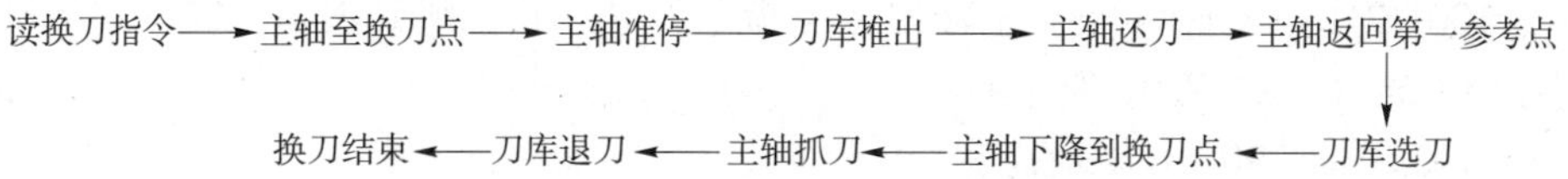

读换刀指令→主轴至换刀点→主轴准停→刀库推出→主轴还刀→主轴返回第一参考点

↓

换刀结束←刀库退刀←主轴抓刀←主轴下降到换刀点←刀库选刀

图 4-18 刀库自动换刀动作过程

（2）换刀流程及思路。

自动换刀需要考虑 T 指令与主轴刀号是否一致、主轴上是否有刀、刀库的刀套号与主轴刀号是否一致等。换刀流程如图 4-19 所示。

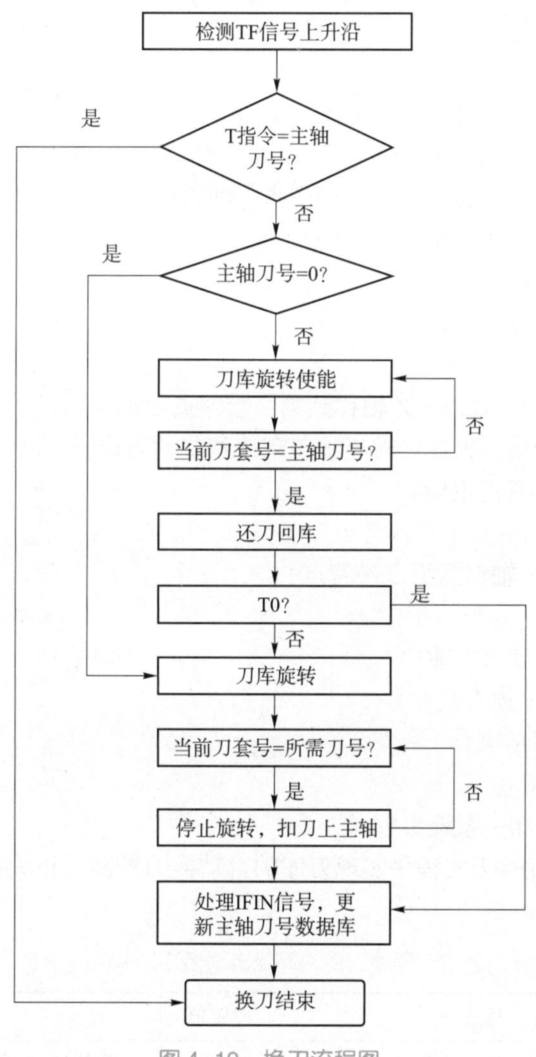

图 4-19 换刀流程图

2. 斗笠式刀库换刀宏程序

（1）编写换刀宏程序。

根据图 4-19 换刀流程图，编制如下“O9001”换刀宏程序。在 EDIT 方式下，按【PROG】程序按钮，输入“O9001”，再按 BG 编辑程序，如下所示。

O9001；（换刀宏程序号）

IF［#1001EQ1］GOTO 40；（主轴刀号与指令刀号一致，跳到 N40，宏变量#1001 对应 PMC 程序中 G54.1；EQ 表示“=”）

#199=#4003；（G90、G91 模态）

#198=#4006；（G20、G21 模态）

IF［#1003EQ1］GOTO 20；（判定主轴是否有刀，没有刀直接跳到 N20，宏变量#1003 对应 PMC 程序中 G54.3）

G21 G91 G30 P2 *Z*0 M19；（*Z* 轴移到第二参考点主轴定向）

M81；（主轴刀号与刀库当前刀号一致性判断，若不一致刀库旋转到与主轴刀号一致为止）

M80；（刀库向前，靠近主轴）

M82；（松刀吹气）

G91 G28 *Z*0；（*Z* 轴移到第一参考点）

IF［#1002EQ1］GOTO 10；（指令是否为 T0？宏变量#1002 对应 PMC 程序中 G54.2）

M83；（在主轴端，刀库旋转至加工程序指定的刀位）

G91 G30 P2 *Z*0；（*Z* 轴移到第二参考点）

N10 M84；（刀具夹紧）

M86；（刀库向后，远离主轴）

GOTO 30；

N20 G21 G91 G28 *Z*0 M19；（*Z* 轴移到第一参考点主轴定向）

M83；（在远离主轴端，刀库旋转至加工程序指定的刀位）

M80；（刀库向前，靠近主轴）

M82；（松刀吹气）

G91 G30 P2 *Z*0；（*Z* 轴移到第二参考点）

M84；（刀具夹紧）

M86；（刀库向后，远离主轴）

N30 G#199G#198；（模态恢复）

N40 M99；（子程序结束）

最后按 BG 编辑结束。

（2）宏程序调用及相关系统参数。

用“M06”指令调用换刀宏程序实现刀库的自动换刀控制，相关系统参数设置如表 4-10 所示。

表 4-10　调用换刀宏程序实现刀库的自动换刀控制的相关系统参数设置

参数号	意义	设置值	说明
6071	用 M 指令调用换刀宏程序 O9001	6	指定 M06 调用宏程序 O9001
3202#4	设置宏程序允许显示、编辑、删除	0	设为“0”时表示允许显示、编辑、删除程序“O9000-O9999”

（3）宏程序中与 PMC 程序 G 地址对应的变量。

宏程序与 PMC 程序直接相关的是宏程序中保存和恢复系统模态、判断跳转条件等内容，因此需要用到系统变量。

1）输入信号变量（G54.1~G54.3）。

宏程序中与 PMC 程序输入信号对应变量说明如表 4-11 所示。

表 4-11 宏程序中与 PMC 程序输入信号对应变量

信号/功能	地址	对应变量
主轴刀号与指令刀号一致	G54.1	#1001
判定主轴是否有刀	G54.3	#1003
指令是否为 T0	G54.2	#1002

表 4-11 中的 G54.1、G54.2、G54.3 是“0”还是“1”由 PMC 程序的运行结果决定。

2）系统模态信息变量（#4003、#4006）。

由于宏程序中使用增量编程，在执行宏程序前必须保护主程序的系统模态，在执行完宏程序后必须恢复主程序的系统模态，因此需要用到系统模态信息变量。主程序中的系统模态主要有公/英制编程和绝对值/增量值编程模态，对应的系统变量为“#4003”（对应系统当前所用的编程坐标模态值 G90（绝对值）/G91(增量值)）和“#4006”（对应系统当前所用的编程单位系统 G20（英制）/G21（公制））。

3. 斗笠式刀库 PMC 程序

（1）主轴定向。

在换刀过程中主轴定向是主轴需要停在某一固定角度，否则刀具与主轴前端定位块会发生碰撞。主轴定向功能主要由 PMC 程序与系统参数设置实现。与主轴定向相关的系统参数及其设置如表 4-12 所示；主轴定向控制功能的资源及其分配如表 4-13 所示；主轴定向的 PMC 参考程序如图 4-20 所示。

表 4-12 与主轴定向相关的系统参数及其设置

参数号	意义	设置值	说明
4077	设置主轴准停位置数据	主轴准停位置数据	保存主轴准停的位置数据
4038	设置主轴定向速度	主轴定向速度	保存主轴定向速度

表 4-13 主轴定向控制功能的资源及其分配

资源类型	信号/功能	地址	说明
输入	主轴定向	X0005.4	操作面板上主轴定向按键
	空气压力	X0008.3	空气压力检测信号
输出	主轴定向灯	Y0005.4	操作面板上主轴定向按键指示灯
内部信号	复位信号	F0001.1	系统处于复位状态（RST 信号）
	手轮选择信号	F0003.1	手轮进给选择确认信号
	JOG 选择信号	F0003.2	手动连续进给选择确认信号
	MDI 选择信号	F0003.3	手动数据输入选择确认信号
	DNC 选择信号	F0003.4	DNC 运行选择确认信号

续表

资源类型	信号/功能	地址	说明
内部信号	自动选择信号	F0003. 5	存储器运行选择确认信号
	主轴定向结束信号	F0045. 7	主轴定向结束
	主轴定向信号	G0070. 6	PMC 向 CNC 发出的控制请求信号

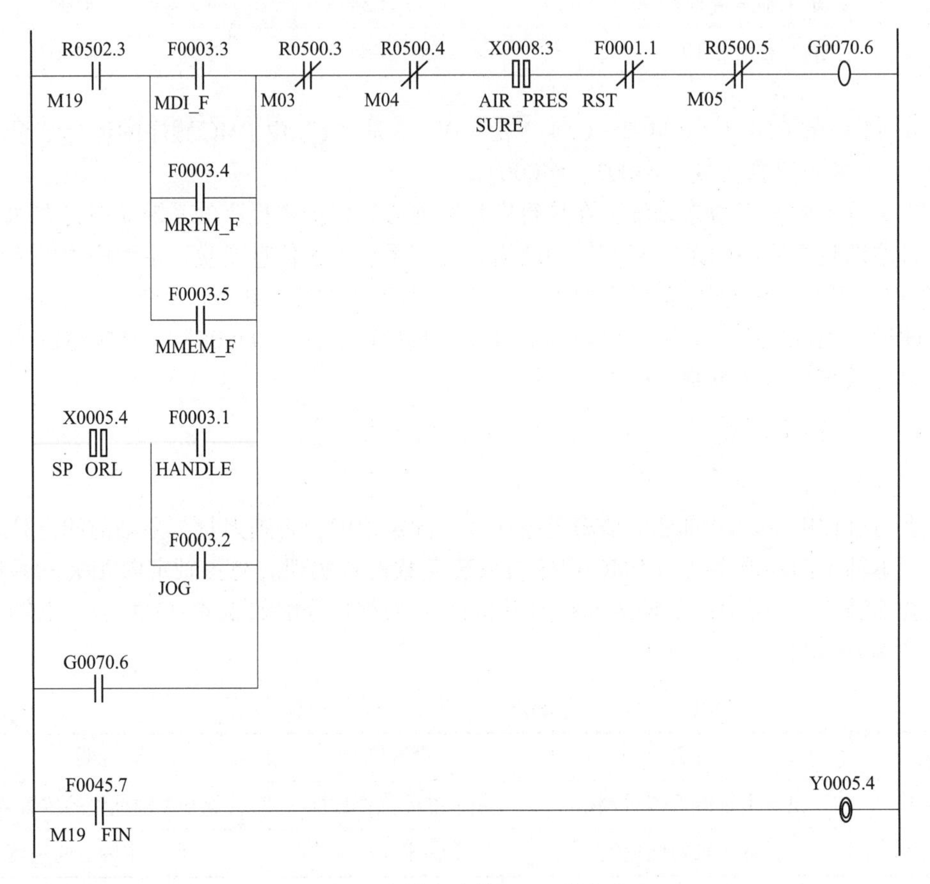

图 4-20　主轴定向的 PMC 参考程序

（2）T 指令等于主轴刀号判定。

在换刀过程中要进行 T 指令是否等于主轴刀号的判定，实现该功能的资源及其分配如表 4-14 所示。T 指令等于主轴刀号判定的 PMC 参考程序如图 4-21 所示。

表 4-14　T 指令等于主轴刀号判定的资源及其分配

资源类型	信号/功能	地址	说明
内部信号	常“1”信号	R9091. 1	PMC 常“1”信号
	复位信号	F0001. 1	系统处于复位状态（RST 信号）
	T 指令选通信号	F0007. 3	CNC 输出给 PMC 的 T 指令选通信号
	T 指令译码寄存器	F0026	CNC 输出给 PMC 的 T 指令译码

续表

资源类型	信号/功能	地址	说明
内部信号	T 指令译码结果信号	R0026	保存 T 指令译码数据
	刀库当前刀号译码结果信号	R0010	保存刀库当前刀号译码数据
	刀库当前刀号寄存器	C0010	可通过 MDI 修改、查询
	T 指令等于主轴刀号信号	G0054. 1	对应变量#1001

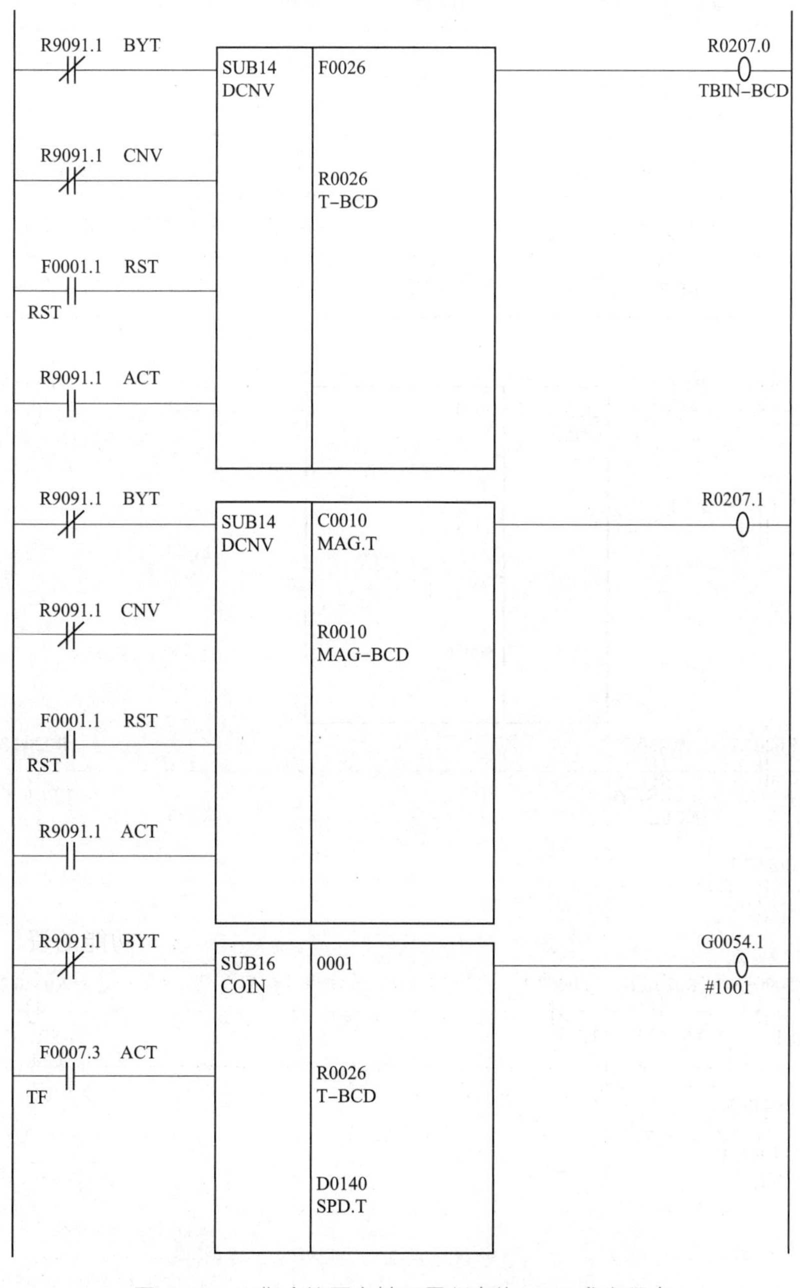

图 4-21　T 指令等于主轴刀号判定的 PMC 参考程序

（3）T 指令取值范围判定。

本项目使用的斗笠式刀库的刀库容量为 12 把刀具。因此，PMC 程序中要有 T 指令取值范围判定，输入刀号不能超过 T12。实现 T 指令取值范围判定的资源及其分配如表 4-15 所示，其 PMC 参考程序如图 4-22 所示。

表 4-15　T 指令取值范围判定的资源及其分配

资源类型	信号/功能	地址	说明
内部信号	常“1”信号	R9091. 1	PMC 常“1”信号
	T 指令错误信号	R0207. 3	R0207. 3 为“1”表示 T 指令超出取值范围，即 $T>12$
	T 指令错误取消信号	R0430. 3	R0430. 3 为“1”表示取消 T 指令错误信号
	复位信号	F0001. 1	系统处于复位状态（RST 信号）
	T 指令选通信号	F0007. 3	NC 输出给 PMC 的 T 指令选通信号

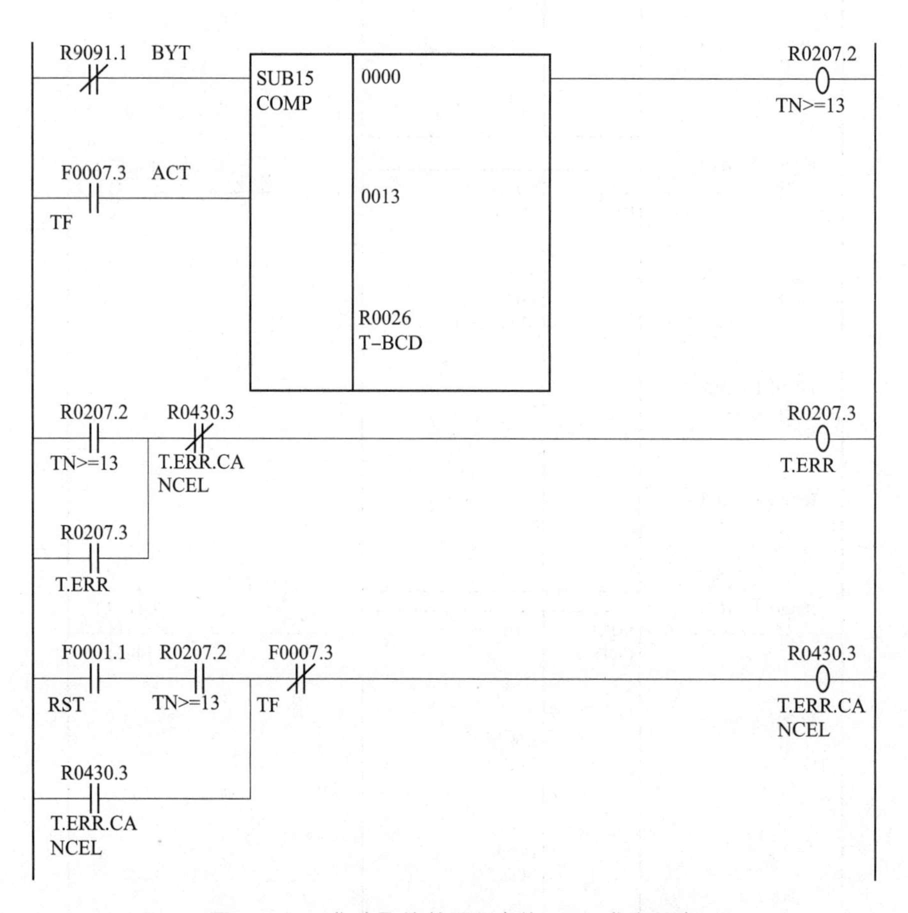

图 4-22　T 指令取值范围判定的 PMC 参考程序

(4) 刀号的判定。

本项目的斗笠式刀库要求进行主轴上是否有刀具的判定、T 指令是否为零的判定及刀库当前刀具换刀点位置的判定。刀号判定的资源及其分配如表 4-16 所示，其 PMC 参考程序如图 4-23 所示。

表 4-16 刀号判定的资源及其分配

资源类型	信号/功能	地址	说明
输出	刀库反转驱动信号	Y0002.6	控制刀库反转
内部信号	常“1”信号	R9091.1	PMC 常“1”信号
	T 指令译码结果信号	R0026	保存 T 指令译码数据
	刀库计数传感器计数下降沿信号	R1000.0	R1000.0 为“1”表示刀库计数传感器计数 1 次
	刀库向后，远离主轴译码信号	R0200.6	M86 信号
	T 指令选通保持信号	R0205.1	R0205.1 为“1”表示 T 指令选通保持
	T 指令选通信号	F0007.3	CNC 输出给 PMC 的 T 指令选通信号
	主轴刀号寄存器（数据表）	D0140	保存主轴当前刀号
	T 指令等于零信号	G0054.2	对应变量#1002
	主轴上无刀信号	G0054.3	对应变量#1003

(5) 刀库按主轴刀号选刀。

当执行换刀指令时，若主轴上有刀，需先将主轴上的刀具返回刀库，刀库再按主轴刀号选刀。刀库按主轴刀号选刀的资源及其分配如表 4-17 所示，其 PMC 参考程序如图 4-24 所示。

表 4-17 刀库按主轴刀号选刀的资源及其分配

资源类型	信号/功能	地址	说明
内部信号	常“1”信号	R9091.1	PMC 常“1”信号
	刀库当前刀号译码结果信号	R0010	保存刀库当前刀号译码数据
	主轴刀号与刀库当前刀号一致性判断译码信号	R0200.1	M81 信号
	刀库当前刀号与主轴刀号判定旋转控制信号	R0205.4	R205.4 为“0”表示刀库需要正转；为“1”表示刀库需要反转
	主轴刀号寄存器（数据表）	D0140	保存主轴当前刀号

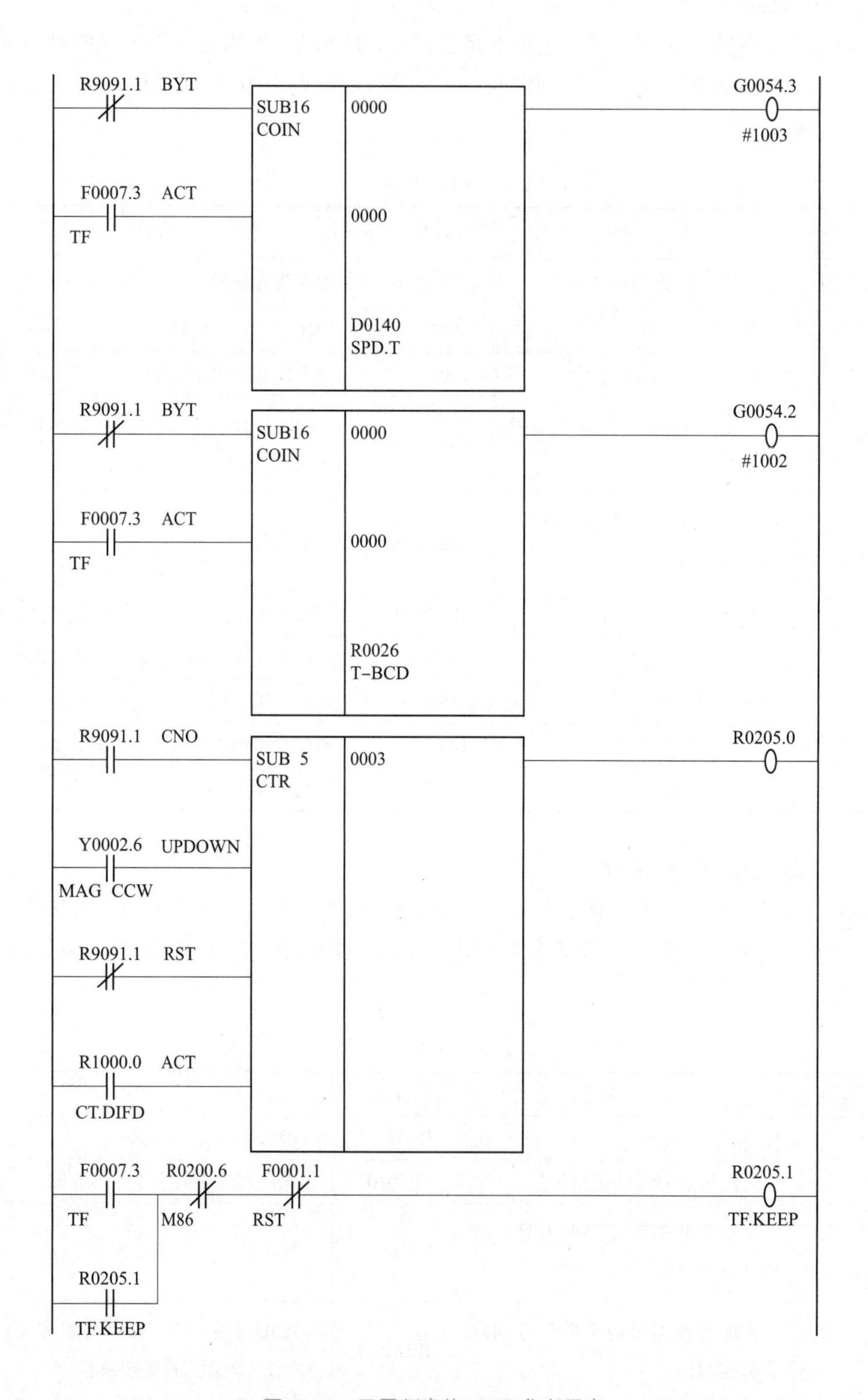

图 4–23 刀号判定的 PMC 参考程序

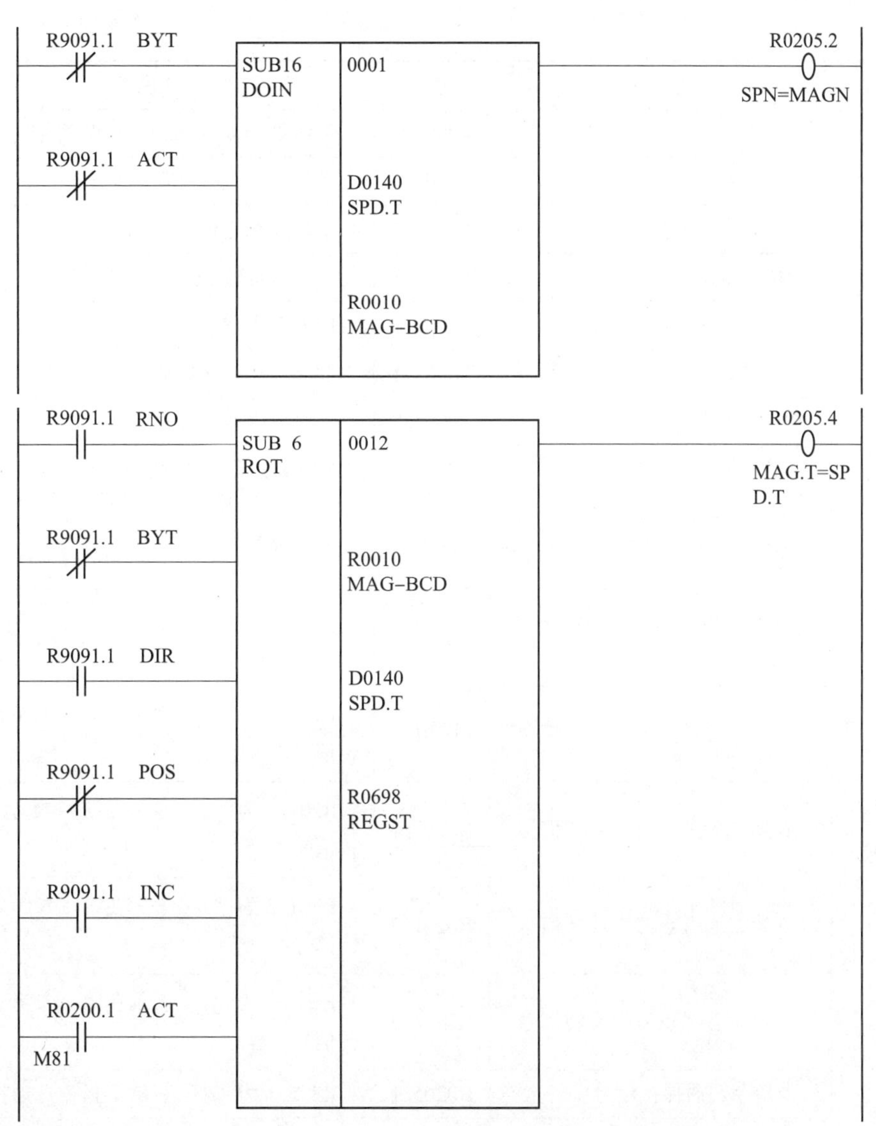

图 4-24　刀库按主轴刀号选刀的 PMC 参考程序

（6）手动刀库正反转。

按下机床操作面板刀库正反转按键，刀库实现正反转，转到下一刀位。手动刀库正反转的资源及其分配如表 4-18 所示，其 PMC 参考程序如图 4-25 所示。

表 4-18　手动刀库正反转的资源及其分配

资源类型	信号/功能	地址	说明
输入	急停信号	X0008. 4	机床紧急停止
输出	刀库反转驱动信号	Y0002. 6	控制刀库反转
	刀库正转驱动信号	Y0002. 7	控制刀库正转
	刀库正转指示灯	Y0000. 7	控制刀库正转指示灯
	刀库反转指示灯	Y0001. 7	控制刀库反转指示灯

续表

资源类型	信号/功能	地址	说明
内部信号	复位信号	F0001.1	系统处于复位状态（RST 信号）
	手轮选择信号	F0003.1	手轮进给选择确认信号
	JOG 选择信号	F0003.2	手动连续进给选择确认信号
	MDI 选择信号	F0003.3	手动数据输入选择确认信号
	DNC 选择信号	F0003.4	DNC 运行选择确认信号
	自动选择信号	F0003.5	存储器运行选择确认信号
	第三轴到达第一参考点	F0094.2	*Z* 轴到达第一参考点确认信号
	第三轴到达第二参考点	F0096.2	*Z* 轴到达第二参考点确认信号
	主轴刀号与刀库当前刀号一致性判断译码信号	R0200.1	M81 信号
	刀库旋转至加工程序指定的刀位	R0200.3	M83 信号
	在第二参考点刀库正转信号	R0201.0	R201.0 为“1”表示在第二参考点刀库正转信号
	在第一参考点刀库正转信号	R0201.1	R201.1 为“1”表示在第一参考点刀库正转信号
	在第二参考点刀库反转信号	R0201.2	R201.2 为“1”表示在第二参考点刀库反转信号
	在第一参考点刀库反转信号	R0201.3	R201.3 为“1”表示在第一参考点刀库反转信号
	T 指令选通保持信号	R0205.1	R205.1 为“1”表示 T 指令选通保持
	主轴刀号与刀库当前刀号一致性判定信号	R0205.2	R205.2 为“1”表示主轴刀号等于刀库当前刀号
	刀库当前刀号与主轴刀号判定旋转控制信号	R0205.4	R205.4 为“0”表示刀库需要正转；为“1”表示刀库需要反转
	刀库当前刀号与 T 指令判定旋转控制信号	R0206.2	R206.2 为“0”表示刀库需要正转；为“1”表示刀库需要反转
	手动刀库正转信号	R0206.3	R206.3 为“1”表示手动刀库正转信号
	手动刀库反转信号	R0206.4	R206.4 为“1”表示手动刀库反转信号
	刀库当前刀号与 T 指令一致性判定信号	R0208.2	R208.2 为“1”表示刀库当前刀号等于 T 指令

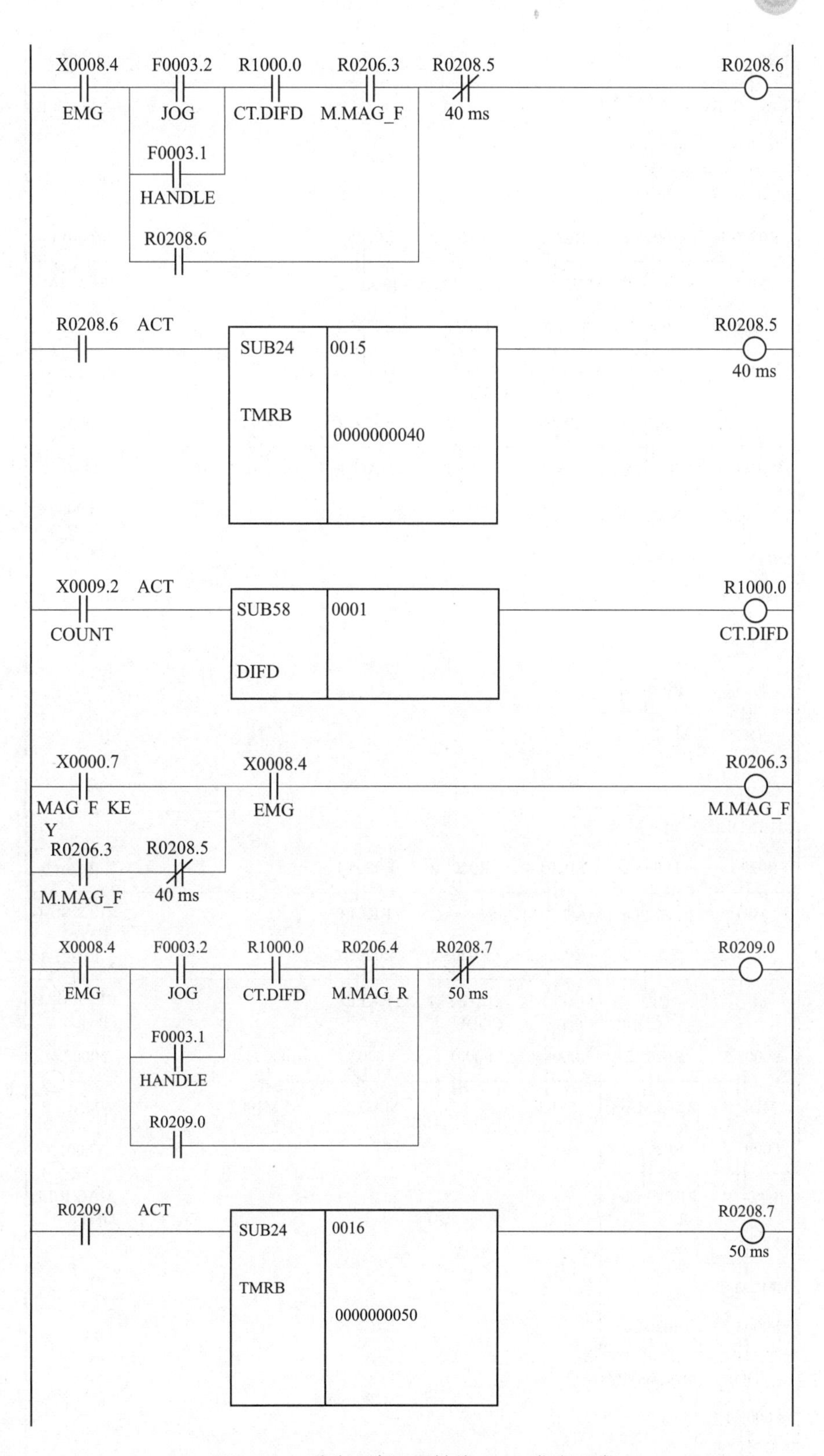

图 4–25　手动刀库正反转的 PMC 参考程序

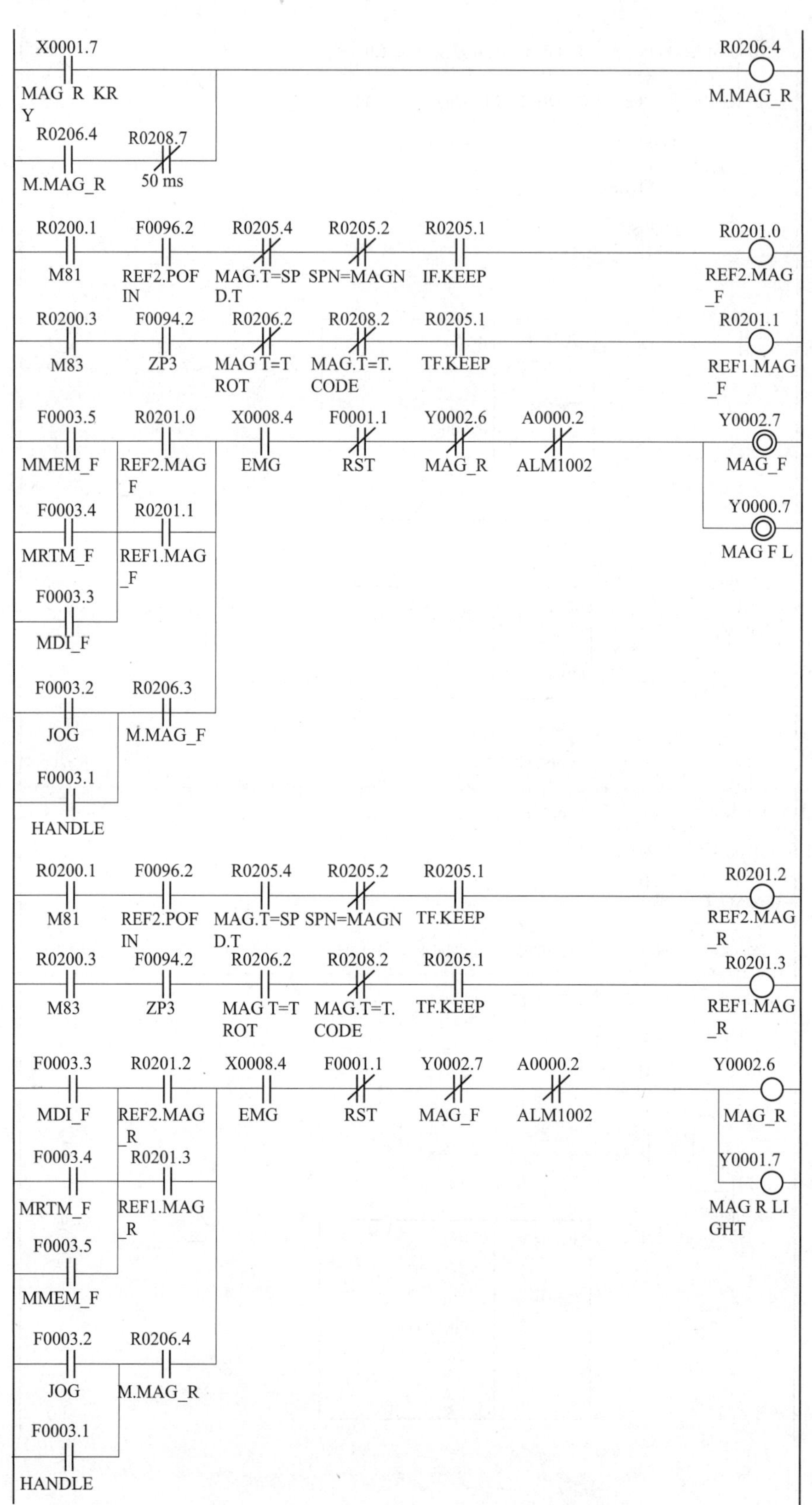

图 4-25　手动刀库正反转的 PMC 参考程序（续）

（7）主轴松刀。

当主轴松刀信号发出后，打刀缸下行推动拉刀机构下行完成松刀动作。主轴松刀的资源及其分配如表 4-19 所示，其 PMC 参考程序如图 4-26 所示。

表 4-19　主轴松刀的资源及其分配

资源类型	信号/功能	地址	说明
输入	手动松刀按钮	X0008. 0	手动松刀信号
	急停信号	X0008. 4	机床紧急停止
输出	打刀信号	Y0003. 0	控制打刀缸下行
	主轴松刀灯	Y0007. 0	手动松刀按钮指示灯信号
内部信号	复位信号	F0001. 1	系统处于复位状态（RST 信号）
	手轮选择信号	F0003. 1	手轮进给选择确认信号
	JOG 选择信号	F0003. 2	手动连续进给选择确认信号
	MDI 选择信号	F0003. 3	手动数据输入选择确认信号
	DNC 选择信号	F0003. 4	DNC 运行选择确认信号
	自动选择信号	F0003. 5	存储器运行选择确认信号
	第三轴到达第二参考点	F0096. 2	*Z* 轴到达第二参考点确认信号
	松刀译码信号	R0200. 2	M82 信号
	刀具夹紧译码信号	R0200. 4	M84 信号

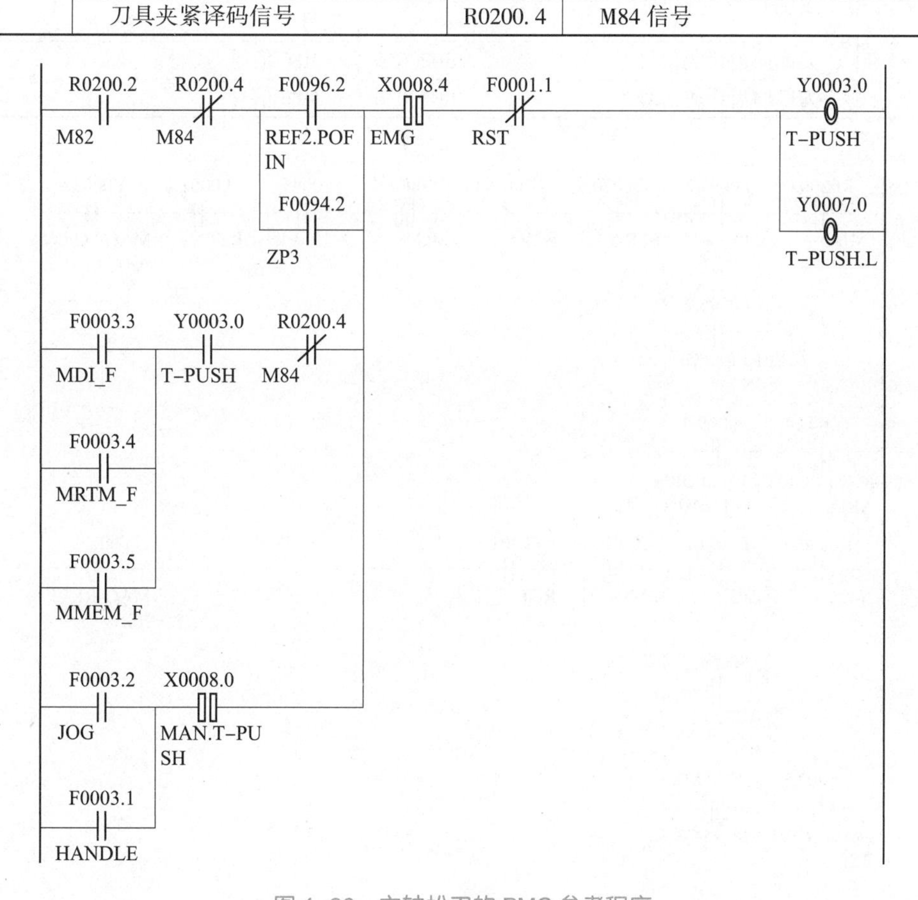

图 4-26　主轴松刀的 PMC 参考程序

(8) 刀库前向后运动。

当机床发出指令换刀时，刀库向前运动接近主轴，换刀完成后刀库向后运动远离主轴。刀库前向后运动信号的资源及其分配如表 4-20 所示，其 PMC 参考程序如图 4-27 所示。

表 4-20 刀库向前、向后运动信号的资源及其分配

资源类型	信号/功能	地址	说明
输入	急停信号	X0008.4	机床紧急停止
	刀库前位信号	X0009.0	刀库前位检测信号
	刀库后位信号	X0009.1	刀库后位检测信号
输出	刀库向前信号	Y0003.4	控制刀库向前
	刀库向后信号	Y0003.5	控制刀库向后
内部信号	复位信号	F0001.1	系统处于复位状态（RST 信号）
	第三轴到达第一参考点	F0094.2	*Z* 轴到达第一参考点确认信号
	第三轴到达第二参考点	F0096.2	*Z* 轴到达第二参考点确认信号
内部信号	主轴定向结束信号	F0045.7	主轴定向结束
	主轴 CW（顺时针）旋转信号	G0070.4	PMC 向 CNC 发出的主轴 CW 旋转信号
	主轴 CCW（逆时针）旋转信号	G0070.5	PMC 向 CNC 发出的主轴 CCW 旋转信号
	刀库向前译码信号	R0200.0	M80 信号
	刀库向后译码信号	R0200.6	M86 信号

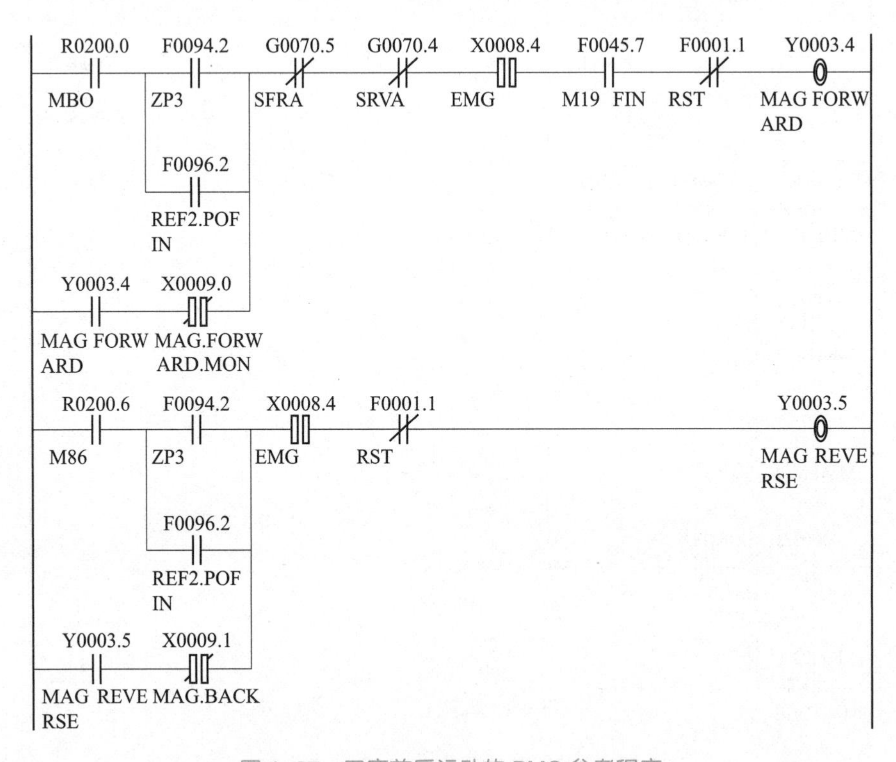

图 4-27 刀库前后运动的 PMC 参考程序

（9）刀库按 T 指令转动。

当执行换刀指令时，主轴上没刀，刀库按 T 指令转动到指定位置，准备换刀。按 T 指令刀库转动的资源及其分配如表 4-21 所示，其 PMC 参考程序如图 4-28 所示。

表 4-21　刀库按 T 指令转动的资源及其分配

资源类型	信号/功能	地址	说明
内部信号	刀库当前刀号寄存器	C0010	可通过 MDI 修改、查询
	刀库当前刀号译码结果信号	R0010	保存刀库当前刀号译码数据
	T 指令译码结果信号	R0026	保存 T 指令译码数据
内部信号	常“1”信号	R9091.1	PMC 常“1”信号
	刀库旋转至加工程序指定的刀位	R0200.3	M83 信号
	刀库当前刀号与 T 指令一致性判定信号	R0208.2	R0208.2 为“1”表示刀库当前刀号等于 T 指令
	刀库当前刀号与 T 指令判定旋转控制信号	R0206.2	R206.2 为“0”表示刀库需要正转；为“1”表示刀库需要反转

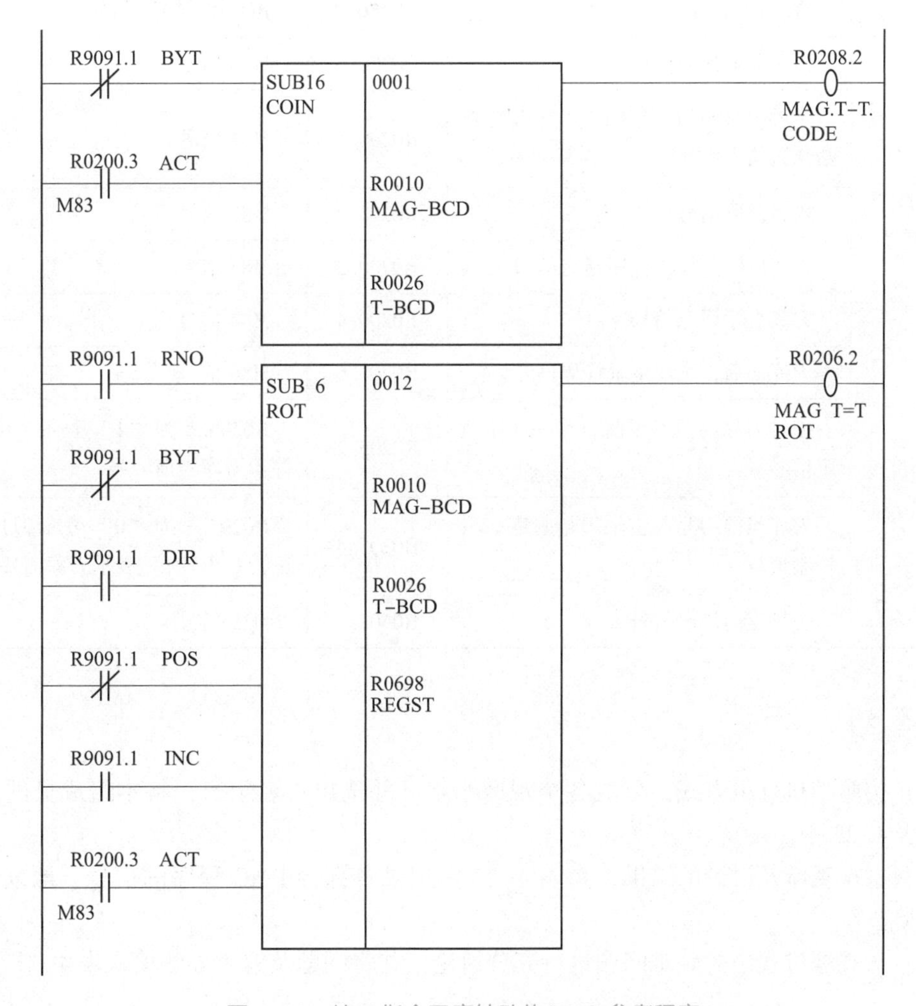

图 4-28　按 T 指令刀库转动的 PMC 参考程序

（10）M、T 指令完成信号。

当换刀完成后，M、T 指令发出完成信号。M、T 指令完成信号的资源及其分配如表 4-22 所示，其 PMC 参考程序如图 4-29 所示。

表 4-22　M、T 指令完成信号的资源及其分配

资源类型	信号/功能	地址	说明
输入	刀具锁紧信号	X0008.2	刀具锁紧检测信号
	刀库前位信号	X0009.0	刀库前位检测信号
	刀库后位信号	X0009.1	刀库后位检测信号
内部信号	辅助功能指令信号	F0010	M 指令译码信号
	主轴定向信号	G0070.6	PMC 向 CNC 发出的控制请求信号
	常“1”信号	R09091.1	PMC 常“1”信号
	刀库向前译码信号	R0200.0	M80 信号
	主轴刀号与刀库当前刀号一致性判断译码信号	R0200.1	M81 信号
	松刀译码信号	R0200.2	M82 信号
	刀库旋转至加工程序指定的刀位	R0200.3	M83 信号
	刀具夹紧译码信号	R0200.4	M84 信号
	刀库向后，远离主轴译码信号	R0200.6	M86 信号
	主轴刀号与刀库当前刀号一致性判定信号	R0205.2	R0205.2 为“1”表示主轴刀号等于刀库当前刀号
	刀库当前刀号与主轴刀号判定旋转控制信号	R0205.4	R0205.4 为“0”表示刀库需要正转；为“1”表示刀库需要反转
	主轴定向译码信号	R0502.3	M19 信号

【任务实施】

（1）按照随机分组方式，分成各个观摩小组，并选出一名组长，将本组成员的个人信息填入表 4-23 中。

（2）根据观摩及讨论的结果，在表 4-23 中记录各部分 PMC 程序的信号、地址、说明及资源类型等。

（3）对观摩过程中的疑难问题进行充分讨论，并将问题及解决办法填入表 4-23 中。

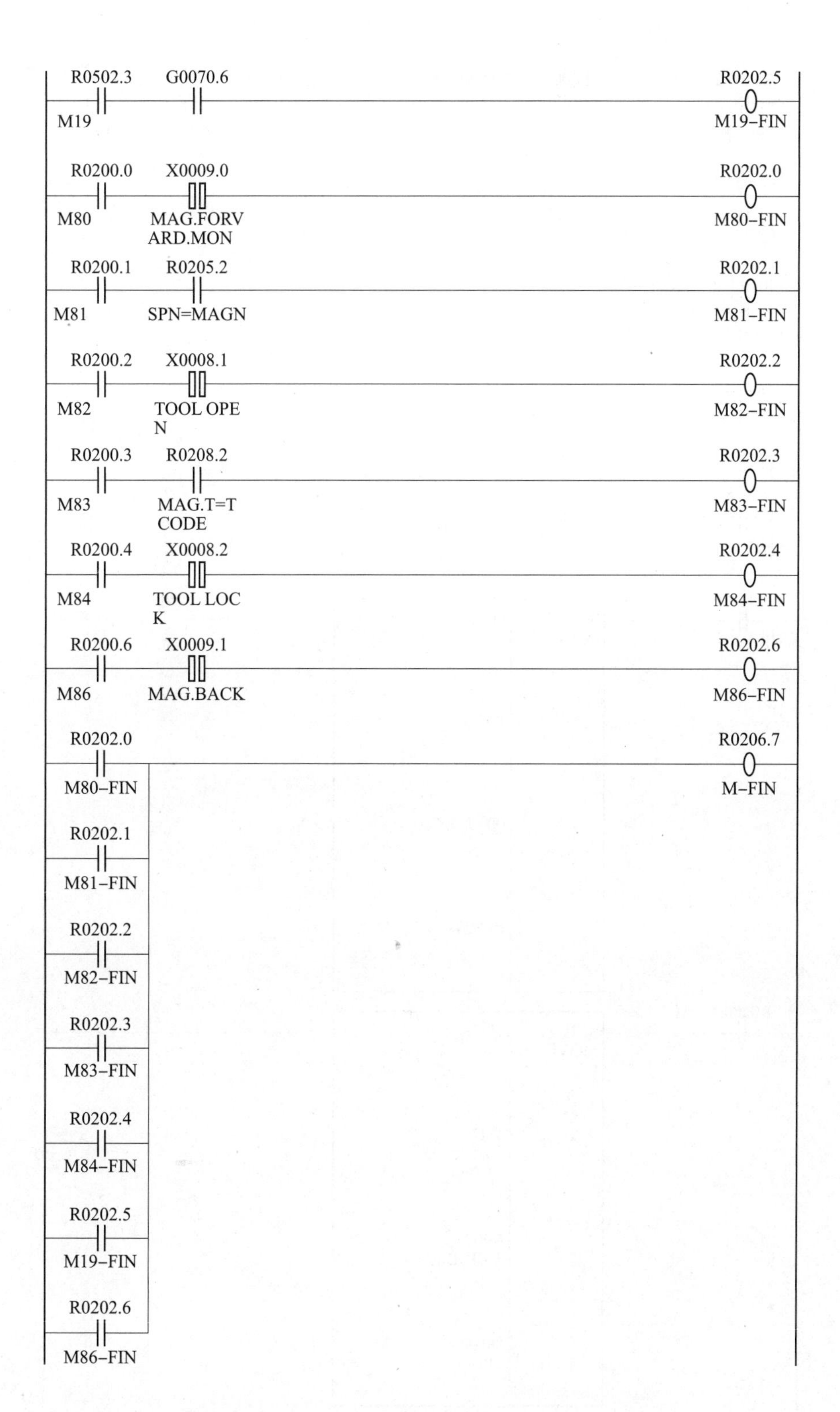

图 4-29　M、T 指令完成信号的 PMC 参考程序

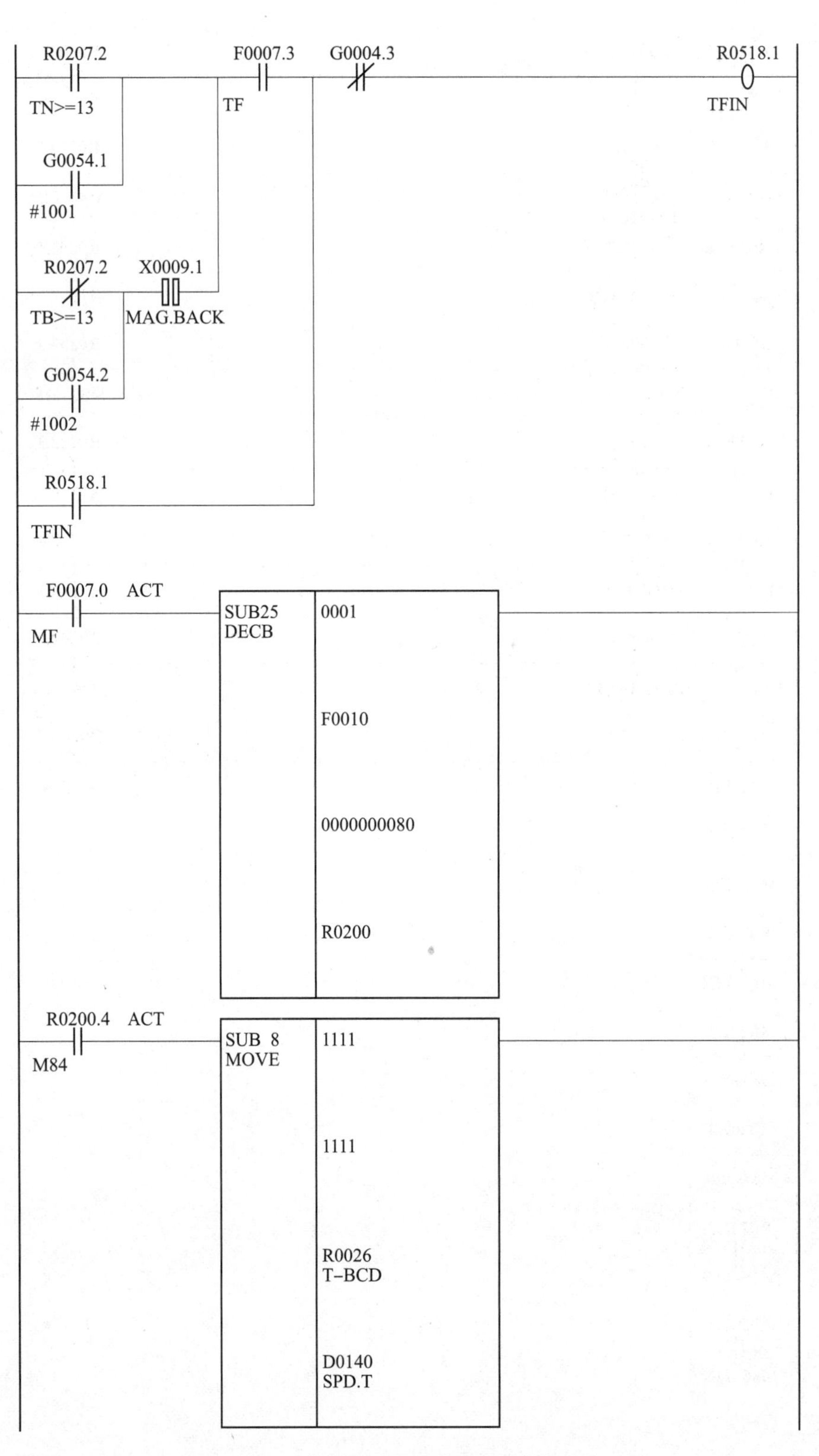

图 4-29　M、T 指令完成信号的 PMC 参考程序（续）

表 4-23　数控系统 PMC 程序任务工单

<table>
<tr><td colspan="6">任务工单</td></tr>
<tr><td colspan="2">姓名：</td><td colspan="2">班级：</td><td colspan="2">学号：</td></tr>
<tr><td colspan="6">所需材料、工量具、设备列表</td></tr>
<tr><td>序号</td><td colspan="2">名称</td><td>数量</td><td>型号或规格</td><td>备注</td></tr>
<tr><td></td><td colspan="2"></td><td></td><td></td><td></td></tr>
<tr><td></td><td colspan="2"></td><td></td><td></td><td></td></tr>
<tr><td></td><td colspan="2"></td><td></td><td></td><td></td></tr>
<tr><td colspan="6">工作过程（观摩）</td></tr>
<tr><td>序号</td><td>信号</td><td>地址</td><td colspan="2">说明</td><td>资源类型</td></tr>
<tr><td></td><td></td><td></td><td colspan="2"></td><td></td></tr>
<tr><td></td><td></td><td></td><td colspan="2"></td><td></td></tr>
<tr><td></td><td></td><td></td><td colspan="2"></td><td></td></tr>
<tr><td></td><td></td><td></td><td colspan="2"></td><td></td></tr>
<tr><td colspan="3">出现问题</td><td colspan="3">解决办法</td></tr>
<tr><td colspan="3"></td><td colspan="3"></td></tr>
</table>

【任务评价】

序号	评价内容	分值	得分
1	能够熟练说出 PMC 的原理及 I/O 地址分配	25	
2	能够熟练掌握 PMC 程序结构	25	
3	能够掌握斗笠式刀库换刀功能 PMC 程序的编写	25	
4	具有有效沟通、表达及团结协作的职业精神	25	

任务四　系统数据传输与备份

【学习目标】

（1）掌握系统数据备份与恢复步骤。

（2）掌握 FANUC LADDER-Ⅲ的 PMC 程序传输步骤。

（3）具有 FANUC 数控系统数据传输与备份的能力。

（4）掌握工程工作方法，培养严谨的工作作风，遵守 5S 管理制度。

【任务描述】

以小组为单位，观摩 FANUC 0i-F 数控系统的系统数据传输与备份功能，并将各种数据传输方法的名称、步骤、主要功能及适用场合等填写到任务工单中。要求观摩过程中务必注意安全，遵守实训场所相关操作规范。

【知识链接】

一、引导界面数据备份与恢复

1. 静态存储器数据备份与恢复

（1）静态存储器（SRAM）的作用。

存储加工程序、刀具补偿量、用户宏变量、CNC 参数、PMC 参数、丝杠螺距误差补偿量等这类数据保存在数控系统主板上的 SRAM 中。当机床断电数据不会丢失，由系统电池供电保持。

（2）SRAM 数据备份或恢复操作步骤。

1）数控机床在关机状态下插入存储卡。

2）在数控系统通电的同时点住显示器右下方的两个软键，进入数控系统引导界面，如图 4-30 所示。

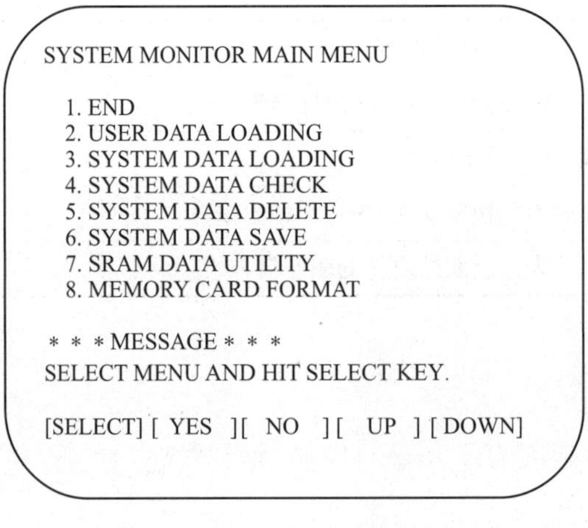

SYSTEM MONITOR MAIN MENU

1. END
2. USER DATA LOADING
3. SYSTEM DATA LOADING
4. SYSTEM DATA CHECK
5. SYSTEM DATA DELETE
6. SYSTEM DATA SAVE
7. SRAM DATA UTILITY
8. MEMORY CARD FORMAT

* * * MESSAGE * * *
SELECT MENU AND HIT SELECT KEY.

[SELECT] [YES][NO][UP] [DOWN]

图 4-30　数控系统引导界面

3）单击【UP】或【DOWN】，把光标移动到“7. SRAM DATA BACKUP”，单击【SELECT】，显示“SRAM DATA BACKUP”界面，如图 4-31 所示。

4）单击【UP】或【DOWN】，进行功能的选择。

SRAM DATA BACKUP

1. SRAM BACKUP　　(CNC→MEMORY CARD)
2. RESTORE SRAM　　(MEMORY CARD→CNC)
3. AUTO BKUP RESTORE　　(F-ROM→CNC)
4. END

* * * MESSAGE * * *
SELECT MENU AND HIT SELECT KEY.

[SELECT] [YES] [NO] [UP] [DOWN]

图 4-31　SRAM DATA BACKUP 界面

使用存储卡备份数据：SRAM BACKUP。

向 SRAM 恢复数据：RESTORE SRAM。

自动备份数据的恢复：AUTO BKUP RESTORE。

（3）完成 SRAM 数据备份或恢复后，将数控机床断电，拔下存储卡。备份数据 SRAM_BAK. 001 文件为机器指令且为打包形式，不能在计算机上打开。

2. PMC 程序备份与恢复

（1）在维修的时候需要修改机床的 PMC，但是在复杂的维修现场有时不能保证记住所修改的内容，因为不熟悉操作以及误改等原因，导致修改错误。为了能够恢复维修前的状态，最好在维修前对机床的参数进行备份。本部分主要介绍关于 PMC 程序备份与恢复的操作步骤。

（2）PMC 程序备份：进入数控系统引导导界面，单击【UP】或【DOWN】，把光标移动到“6. SYSTEM DATA SAVE”，单击【SELECT】，显示 SYSTEM DATA SAVE 界面，如图 4-32 所示。

（3）PMC 程序恢复界面如图 4-33 所示。

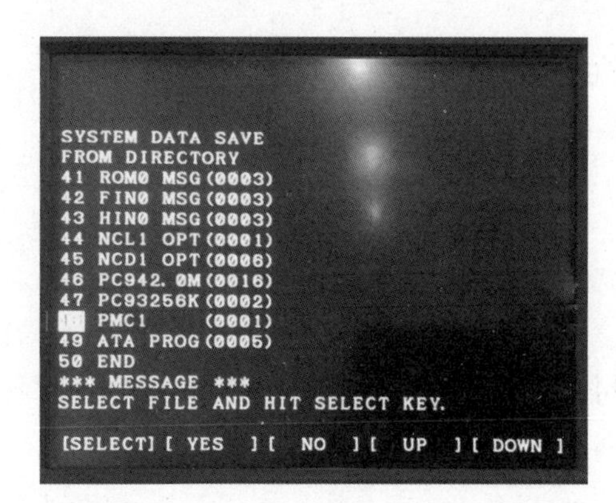

图 4-32　SYSTEM DATA SAVE 界面

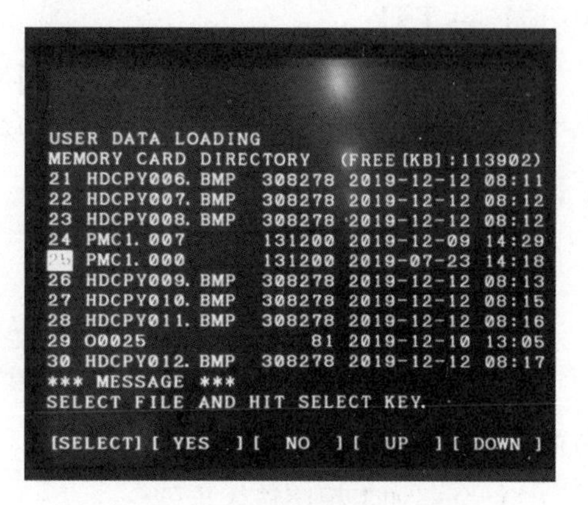

图 4-33　PMC 程序恢复界面

二、工作界面数据备份

1. 系统参数备份

选择 MDI 模式：

（1）打开系统参数写保护；

（2）单击【SYSTEM】—【参数】，将 20 号参数设为 4；

（3）选择 EDIT 模式，依次单击【SYSTEM】—【参数】—【操作】—【>】—【FOUTPT】—【执行】—【覆盖】，等待完成后在 PCMCIA 卡中自动生成“CNC - PARA. TXT”文件。

2. PMC 参数备份

选择 EDIT 模式。依次单击【SYSTEM】—【>】3 次—【PMCMNT】—【I/O】—光标【存储卡、写、参数、文件名】。依次单击【操作】—【文件名】，自动生成“PMC1_ PRM. 000”—【执行】。

3. 加工程序备份（从机床到存储卡）

选择 EDIT 模式。依次单击【PROG】—【目录】—【操作】—【>】—【设备】—【CNCMEM】。

方法一：输入全部文件名，如 O1000，然后单击【FOUTPT】—【执行】。

方法二：输入要备份的文件名，如 O1000 然后单击【P 设定】。将输入存到卡内的文件名，如 O123，然后单击【F 设定】—【执行】，将文件改名后保存到存储卡，如将机床 O1000 存到卡内的文件名改为 O123。

4. 加工程序输入

加工程序输入过程：将机床操作面板上钥匙开关由“0”旋至“1”。

选择 EDIT 模式。依次单击【PROG】—【目录】—【操作】—【>】—【设备】—【M-卡】—【>】。

方法一：“输入程序名”—【F INPT】—【执行】。

方法二：单击【F INPT】—输入卡中程序名—【F 设定】—“输入机床侧程序名”—【P 设定】—【执行】。

5. 全部加工程序输入/输出

（1）全部加工程序输出步骤。

选择 EDIT 模式。依次单击【PROG】—【目录】—【操作】—“输入 PROGRAM. ALL”—【FOUTPT】—【执行】。

（2）全部加工程序输入步骤。

选择 EDI 模式依次单击【PROG】—【目录】—【操作】—”输入 O9999”—【F INPT】—【执行】。

6. 所有 I/O 界面数据输入/输出

选择 EDIT 模式。依次单击【SYSTEM】—【>】3 次—【所有 I/O】，选择进入如图 4-34 所示所有 I/O 界面，可输入输出加工程序、参数、刀偏、宏程序、螺补、工件坐标系等数据。

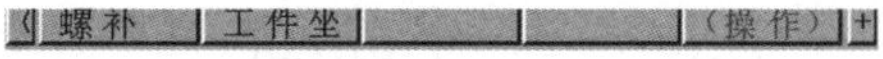

螺补
工件坐
（操作）
+

图 4-34　所有 I/O 界面

（1）螺补数据输出。

选择【螺补】—【操作】进入参数输入/输出界面。

选择【FOUTPT】，如果直接选择执行则输出的文件名为系统默认名称 PITCH. TXT，如图 4-35 所示。

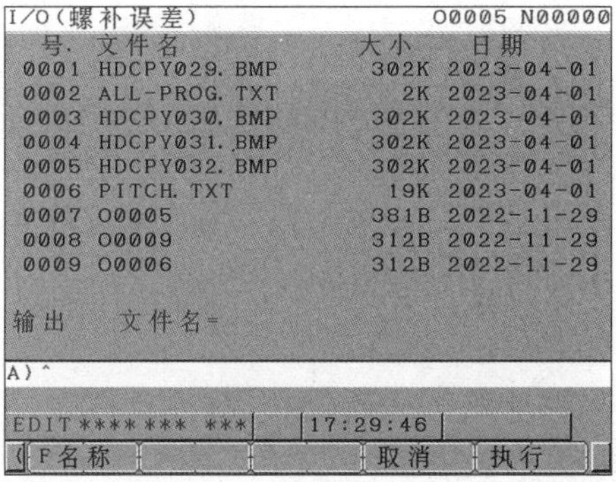

I/O（螺补误差）　O0005 N00000
号.　文件名　大小　日期
0001 HDCPY029. BMP　302K 2023-04-01
0002 ALL-PROG. TXT　2K 2023-04-01
0003 HDCPY030. BMP　302K 2023-04-01
0004 HDCPY031. BMP　302K 2023-04-01
0005 HDCPY032. BMP　302K 2023-04-01
0006 PITCH. TXT　19K 2023-04-01
0007 O0005　381B 2022-11-29
0008 O0009　312B 2022-11-29
0009 O0006　312B 2022-11-29
输出　文件名=
A)^
EDIT **** *** ***　17:29:46
F名称　取消　执行

图 4-35　选择【FOUTPT】

（2）F 输入方式螺补数据输入。

选择 EDIT 模式。依次单击【SYSTEM】—【>】—【所有 I/O】，进入所有 I/O 界面。选择【螺补】—【操作】—【F INPT】—“输入文件号：6”—【F 设定】—【执行】，如图 4-36 所示。

（3）N 输入方式螺补数据输入。

选择 EDIT 模式。依次单击【SYSTEM】—【>】2—【所有 I/O】，进入所有 I/O 界面。选择【参数】—【操作】—【N INPT】—“输入文件名 PITCH. TXT”—【F 名称】—【执行】，如图 4-37 所示。

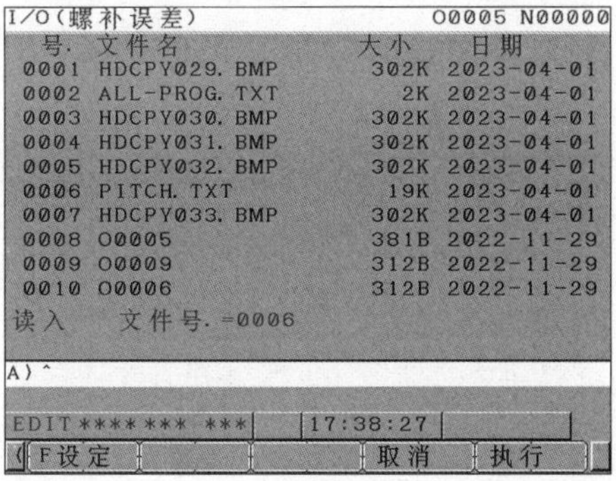

I/O（螺补误差）　O0005 N00000
号.　文件名　大小　日期
0001 HDCPY029. BMP　302K 2023-04-01
0002 ALL-PROG. TXT　2K 2023-04-01
0003 HDCPY030. BMP　302K 2023-04-01
0004 HDCPY031. BMP　302K 2023-04-01
0005 HDCPY032. BMP　302K 2023-04-01
0006 PITCH. TXT　19K 2023-04-01
0007 HDCPY033. BMP　302K 2023-04-01
0008 O0005　381B 2022-11-29
0009 O0009　312B 2022-11-29
0010 O0006　312B 2022-11-29
读入　文件号.=0006
A)^
EDIT **** *** ***　17:38:27
F设定　取消　执行

图 4-36　F 输入方式

I/O（螺补误差）　O0005 N00000
号.　文件名　大小　日期
0001 HDCPY029. BMP　302K 2023-04-01
0002 ALL-PROG. TXT　2K 2023-04-01
0003 HDCPY030. BMP　302K 2023-04-01
0004 HDCPY031. BMP　302K 2023-04-01
0005 HDCPY032. BMP　302K 2023-04-01
0006 PITCH. TXT　19K 2023-04-01
0007 HDCPY033. BMP　302K 2023-04-01
0008 O0005　381B 2022-11-29
0009 O0009　312B 2022-11-29
0010 O0006　312B 2022-11-29
读入　文件名=PITCH. TXT
A)_
EDIT **** *** ***　17:43:31
F名称　取消　执行

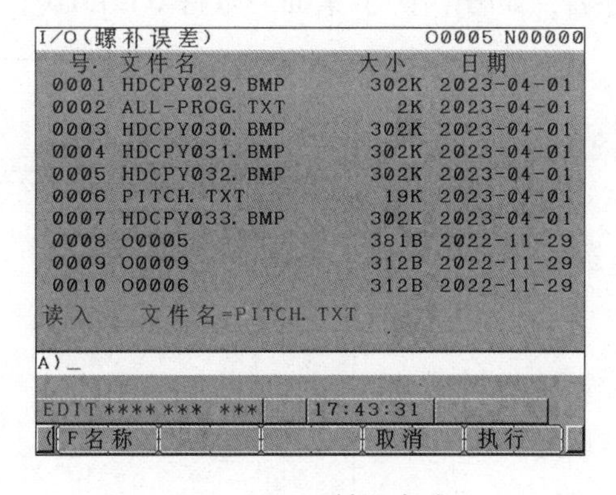

图 4-37　N 输入方式

三、FANUC LADDER-Ⅲ的 PMC 程序传输

1. 使用 LADDER-Ⅲ导入 PMC 程序

通过存储卡备份的 PMC 梯形图称为存储卡格式的 PMC 程序。由于其为机器语言格式，FAPT LADDER-Ⅲ不能直接识别和读取并进行修改和编辑，所以必须通过导入方式进行格式

转换。

导入程序的步骤如下。

（1）运行 FAPT LADDER-Ⅲ，新建一个类型与备份的 M-CARD 格式的 PMC 程序类型相同的空文件。

（2）选择【文件】菜单中的【导入】，系统会弹出【选择导入文件的类型】对话框，如图 4-38 所示。根据提示选择【存储卡格式文件】格式，然后单击【下一步】按钮。

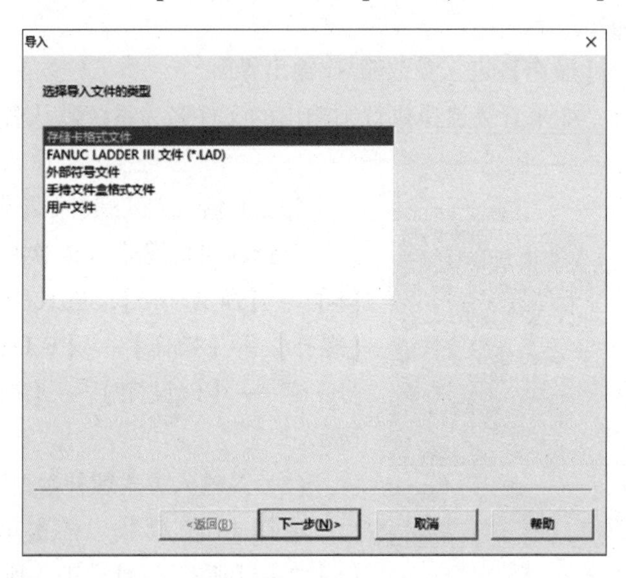

图 4-38　【选择导入文件的类型】对话框

（3）弹出【指定导入的文件名】对话框，如图 4-39 所示。在相应的路径选择要导入的文件名，如图中 D：\ 桌面 \ 教材 \ 1. LAD ，单击【完成】按钮，完成文件导入。

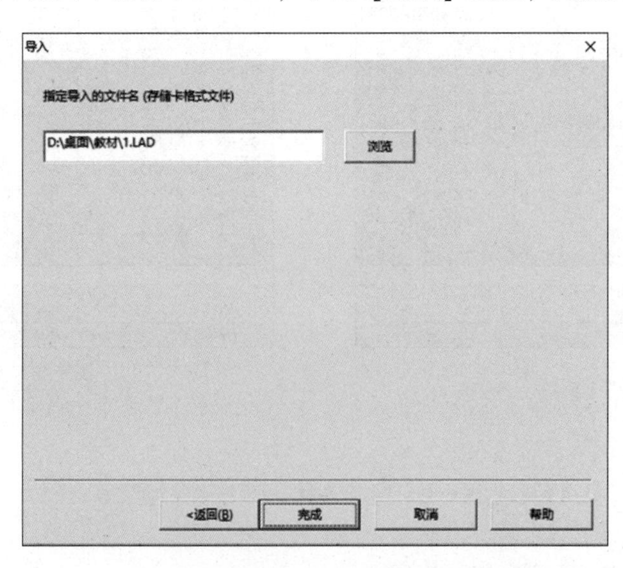

图 4-39　【指定导入的文件名】对话框

2. 使用 LADDER-Ⅲ软件导出 PMC 程序

同样，在计算机上编辑好的 PMC 程序不能直接存储到 M-CARD 上，也必须通过格式转换，然后才能装载到 CNC 系统中。导出程序的操作即完成此格式转换，其操作步骤如下。

（1）在 FAPT LADDER-Ⅲ中打开要转换的 PMC 程序。

（2）在【工具】菜单选择【编译】，将该程序进行编译。如果没有提示有错误，则编译成功；如果提示有错误，退出并修改后重新编译，然后保存。

（3）选择【文件】菜单中的【导出】选项。

（4）在选择【导出】后，在弹出【选择导出文件类型】对话框中提示选择输出的文件类型，选择【储存卡格式文件】格式，然后单击【下一步】。

（5）在弹出【指定导出文件名】对话框中选择文件路径，输入需要导出的文件名。单击【保存】按钮，再次弹出【导出】对话框。单击【完成】，出现【导出完成】信息框（图 4-40）后，单击【确定】按钮，完成整个导出过程。

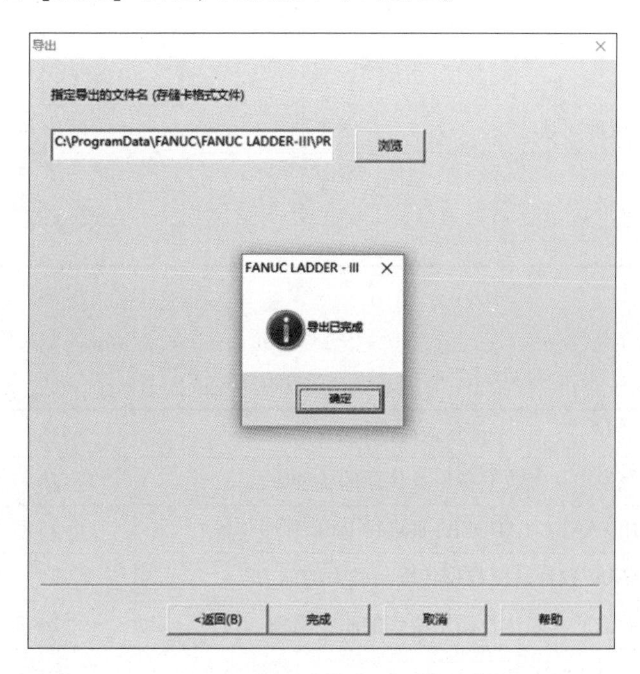

图 4-40 【指定导出的文件名】对话框

【任务实施】

（1）按照随机分组方式，分成各个观摩小组，并选出一名组长，将本组成员的个人信息填入表 4-24 中。

（2）根据观摩及讨论的结果，在表 4-24 中记录各种数据传输方法的名称、步骤、主要功能及适用场合等。

（3）对观摩过程中的疑难问题进行充分讨论，并将问题及解决办法填入表 4-24 中。

表4-24 数控系统数据传输与备份任务工单

<table>
<tr><td colspan="6">任务工单</td></tr>
<tr><td colspan="2">姓名：</td><td colspan="2">班级：</td><td colspan="2">学号：</td></tr>
<tr><td colspan="6">所需材料、工量具、设备列表</td></tr>
<tr><td>序号</td><td colspan="2">名称</td><td>数量</td><td>型号或规格</td><td>备注</td></tr>
<tr><td></td><td colspan="2"></td><td></td><td></td><td></td></tr>
<tr><td></td><td colspan="2"></td><td></td><td></td><td></td></tr>
<tr><td></td><td colspan="2"></td><td></td><td></td><td></td></tr>
<tr><td colspan="6">工作过程（观摩）</td></tr>
<tr><td>序号</td><td>名称</td><td>步骤</td><td colspan="2">主要功能</td><td>适用场合</td></tr>
<tr><td></td><td></td><td></td><td colspan="2"></td><td></td></tr>
<tr><td></td><td></td><td></td><td colspan="2"></td><td></td></tr>
<tr><td></td><td></td><td></td><td colspan="2"></td><td></td></tr>
<tr><td></td><td></td><td></td><td colspan="2"></td><td></td></tr>
<tr><td colspan="3">出现问题</td><td colspan="3">解决办法</td></tr>
<tr><td colspan="3"></td><td colspan="3"></td></tr>
</table>

【任务评价】

序号	评价内容	分值	得分
1	能够熟练说出系统数据备份及还原的各种方法	25	
2	能够使用 FANUC LADDER-Ⅲ进行 PMC 程序传输	25	
3	具备 FANUC 数控系统数据传输与备份的能力	25	
4	具有有效沟通、表达及团结协作的职业精神	25	

复习思考题

（1）简述 FANUC βi-B 伺服接口定义。

（2）简述 FANUC 0i-F 数控系统的主板各接口连接定义。

（3）简述 FANUC 0i-F 数控系统参数的表达方式。

（4）简述系统参数初始化操作步骤。

（5）简述 PMC 的第一级程序内容及编写要求。

（6）简述立式加工中心斗笠式刀库的换刀流程。

（7）简述引导界面的 PMC 程序和 SRAM 数据备份与恢复步骤。

（8）简述在所有 I/O 界面能输入、输出的数据。

（9）简述 FANUC LADDER-Ⅲ的 PMC 程序传输步骤。

模块五　伺服系统故障诊断与维修

如果说 CNC 装置是数控机床的“大脑”，是发布“命令”的指挥机构，那么，伺服系统就是数控设备的“四肢”，是一种“执行机构”，它忠实而准确地执行由 CNC 装置发来的运动命令。

伺服系统概述

伺服系统又叫随动系统，专指被控变量（系统的输出变量）是机械位移（或转角）、速度或加速度的反馈控制系统，其作用是使输出的机械位移（或转角）准确地跟踪输入的位移（或转角）。伺服系统是数控机床的重要组成部分，通常是指各运动坐标轴的进给伺服系统。伺服系统由伺服电机、伺服驱动装置、位置检测装置等组成。它是数控系统和机床机械传动部件间的连接环节。在数控机床中，伺服系统接收来自数控装置的进给脉冲信号，经过伺服驱动装置进行电压、功率放大，由伺服电机驱动机床运动部件实现运动，并保证动作的快速性和准确性。它把数控系统插补运算生成的位置指令，精确地变换为机床移动部件的位移，直接反映了机床坐标轴跟踪运动指令和实际定位的性能。伺服系统的高性能在很大程度上决定了数控机床的高效率、高精度，是数控机床的重要组成部分。它包含了机械传动、电气驱动、检测、自动控制等方面的内容，涉及强电与弱电控制。

数控机床的伺服系统包括进给伺服系统和主轴伺服（驱动）系统。进给伺服系统是以机械位移（位置控制）为直接控制目标的自动控制系统，用来保证加工轮廓。主轴伺服系统以速度控制为主，一般只满足主轴调速及正反转功能，提供切削过程中需要的转矩和功率。但当要求机床有螺纹加工、准停和恒线速度加工等功能时，还需要对主轴提出相应的位置控制要求。

本模块主要介绍数控机床伺服系统、数控机床主轴伺服系统与进给伺服系统常见故障诊断与维修。

任务一　数控机床伺服系统

【学习目标】

（1）了解数控机床伺服系统的分类。

（2）了解数控机床伺服系统的组成。

（3）了解数控机床对伺服系统的要求。

（4）掌握数控机床伺服进给系统位置控制的基本原理。

（5）了解数控机床伺服进给系统速度控制的驱动装置。

（6）具有伺服系统控制工作原理解读的能力。

（7）掌握工程工作方法，培养严谨的工作作风，遵守 5S 管理制度。

【任务描述】

以小组为单位，观摩 YL569 型数控铣床的伺服系统，并将数控机床各伺服系统的名称、分类及组成、主要功能、适用场合等填写到任务工单中。要求观摩过程中务必注意安全，遵守实训场所相关操作规范。

【知识链接】

一、伺服系统的分类和组成

伺服系统的分类可以采用多种方式，一般按调节控制方式来划分。伺服系统按调节控制方式分类通常可分为开环伺服系统、闭环伺服系统和半闭环伺服系统。

（1）开环伺服系统。

开环伺服系统通常使用步进电机，进给脉冲经功率放大后直接控制步进电机。在步进电机轴上或工作台上没有速度或位置检测装置，没有反馈信号，由数控装置输出脉冲的频率控制步进电机的速度，由输出脉冲的数量控制工作台的位置。图 5-1 所示为开环伺服系统。

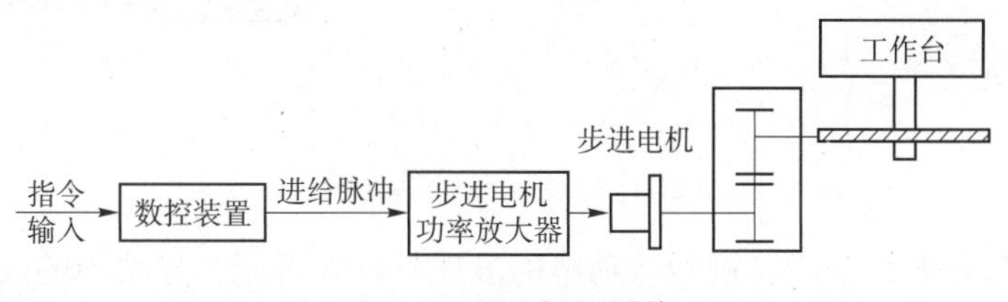

图 5-1　开环伺服系统

开环伺服系统的结构简单、运行平稳、易于控制、价格低廉、使用和维修简单，但精度不高、低速不稳定、高速扭矩小，一般用于经济型数控机床及普通的机床改造。

（2）闭环伺服系统。

闭环和半闭环伺服系统通常使用直流伺服电机或交流伺服电机，由速度检测元件和位置检测元件获取被控制对象，如工作台的速度和位置的反馈信号，并用其来调节伺服电机的速度和位置。

闭环伺服系统的位置检测装置直接安装在机床移动部件上。例如，工作台上安装的直线位置检测装置，该装置将检测到的实际位置反馈到数控系统中，进行位置比较计算，产生控制输出。图 5-2 所示为闭环伺服系统。

由于闭环伺服系统的位置检测包含了传动链的全部误差，因此可以达到很高的控制精度。但不能认为闭环伺服系统可以降低对传动机构的要求，因为传动机构会影响系统的动态特性，给调试和稳定带来困难，导致调整环路时必须要降低位置增益，会对跟随误差与轮廓加工误差产生不利影响。所以采用闭环方式时必须增大机床的刚性，改善滑动面的摩擦特

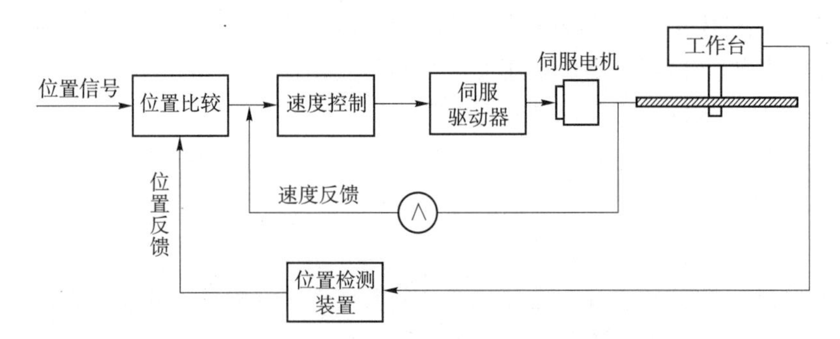

图 5-2 闭环伺服系统

性，减小传动间隙，这样才有可能提高位置增益。闭环伺服系统主要应用于精度要求高的大型数控机床。

（3）半闭环伺服系统。

半闭环伺服系统的位置检测装置一般安装在电机轴上或滚珠丝杠轴端，通过测量角位移间接地测量出移动部件的直线位移，然后反馈到数控系统中，与系统中的位置指令值进行比较，用比较后的差值控制移动部件移动，直到差值消除时停止移动。半闭环伺服系统如图 5-3 所示。

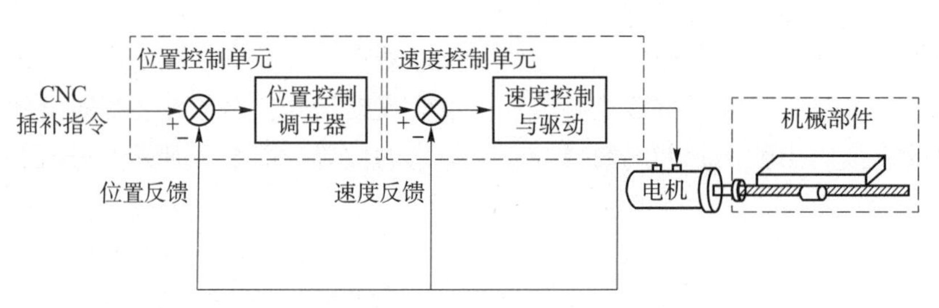

图 5-3 半闭环伺服系统

由于在半闭环伺服系统中，进给传动链的滚珠丝杠螺母副、导轨副的误差不全包括在位置反馈中，所以传动机构的误差仍然会影响移动部件的位置精度。但由于反馈过程中不稳定因素的减少，系统容易达到较高的位置增益而不发生振荡，其快速性好、动态精度高，目前应用比较广泛。至于传动链误差，如反向间隙、丝杠螺距累积误差可通过数控系统的参数设置来进行补偿，以提高机床的定位精度。

二、位置控制

位置控制和速度控制是数控机床进给伺服系统的重要环节。位置控制装置是将计算机插补运算得到的位置指令与位置检测装置反馈来的机床坐标轴的实际位置相比较，形成位置偏差，经变换得到速度指令电压。速度控制装置根据位置控制装置输出的速度电压信号和速度检测装置反馈的实际转速对伺服电机进行控制，以驱动机床传动部件。因为速度控制装置是伺服系统中的功率放大部分，所以速度控制装置也称驱动装置或伺服放大器。

位置控制的基本原理如下。

位置控制是一个闭环或半闭环系统。移动部件的位置可由检测元件检测到，并送给计算机作比较，因而机床移动部件的定位精度较高。

图 5-4 所示为位置控制系统，由图可知该系统主要由位置控制、速度控制和位置检测三部分组成。位置控制装置的作用是将插补计算得出的瞬时位置指令值 P_s 和检测出的实际位置相比较，产生位置偏移量 ΔP，再把 ΔP 变换为瞬时速度指令电压 U_{sn}；速度控制装置的作用是将瞬时速度指令电压 U_{sn} 和检测的速度电压 U_{fn} 比较后放大为驱动伺服电机的电枢电压；位置传感器的作用是把位置检测元件检测到的信号转换为与指令位置量级相同的数字量 P_f，供位置控制环节比较使用。

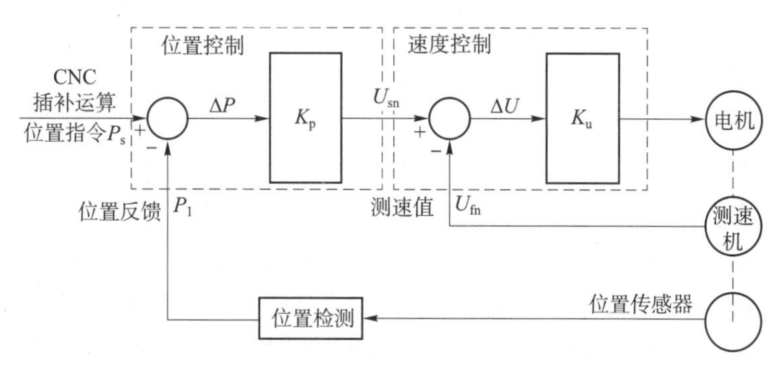

图 5-4　位置控制系统

三、速度控制

速度控制装置也称驱动装置，数控机床中的驱动装置因驱动电机的不同而不同。交流伺服电机的驱动装置有它控变频控制和自控变频控制。直流主轴电机的驱动装置由交流晶闸管控制，交流主轴电机的驱动装置有通用变频控制和矢量控制。驱动装置中常用的功率器件有大功率晶体管、功率场效应晶体管、绝缘门极晶体管、普通晶闸管和可关断晶闸管等，这些功率器件在驱动装置中的作用就是将控制信号进行功率放大。

进给用交流伺服电机多采用三相交流永磁同步电机。常采用的调速方法是变频调速。变频调速系统可以分为它控变频和自控变频两大类。它控变频调速系统用独立的变频装置给电机提供变压变频电源；自控变频调速系统用电机轴上所带的转子位置检测器来控制变频。

（1）频率调制原理。

工业用电的频率是固定的 50 Hz，必须采用变频的方法改变电机供电频率。常用的方法有两种：直接的交-交变频和间接的交-直-交变频。

交-交变频是用晶闸管整流器把工频交流电直接变成频率较低的脉动交流电，正组输出正电压脉波，反组输出负电压脉波，这个脉动交流电的基波电压就是所需变频电压。但这种方法得到的交流电波动较大。

交-直-交变频（图 5-5）是先把交流电整流成直流电，然后把直流电压变成矩形脉冲波电压，这个矩形脉冲波的基波就是所要的变频电压。因交-直-交变频所得交流电波动小、调频范围宽、调节线性度好，所以数控机床上经常采用这种方法。在交-直-交变频中有中间直流电压可调式 PWM 逆变器和中间直流电压固定式 PWM 逆变器。根据中间直流电路上的储能元件是大电容或大电感分为电压型 PWM 逆变器和电流型 PWM 逆变器。电压固定的 PWM 逆变器是典型的交-直-交逆变器。

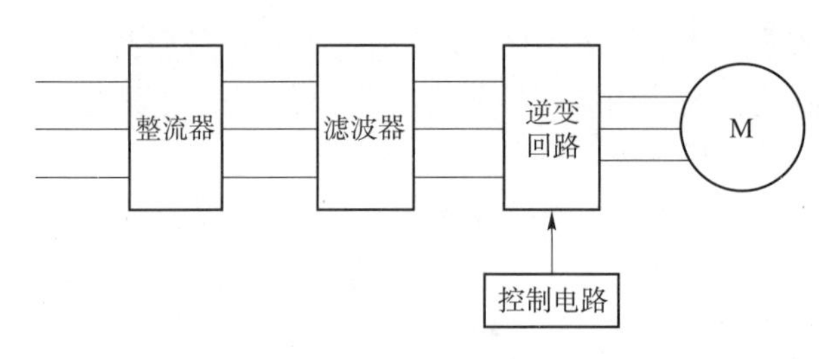

图 5-5　交-直-交逆变器的组成

（2）SPWM 原理。

正弦波脉宽调制 Sine SPWM，是 PWM 的一种。SPWM 变频器不仅适用于交流永磁式伺服电机，也适用于交流感应式伺服电机。SPWM 采用正弦规律脉宽调制原理，具有功率因数高、输出波形好等优点，因而在交流调速系统中获得了广泛应用。

图 5-6 所示为模拟式单相 SPWM 波调制原理。在 SPWM 中，输出电压是由三角载波调制的正弦电压得到的。SPWM 的输出电压 U_0 是一个幅值相等、宽度不等的方波信号。其各脉冲的面积与正弦波下的面积成比例，所以脉宽基本上按正弦分布，其基波是等效正弦波。用这个输出脉冲信号经功率放大后作为交流伺服电机的相电压（电流），改变正弦基波的频率就可改变电机相电压（电流）的频率，实现调频调速的目的。

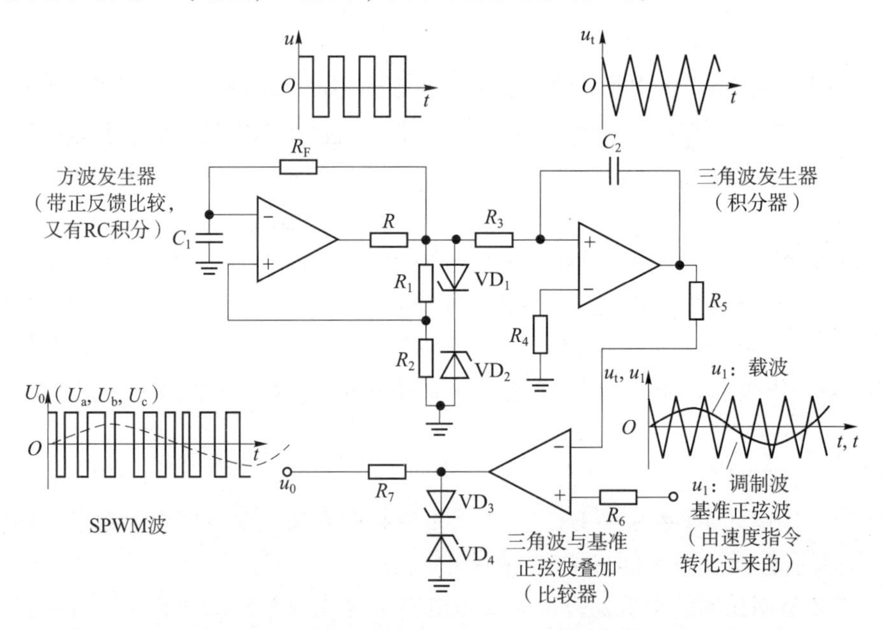

图 5-6　模拟式单相 SPWM 波调制原理

除了模拟式 SPWM 方法外，数字控制 SPWM 应用越来越广泛，其通常采用的方法有：①微机存储事先算好的 SPWM 数据，由指令调出，或通过系统实时生成；②专用集成芯片；③单片机微处理器直接带有 SPWM 信号产生功能。

【任务实施】

（1）按照随机分组方式，分成各个观摩小组，并选出一名组长，将本组成员的个人信

息填入表 5-1 中。

（2）根据观摩及讨论的结果，在表 5-1 中记录数控机床各伺服系统的名称、分类及组成、主要功能、适用场合等。

（3）对观摩过程中的疑难问题进行充分讨论，并将问题及解决办法填入表 5-1 中。

表 5-1　数控机床伺服系统任务工单

<table>
<tr><td colspan="6">任务工单</td></tr>
<tr><td colspan="2">姓名：</td><td colspan="2">班级：</td><td colspan="2">学号：</td></tr>
<tr><td colspan="6">所需材料、工量具、设备列表</td></tr>
<tr><td>序号</td><td colspan="2">名称</td><td>数量</td><td>型号或规格</td><td>备注</td></tr>
<tr><td></td><td colspan="2"></td><td></td><td></td><td></td></tr>
<tr><td></td><td colspan="2"></td><td></td><td></td><td></td></tr>
<tr><td></td><td colspan="2"></td><td></td><td></td><td></td></tr>
<tr><td colspan="6">工作过程（观摩）</td></tr>
<tr><td>序号</td><td>名称</td><td>分类及组成</td><td colspan="2">主要功能</td><td>适用场合</td></tr>
<tr><td></td><td></td><td></td><td colspan="2"></td><td></td></tr>
<tr><td></td><td></td><td></td><td colspan="2"></td><td></td></tr>
<tr><td></td><td></td><td></td><td colspan="2"></td><td></td></tr>
<tr><td></td><td></td><td></td><td colspan="2"></td><td></td></tr>
<tr><td colspan="3">出现问题</td><td colspan="3">解决办法</td></tr>
<tr><td colspan="3"></td><td colspan="3"></td></tr>
</table>

【任务评价】

序号	评价内容	分值	得分
1	能够熟练说出数控机床伺服系统的分类及组成	25	
2	能够掌握数控机床进给伺服系统位置控制的基本原理	25	
3	具有伺服系统控制工作原理解读的能力	25	
4	具有有效沟通、表达及团结协作的职业精神	25	

任务二 主轴驱动系统常见故障诊断与维修

【学习目标】

(1) 掌握 FANUC 数控系统主轴监控界面的操作。
(2) 了解 FANUC 数控系统串行主轴报警信息的含义。
(3) 掌握 FANUC 主轴驱动系统维修的基本知识。
(4) 掌握查阅系统诊断界面中主轴的诊断号以及含义。
(5) 具有自主分析主轴驱动系统故障和排除主轴驱动系统故障的能力。
(6) 掌握工程工作方法，培养严谨的工作作风，遵守 5S 管理制度。

【任务描述】

以小组为单位，观摩数控机床的 FANUC 数控系统主轴诊断和维护界面，并将 FANUC 数控系统主轴诊断和维护界面各信息的名称、信号符号、主要功能及适用场合等填写到任务工单中。要求观摩过程中务必注意安全，遵守实训场所相关操作规范。

【知识链接】

一、FANUC 数控系统主轴诊断和维护界面

1. FANUC 串行主轴提供维护和维修手段

FANUC 串行主轴为了维护和维修方便，提供了多方面的维护和维修手段，从系统诊断 400 开始就提供了与主轴有关的诊断信息，在主轴放大器七段 LED 数码管上也显示了运行状态。主轴监控界面如图 5-7 所示。

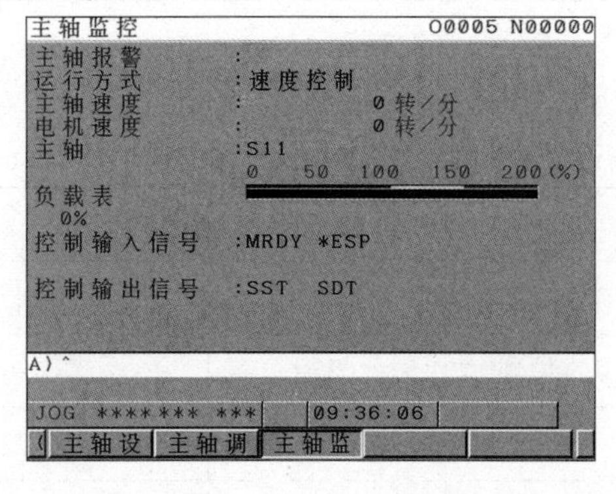

图 5-7 主轴监控界面

在图 5-7 所示界面中，可以选择主轴设定界面、主轴调整界面、主轴监控界面。主轴监控界面提供了丰富的维护和维修信息，为维护和维修带来了极大的方便。现代数控机床要充分利用数控系统提供的丰富信息进行故障诊断和维修。

2. FANUC 主轴监控界面

在 FANUC 主轴监控界面中有监控信息，如图 5-7 所示。不同的运行方式有不同的参数调整和不同的监视内容。

(1) 主轴报警。

“主轴报警”信息栏提供了当主轴报警时即时显示的主轴以及主轴电机等的主轴报警信息。主轴报警信息多达 63 种，部分主轴报警信息见表 5-2。

表 5-2　部分主轴报警信息一览表

报警号：报警信息	报警号：报警信息	报警号：报警信息
1：电机过热	29：短暂过载	61：半侧和全侧位置返回误差报警
2：速度偏差过大	30：输入电路过电流	65：磁极确定动作时的移动量异常
3：直流链路熔断器熔断	31：电机受到限制	66：主轴放大器间通信报警
4：输入熔断器熔断	32：用于传输的 RAM 异常	72：电机速度判定不一致
6：温度传感器断线	33：直流链路充电异常	73：电机传感器断线
7：超速	34：参数设定异常	80：通信的下一个主轴放大器异常
9：主回路过载	41：位置编码器一转信号检测错误	82：尚未检测出电机传感器一转信号
11：直流链路过电压	42：尚未检测出位置编码器一转信号	83：电机传感器信号异常
12：直流链路过电流	43：差速控制用位置编码器信号断线	84：主轴传感器断线
15：输出切换报警	46：螺纹切削用位置传感器一转信号检测错误	85：主轴传感器一转信号检测错误
16：RAM 异常	47：位置编码器信号异常	87：主轴传感器信号异常
19：U 相电流偏置过大	51：变频器直流链路过电压	110：放大器间通信异常
20：V 相电流偏置过大	52：ITP 信号异常 I	111：变频器控制电源低电压
21：位置传感器极性设定错误	56：内部散热风扇停止	112：变频器再生电流过大
24：传输数据异常或停止	57：变频器减速电力过大	120：通信数据报警
27：位置编码器断线	58：变频器主回路过载	137：设备通信异常

在维修时，通过主轴参数调整监控界面，可以方便又直观地了解主轴放大器、主轴电机、主轴传感器反馈等相关故障诊断信息，要充分利用主轴监控界面提供的故障诊断信息。

（2）运行方式。

“运行方式”信息栏提供了当前主轴的运行方式。FANUC 主轴运行方式比较丰富和灵活，主要有速度控制、主轴定向、同步控制、刚性攻螺纹、主轴 CS 轮廓控制、主轴定位控制（T 系列）。不是每种主轴都有六种运行方式，这主要取决于机床制造厂家是否二次开发了用户需要的运行方式。此外，有的运行方式还需要数控系统具备相应的软件选项和主轴电机具备实现功能的硬件。

（3）主轴控制输入信号。

编制 PMC 程序使主轴实现相关功能时，经常把逻辑处理结果输出到 PMC 的 G 地址，最终实现主轴功能。例如，要使第 1 主轴正转，需要编制包含 M03 的加工程序，经过梯形图

逻辑处理输出到 G70.5，而 FANUC 规定 G70.5 地址信号用符号表示就是 SFRA，即只要第 1 主轴处于正转状态，就能在“控制输入信号”栏看到“SFR”，在“主轴”栏看到“S1”。常用的主轴控制输入信号一览表如表 5-3 所示。

表 5-3 常用的主轴控制输入信号一览表

信号符号：信号含义	信号符号：信号含义
TLML：转矩限制信号（低）	*ESP：急停（负逻辑）信号
TLMH：转矩限制信号（高）	SOCN：软启动/停止信号
CTH1：齿轮信号 1	RSL：输出切换请求信号
CTH2：齿轮信号 2	RCH：动力线状态确认信号
SRV：主轴反转信号	INDX：定向停止位置变更信号
SFR：主轴正转信号	ROTA：定向停止位置旋转方向信号
ORCM：主轴定向信号	NRRO：定向停止位置快捷信号
MRDY：机械准备就绪信号	INTG：速度积分控制信号
ARST：报警复位信号	DEFM：差速方式指令信号

若主轴某一功能没有实现，可以在图 5-7 主轴监控界面的“控制输入信号”栏检查有无信号显示。若有信号显示，就不需要到梯形图中分析程序了；若某一实现功能的信号没显示，还必须借助梯形图来分析逻辑关系。

（4）主轴控制输出信号。

主轴控制输出信号理解思路与主轴控制输入信号一样，当主轴控制处于某个状态时，由 CNC 把相关的状态输出至 PMC 的 F 存储区，使维修人员直观地了解主轴目前所处的控制状态。例如，当第 1 主轴速度达到运行转速时，CNC 就输出速度到达信号，信号地址是 F45.3，FANUC 定义的符号是 SARA，在“控制输出信号”栏可以看到“SAR”，在“主轴”栏看以到常用的主轴控制输出信号，如表 5-4 所示。

表 5-4 常用的主轴控制输出信号一览表

信号符号：信号含义	信号符号：信号含义
ALM：报警信号	LDT2：负载检测信号 2
SST：速度零信号	TLM5：转矩限制信号
SDT：速度检测信号	ORAR：定向结束信号
SAR：速度到达信号	SRCHP：输出切换信号
LDT1：负载检测信号 1	RCFN：输出切换结束信号

在维修主轴时，可以从主轴监控界面的“控制输出信号”栏了解目前主轴运行状态。

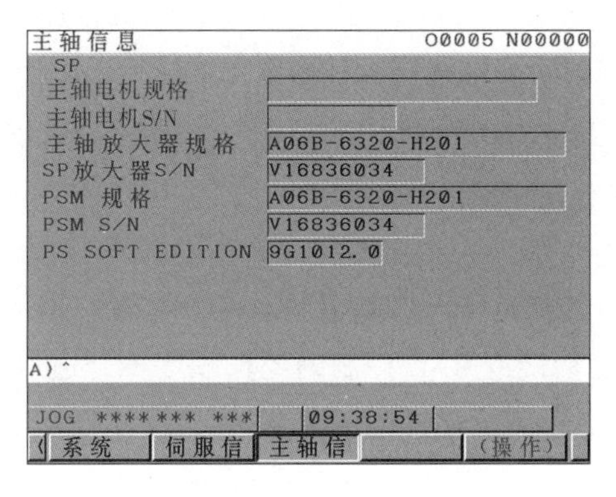

图 5-8　主轴信息界面

3. FANUC 主轴维护界面

为了方便用户维修和更换备件，FANUC 数控系统在主轴放大器首次启动时，自动地从连接的主轴放大器读出并记录 ID 信息。由于主轴放大器连接的主轴电机不同，主轴电机信息不被自动读出。主轴信息界面如图 5-8 所示。

从图 5-8 可以看出，“主轴放大器规格”“PSM 规格”等处的序号就是向 FANUC 公司订货的各部件的订货号。有了订货号，维修人员订购备件就很方便。CNC 启动后，若存储的主轴放大器等信息与实物不符，不同之处用“ * ”标识。

二、FANUC 串行主轴驱动系统故障诊断与维修

1. FANUC 主轴驱动系统

FANUC 主轴驱动系统分串行主轴驱动系统和模拟主轴驱动系统两种。

(1) 串行主轴驱动系统。

串行主轴驱动系统选用 FANUC 主轴放大器与 FANUC 主轴电机。CNC 与主轴放大器之间通过串行总线进行数据交换。

(2) 模拟主轴驱动系统。

模拟主轴驱动系统是由 FANUC 数控系统输出 0~±10 V 模拟电压，输入具有模拟接口的主轴放大器，实现对主轴电机的速度控制。目前 FANUC 的主轴放大器都是串行接口，没有具有模拟接口的主轴放大器。

2. 主轴放大器上的主轴报警和错误代码显示

FANUC 数控系统除在 CNC 上提供了涉及串行主轴报警的多达 130 余项报警信息外，还在主轴放大器上提供了报警和错误信息。FANUC 在 βi 系列主轴放大器上利用图 5-9 所示的七段 LED 数码管和指示灯来反映主轴放大器以及主轴电机运行情况和故障状态等信息。在维修主轴放大器和主轴电机时，要充分利用系统和主轴放大器本体上提供的报警和错误信息进行故障诊断。

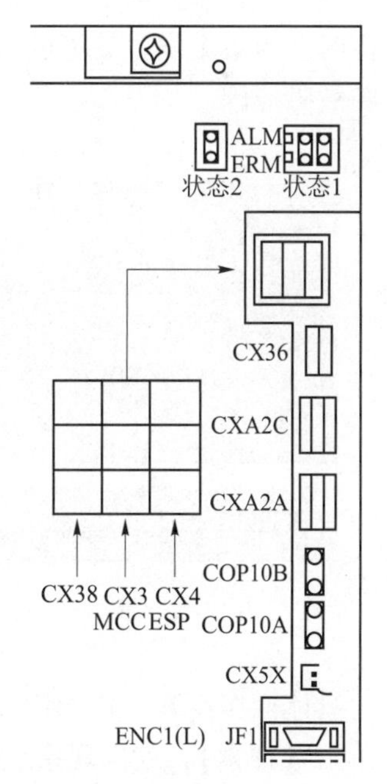

图 5-9　βi 主轴放大器上的七段 LED 数码管和指示灯

下面主要介绍 βi 主轴放大器主轴报警和错误信息显示。

虽然 βi 主轴放大器和 βi 伺服放大器是一体化设计的，但涉及主轴的报警和错误信息是与伺服放大器分开的。如图 5-9 所示，“状态 1”用于主轴报警和错误信息显示，“状态 2”用于伺服放大器报警和错误信息显示。βi 主轴放大器也有和 αi 主轴放大器模块一样的显示主轴报警和错误信息的七段 LED 数码管和指示灯。

（1）βi 主轴放大器报警信息显示。

当 βi 主轴放大器有报警时，“状态 1”旁边的红色指示灯亮，同时两个七段 LED 数码管显示报警信息，维修方法同 αi 主轴放大器模块。主轴报警信息以及显示屏上显示的报警信息与 αi 主轴放大器模块的报警信息和错误信息基本上是相同的。

（2）βi 主轴放大器错误信息显示。

当 βi 主轴放大器有错误信息时，“状态 1”旁边的黄色指示灯亮，同时两个七段 LED 数码管提示错误信息，主要内容仍然是发生了参数设定错误或 CNC 控制的逻辑顺序不合适等，维修方法同 αi 主轴放大器模块。主轴报警信息以及显示屏上显示的错误信息与 αi 主轴放大器模块的报警信息和错误信息基本是相同的。

3. 主轴调整界面和主轴监控界面

主轴调整界面和主轴监控界面提供了维修诊断信息，如图 5-10 和图 5-7 所示。在维修时，可以结合主轴调整界面和主轴监控界面维修诊断信息进行故障诊断和分析。

4. 系统诊断界面提供主轴维修信息

FANUC 系统除了提供 130 余项涉及主轴的报警信息和主轴监控界面外，还在系统诊断界面提供了主轴故障诊断信息。在系统诊断界面相关的诊断号具体指出主轴实际运行数据情况，包括主轴转速、主轴负载显示、刚性攻螺纹的位置指令以及误差、在线主轴位置、编码器信息等，还显示了错误和报警代码以及原因等。

涉及主轴报警的系统诊断号从 400 开始。涉及主轴系统的诊断界面如图 5-11 所示。在维修时，当主轴发生故障或错误时，可以对照界面，根据 CNC 提供的具体诊断信息，综合判断故障原因。

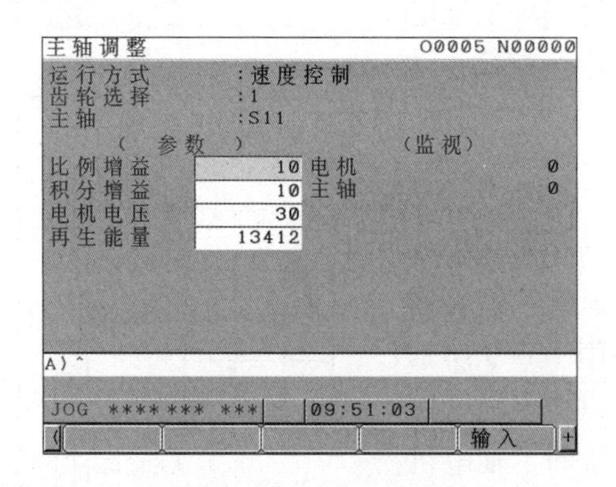

图 5-10 主轴调整界面

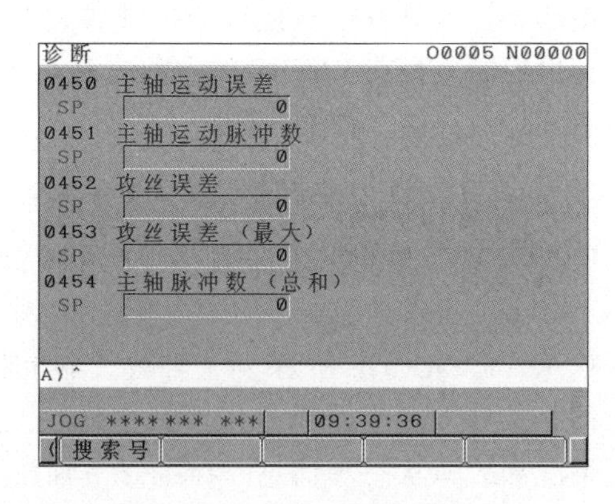

图 5-11 主轴系统的诊断界面

具体操作步骤：在 MDI 方式下，多次单击【SYSTEM】，单击【诊断】，输入“400”，再单击【搜索】，就出现系统 400 诊断界面，再依次检查其他诊断号即可。

【任务实施】

（1）按照随机分组方式，分成各个观摩小组，并选出一名组长，将本组成员的个人信息填入表 5-5 中。

（2）根据观摩及讨论的结果，在表 5-5 中记录 FANUC 数控系统主轴诊断和维护界面各信息的名称、信号符号、主要功能及适用场合等。

（3）对观摩过程中的疑难问题进行充分讨论，并将问题及解决办法填入表 5-5 中。

表 5-5　FANUC 数控系统主轴诊断和维护界面观摩任务工单

任务工单				
姓名：	班级：		学号：	
所需材料、工量具、设备列表				
序号	名称	数量	型号或规格	备注
工作过程（观摩）				
序号	信息名称	信号符号	主要功能	适用场合
出现问题			解决办法	

【任务评价】

序号	评价内容	分值	得分
1	能够熟练说出 FANUC 数控系统主轴监控界面的操作	25	
2	能够掌握 FANUC 主轴驱动系统维修的基本知识	25	
3	具有自主分析主轴驱动系统故障和排除主轴驱动系统故障的能力	25	
4	具有有效沟通、表达及团结协作的职业精神	25	

任务三　进给伺服系统常见故障诊断与维修

【学习目标】

(1) 了解数控系统伺服参数调整的操作。
(2) 了解伺服参数调整界面的含义。
(3) 掌握常见的伺服参数的调整。
(4) 了解伺服系统出现故障时的表现形式。
(5) 掌握伺服放大器故障种类和故障节点。
(6) 具有自主分析进给伺服系统故障和排除进给伺服系统故障的能力。
(7) 掌握工程工作方法，培养严谨的工作作风，遵守5S管理制度。

【任务描述】

以小组为单位，观摩数控系统伺服参数调整界面，并将系统伺服参数调整界面各信息的参数名称、号码、含义及主要功能等填写到任务工单中。要求观摩过程中务必注意安全，遵守实训场所相关操作规范。

【知识链接】

一、伺服参数调整界面

1. 伺服参数调整的意义

数控系统经过硬件连接正常通电后，必须进行伺服参数初始化以及光缆初始化。但是，要使机床能真正满足用户正常操作及使用要求，还有好多参数需要调整。比如，位置环增益是否满足本机床需要，伺服响应时间是否满足本机床工作需要，机床运行有无振动、爬行，机床运行速度是否满足需求等。加工工件时，机床运行速度以及每一个轴运行速度需求都不一样，伺服参数都需要调整；伺服轴进给速度不一样，相应轴的跟随误差也不一样，每台设备的丝杠间隙也不一样等，这些也都需要调整才能满足机床实际需求。

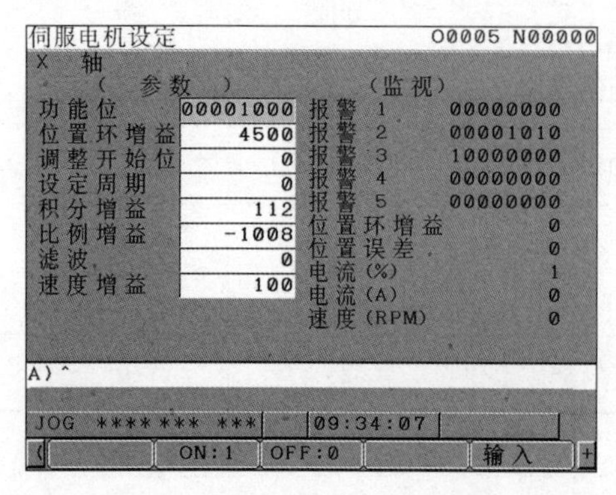

图 5-12　伺服参数调整界面

2. 伺服参数调整界面的含义

FANUC数控系统为了使用户能直观了解伺服电机运行情况以及运行中是否有报警情况等，专门设计了伺服参数调整界面，如图5-12所示。

参数含义情况如下：

1）功能位：对应参数 2003。

2）位置环增益：对应参数 1825，一般默认设置为 3000，也可以根据机床情况调整此参数。

3）调整开始位：在伺服自动调整功能中使用。

4）设定周期：在伺服自动调整功能中使用。

5）积分增益：对应参数 2043。

6）比例增益：对应参数 2044。

7）滤波：对应参数 2067。

8）速度增益：设定值= [（参数 2021/256）+1]×100，其中参数 2021 为负载量比。

9）报警 1：对应诊断号 0200，具体含义见下文。

10）报警 2：对应诊断号 0201，具体含义见下文。

11）报警 3：对应诊断号 0202，具体含义见下文。

12）报警 4：对应诊断号 0203，具体含义见下文。

13）报警 5：对应诊断号 0204，具体含义见下文。

14）位置环增益：表示实际环路增益。在此能看到伺服初始化和调整以后的实际值。

15）位置误差：表示实际位置误差值（对应诊断号 0300）。

16）电流（%）：以相对于伺服电机额定值的百分比表示电流值。

17）电流（A）：以峰值表示实际电流。

18）速度（RPM）：表示该界面中所示进给轴的伺服电机实际转速。

在图 5-12 所示界面中，可以通过单击【+】查看其他轴同样参数的值。伺服参数调整界面对日常维护是非常重要的。界面中 1)~8）项用户可以根据需要进行修改，但从维修角度来讲，一般无须修改。界面中 9)~13）项报警 1~报警 5 的含义，要查阅维修资料才能理解。鉴于现在系统诊断信息细化，当有故障报警时，已经在界面中显示报警原因了，因此可以把此报警信息作为维修综合参考因素。界面中 14)~18）项可以实时显示伺服电机运行状态。

3. 伺服参数调整界面中维修用检测内容

在伺服参数调整界面中，9)~13）项为五组报警，当 FANUC 数控系统有故障报警，且具体报警原因不清楚时，可以先检查一下伺服参数调整界面的 9)~13）项有无为 1 的报警，再检查诊断号 0200~0204 每一位的状态变化。诊断号 0200~0204 各位的含义如表 5-6 所示。

表 5-6　诊断号 0200~0204 各位的含义

诊断号	#7	#6	#5	#4	#3	#2	#1	#0
0200	OVL	LV	OVC	HCA	HVA	DCA	FBA	OFA
0201	ALD	PCR	—	EXP	—	—	—	—
0202	—	CSA	BLA	PHA	RCA	BZA	CKA	SPH
0203	DTE	CRC	STB	PRM	—	—	—	—
0204	—	OFS	MCC	LDA	PMS	—	—	—

在表 5-6 中，诊断号中的每一位都用报警英文缩写来表示，具体含义可查阅 FANUC Series 0i-MODE F 维修说明书（B-64605CM/01），或参考下文介绍。

（1）诊断号 0200。

诊断号 0200 各位的含义如下所述。

0200#0:OFA，溢流报警；

0200#1:FBA，反馈电缆断线报警；

0200#2:DCA，放电报警；

0200#3:HVA，过电压报警；

0200#4:HCA，异常电流报警；

0200#5:OVC，过电流报警；

0200#6:LV，不足电压报警；

0200#7:OVL，过载报警。

（2）诊断号 0201。

诊断号 0201 各位的含义如下所述。

1）诊断号 0201#4 和 0201#7 状态位含义如表 5-7 所示。

表 5-7　诊断号 0201#4 和 0201#7 状态位含义

类型	0201#7（ALD）	0201#4（EXP）	含义
过载报警	0	—	伺服电机过热
	1	—	伺服放大器过热
断线报警	1	0	内置脉冲编码器反馈电缆断线（硬件）
	1	1	外置脉冲编码器反馈电缆断线（硬件）
	0	0	脉冲编码器反馈电缆断线（软件）

2）诊断号 0201#6：PCR。手动返回参考点时，捕捉到了位置检测器的一转信号。由于已经建立起了用于手动返回参考点的栅格，所以可以手动返回参考点。此位在没有开始手动返回参考点方式的动作时没有意义。

（3）诊断号 0202。

诊断号 0202 各位的含义如下所述。

0202#0:SPH，串行脉冲编码器或反馈电缆异常或反馈脉冲信号的计数不正确；

0202#1:CKA，串行脉冲编码器异常，内部块停止工作；

0202#2:BZA，电池电压降为 0，应更换电池，并设定参考点；

0202#3:RCA，串行脉冲编码器异常，转速的计数不正确；

0202#4:PHA，串行脉冲编码器或反馈电缆异常，反馈脉冲信号的计数不正确；

0202#5:BLA，电池电压下降（警告）；

0202#6:CSA，串行脉冲编码器的硬件异常。

（4）诊断号 0203。

诊断号 0203 各位的含义如下所述。

0203#4:PRM，数字伺服一侧检测出参数非法。这时需查阅诊断号 0352 中所描述的原因

和对策；

0203#5:STB，串行脉冲编码器通信异常，传输过来的数据有误；

0203#6:CRC，串行脉冲编码器通信异常，传输过来的数据有误；

0203#7:DTE，串行脉冲编码器通信异常，没有通信的响应。

（5）诊断号 0204。

诊断号 0204 各位的含义如下所述。

0204#3:PMS，由于串行脉冲编码器 C 或者反馈电缆异常，反馈不正确；

0204#4:LDA，串行脉冲编码器的 LED 异常；

0204#5:MCC，伺服放大器中的电磁开关触点化；

0204#6:OFS，数字伺服的电流值的 A/D 变换异常。

多次单击【SYSTEM】，单击【诊断】，就进入诊断界面，输入“200”，单击【搜索】，就进入诊断号 0200 诊断界面。

二、伺服系统故障诊断维修

1. 概述

当伺服系统出现故障时，通常有 3 种表现形式。①在显示屏上或操作面板上显示报警内容或报警信息。②在进给伺服驱动单元上用报警灯或数码管显示故障。③运动不正常，但无任何报警。

对于①和②两种形式，因为某些报警的含义比较明确，可查阅《βi 系列交流伺服电机/主轴电机/伺服放大器维修说明书（B-65325CM）》。对于③这种形式，需根据进给伺服系统的各个环节控制过程，参考如图 5-13 所示 βi 系列伺服放大器故障节点，查阅《βi 系列交流伺服电机/主轴电机/伺服放大器维修说明书（B-65325CM）》，利用数控系统提供的伺服诊断信息，逐步检查、排除，直至查到真正的原因。

2. 伺服放大器日常维修基本知识

（1）维修类别。

伺服放大器电路的维修分板级维修和片级维修，相应地，伺服放大器故障诊断分电路板级诊断维修和芯片级诊断维修。实际上对最终用户而言，主要进行模块或电路板级维修；对伺服放大器制造商而言，主要进行片级维修，现在高密度的电路板也逐步进行板级维修。现场维修时，一般都是模块或电路板级维修，也就是快速进行故障诊断，再进行电路板或模块的更换处理。

（2）维修理念。

伺服放大器是硬件集成度很高的控制部件，既包括硬件电路，也包括软件算法，因此对伺服放大器的维修不是简单通过万用表或普通仪器就可以的，而要通过伺服放大器制造厂家提供维修帮助来判断故障存在的地方。现在的伺服放大器都有故障显示以及丰富的故障诊断软件来帮助用户进行故障定位。FANUC 伺服放大器有丰富的故障报警功能，为用户维修伺服放大器提供了极大的便利。

3. 伺服放大器故障种类和故障节点

虽然伺服放大器是一个技术密集型的产品，涉及很多自动控制理论以及电子电路，但作

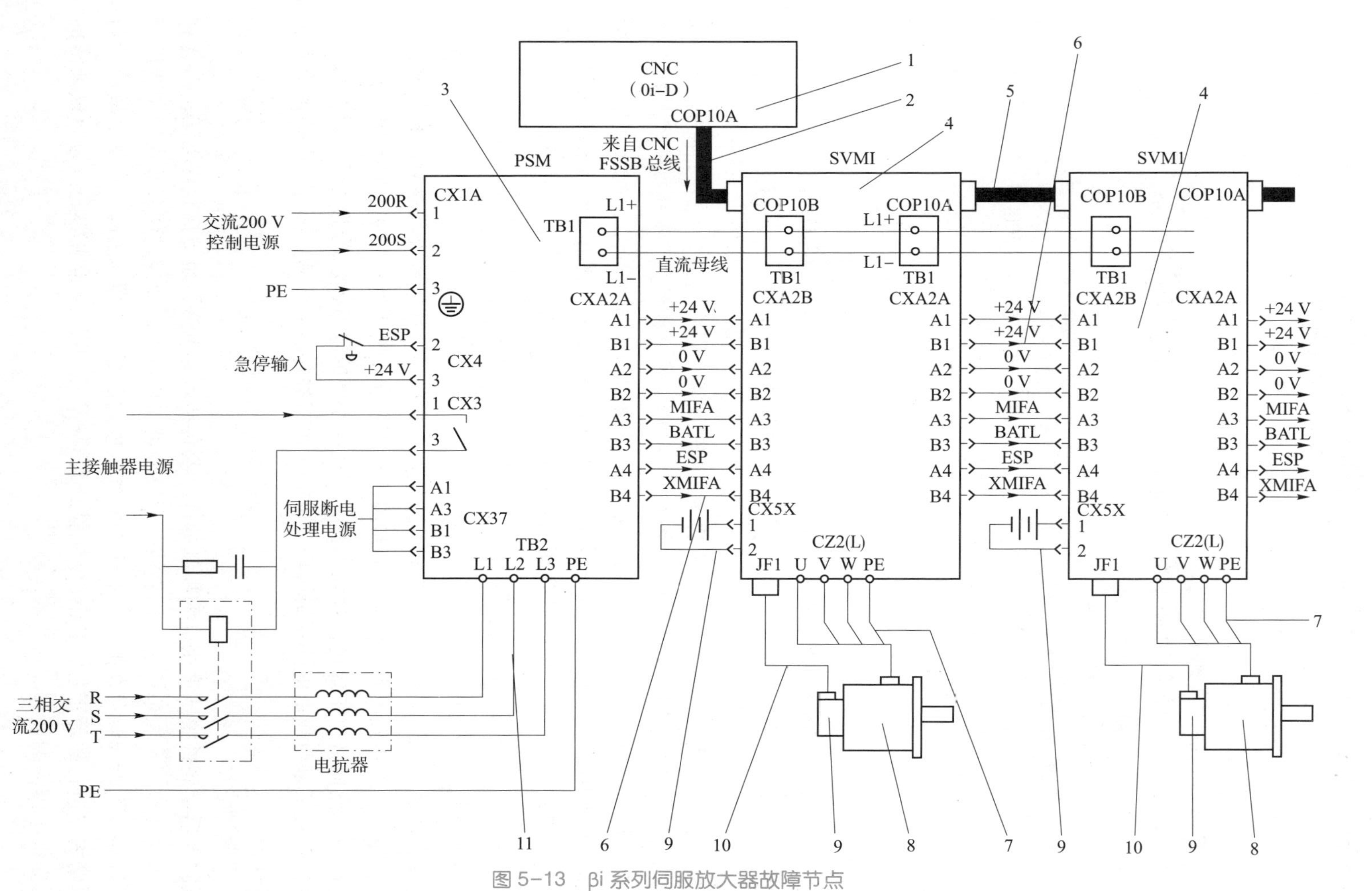

图 5-13 βi 系列伺服放大器故障节点

1—CNC 数控系统；2—FSSB 总线；3—电源模块；4—主轴伺服模块；5—FSSB 总线；6—伺服模块接口；7—电机动力电缆；8—伺服电机；9—电机编码器；10—电机编码器反馈电缆；11—电源模块电缆

为使用和维修用户，只要熟悉伺服放大器电路组成、模块硬件连接、伺服放大器本体接口功能，充分利用系统和伺服放大器本身提供的丰富的故障诊断手段，就能快速进行故障定位和维修。按伺服模块的组成，可以把故障分为 CNC（轴卡）故障、伺服放大器模块故障、伺服电机故障、编码器故障（含电池）、外围连接故障等。

βi 系列伺服放大器故障节点如图 5-13 所示，具体内容如表 5-8 所示。具体诊断和判别 βi 系列伺服放大器故障节点，也需要利用 FANUC 数控系统软件和伺服诊断软件进行综合判断。

表 5-8　βi 系列伺服放大器故障节点

序号	故障节点	详细故障可能点	主要措施
1	CNC（轴卡）故障	CNC 或轴卡	更换 CNC 或轴卡
2	光缆通信故障	CNC 光缆通信接口、光缆连接、伺服模块（第一个）光缆通信接口	更换轴卡、光缆、伺服模块或光缆通信板
3	伺服放大器故障	伺服放大器本体各部分（控制和动力）	更换伺服模块、动力或控制印制电路板
4	下一个光缆通信故障	上一个光缆通信接口、光缆、下一个光缆通信接口	更换上一个光缆通信板、光缆、下一个光缆通信板或下一个伺服模块
5	控制信号互连线故障	电源模块控制线输出接口、控制线、伺服模块控制线接口	更换电源模块、电源模块部件、控制线、伺服模块部件
6	能耗制动模块故障	能耗制动模块电阻或过热检测	更换能耗制动电阻
7	电机动力电缆故障	动力电缆短路和断路	更换动力电缆
8	伺服电机故障	伺服电机本体故障	更换伺服电机
9	编码器故障（含电池）	编码器本体、电池	更换编码器、电池
10	反馈电缆故障	伺服模块控制印制电路板、反馈电缆断路/短路、编码器	更换伺服模块控制印制电路板、反馈电缆、编码器等
11	模块外围物理连接	外围控制电源（单相 200 V）、动力电源（三相 200 V）、急停回路、MCC 控制回路等	更换外围电气部件

4. 系统提供伺服放大器及伺服电机维修信息

现代伺服放大器等都是全数字系统，集成度相当高，基本不能用常规工具来检查。现代数控维修主要依靠伺服放大器本身提供的丰富的诊断功能，数控维修人员要有现代数控维修理念，要充分利用数控系统提供的诊断功能和信息来帮助维修。

当 FANUC 进给伺服驱动系统部件有故障时，除在数控系统显示界面及时出现故障报警

信息外，FANUC 数控系统还在伺服参数调整界面提供了报警 1~5 共 40 位报警信息位，在伺服参数调整界面部分做了详细介绍（多次单击【SYSTEM】—【+】—【SV 调整】，出现如图 5-12 所示界面）。同时还在系统诊断界面中从诊断号 0200 开始有涉及伺服的报警信息内容（单击【SYSTEM】—单击【+】—【检断】—输入诊断号，单击【搜索】可出现系统界面）。

由于现在数控系统提供了丰富的报警信息，在故障报警信息不是十分具体和明确时，可以利用系统诊断号和伺服参数调整界面报警信息位综合加以判断。

FANUC 数控系统提供了七十余种与伺服报警有关的信息，具体可以参考相关维修说明书（B-64305CM）。

【任务实施】

（1）按照随机分组方式，分成各个观摩小组，并选出一名组长，将本组成员的个人信息填入表 5-9 中。

（2）根据观摩及讨论的结果，在表 5-9 中记录系统伺服参数调整界面各信息的参数名称、号码、含义及主要功能等。

（3）对观摩过程中的疑难问题进行充分讨论，并将问题及解决办法填入表 5-9 中。

表 5-9　系统伺服参数调整界面任务工单

<table>
<tr><td colspan="5">任务工单</td></tr>
<tr><td colspan="2">姓名：</td><td colspan="2">班级：</td><td>学号：</td></tr>
<tr><td colspan="5">所需材料、工量具、设备列表</td></tr>
<tr><td>序号</td><td>名称</td><td>数量</td><td>型号或规格</td><td>备注</td></tr>
<tr><td></td><td></td><td></td><td></td><td></td></tr>
<tr><td></td><td></td><td></td><td></td><td></td></tr>
<tr><td></td><td></td><td></td><td></td><td></td></tr>
<tr><td colspan="5">工作过程（观摩）</td></tr>
<tr><td>序号</td><td>名称</td><td>号码</td><td>含义</td><td>主要功能</td></tr>
<tr><td></td><td></td><td></td><td></td><td></td></tr>
<tr><td></td><td></td><td></td><td></td><td></td></tr>
<tr><td></td><td></td><td></td><td></td><td></td></tr>
<tr><td></td><td></td><td></td><td></td><td></td></tr>
<tr><td colspan="3">出现问题</td><td colspan="2">解决办法</td></tr>
<tr><td colspan="3"></td><td colspan="2"></td></tr>
</table>

【任务评价】

序号	评价内容	分值	得分
1	能够熟练说出数控系统伺服调整界面的操作	25	
2	能够掌握伺服参数调整界面的含义	25	
3	具有自主分析进给伺服系统故障和排除进给伺服系统故障的能力	25	
4	具有有效沟通、表达及团结协作的职业精神	25	

复习思考题

（1）简述数控机床伺服系统的分类与组成。
（2）数控机床对伺服系统的要求有哪些？
（3）简述串行主轴驱动系统与模拟主轴驱动系统的区别。
（4）如何进入主轴调整界面？
（5）常见的伺服参数有哪些？
（6）伺服系统故障时的表现形式有哪些？

模块六　数控机床机械装调与检测

数控机床机械部分是数控机床的主体部分，是完成各种切削加工的机械结构。来自数控装置的各种运动和动作指令都必须由机床本体转换成真实的、准确的机械运动和动作，才能实现数控机床的功能，其精度的好坏直接影响加工零件的质量。

数控机床的机械结构始终处在一个逐步发展变化的过程中，早期的数控机床主要是对普通机床的进给系统进行了革新和改造，其外形和结构与普通机床基本相同。目前简易型数控机床与早期数控机床在机械结构上有许多相似之处，其原因正是这些产品也是在普通机床的总体结构基础上经局部改进发展而来的。随着 CNC、计算机、自动控制、信息处理、传感检测、动力元件、液压、气动等技术和新材料的飞速发展，以及对高精度和高效率需要的不断适应，数控机床的机械结构已从初期对普通机床局部结构的改进阶段，逐步发展到形成数控机床的独特机械结构阶段。

本模块在分析数控机床机械结构的同时，针对数控机床主传动系统、进给系统、换刀系统等进行全面的剖析，同时针对数控机床几何精度检测的具体内容进行详细的讲解。

任务一　数控机床机械系统组成

【学习目标】

（1）了解数控机床的机械结构。

（2）掌握数控机床主传动系统的结构及要求。

（3）掌握数控机床进给系统的结构。

（4）掌握数控机床换刀系统的结构。

（5）具备自主分析问题和解决问题的能力。

（6）掌握工程工作方法，培养严谨的工作作风，遵守 5S 管理制度。

【任务描述】

以小组为单位，观摩 YL569 型数控铣床的机械系统单元，并将机床各机械结构名称、结构特征、主要功能及适用场合等填写到任务工单中。要求观摩过程中务必注意安全，遵守实训场所相关操作规范。

【知识链接】

一、数控机床机械结构的基本组成

数控机床机械结构主要由以下几部分组成。

（1）主传动系统，其功用是实现主运动。

（2）进给系统，其功用是实现进给运动。

（3）机床基础件（又称机床大件），通常指床身、底座、立柱、滑座和工作台等，它们是整台机床的基础和框架，其功用是支撑机床本体的其他零部件，并保证这些零件在工作时固定在基础件上，并且在导轨上运动。

（4）实现某些零件动作和辅助功能的系统和装置，如液压、气动、润滑、冷却、防护和排屑等装置。

（5）刀库、刀架和自动换刀装置。

（6）自动托盘交换装置，如双工位托盘自动交换装置。

（7）实现工件回转、分度定位的装置和附件，如回转工作台。

（8）特殊功能装置，如刀具破损检测、精度检测和监控装置等。

二、数控机床的主传动系统及主轴部件

1. 主传动系统的要求

主传动系统是指数控机床主轴的运动传动系统。数控机床主轴的运动是机床的成形运动之一，它的精度决定了零件的加工精度。数控机床的主轴传动系统必须满足如下要求。

（1）具有较大的调速范围并实现无级调速。为了保证数控机床加工时能选用合理的切削用量，从而获得较高的生产率以及较好的加工精度和表面质量，要求其必须具有较大的调速范围。对于加工中心，为了适应各种工序和各种加工材料的要求，主轴系统的调速范围还应进一步扩大。

（2）具有较高的精度与刚度、传动平稳、噪音低。数控机床加工精度的提高与主轴系统的精度密切相关。为提高传动件的制造精度与刚度，齿轮齿面应采用高频感应加热淬火工艺以增加耐磨性。最后一级应采用斜齿轮传动，使传动更平稳；应采用精度高的轴承及合理的支承跨距，以提高主轴组件的刚性。

（3）良好的抗振性和热稳定性。数控机床加工时，可能由于断续切削、加工余量不均匀、运动部件不平衡以及切削过程中的自振等原因引起冲击力和交变力，使主轴产生振动，影响加工精度和表面粗糙度，严重时甚至可能破坏刀具和主轴系统中的零件，使其无法工作。主轴系统发热使其中所有零件产生热变形，降低传动效率，破坏零件之间的相对位置精度和运动精度，从而造成加工误差。因此，主轴组件要有较高的固有频率、较好的动平衡，且要保持合适的配合间隙，并要进行循环润滑。

2. 主轴传动的配置方式

数控机床主轴传动有 3 种配置方式，如图 6-1 所示。

（1）带变速齿轮的主轴传动（图 6-1（a））。这是大中型机床采用较多的配置方式，滑移齿轮多采用液压拨叉或直接由液压缸带动齿轮来实现。

（2）通过带传动的主轴传动（图 6-1（b））。这种配置方式主要用于小型机床，可以避免齿轮传动引起的振动和噪声，而且只能用在低扭矩情况下。

（3）由调速电机直接驱动的主轴传动（图 6-1（c））。这种传动方式非常好，简化了机构，提高了主轴的刚度，但输出扭矩比较小。

3. 主轴轴承的主要配置方式

数控机床主轴轴承的配置方式主要有 3 种，如图 6-2 所示。

（1）前支承采用双列短圆柱滚子轴承和 60°角接触双列向心推力球轴承组合，后支承采用成对向心推力球轴承（图 6-2（a））。此配置可提高主轴的综合刚度，可满足强力切削的要求。

（2）前支承采用高精度向心推力球轴承（图 6-2（b））。向心推力球轴承有良好的高速性，但它的承载能力小，适用于高速、轻载、精密的数控机床主轴。

（3）采用双列和单列圆锥滚子轴承（图 6-2（c））。这种轴承径向和轴向刚度高，能承受重载荷，尤其是可承受较强的动载荷。它的安装、调整性能好，但限制主轴转速和精度，所以用于中等精度、低速、重载的数控机床的主轴。

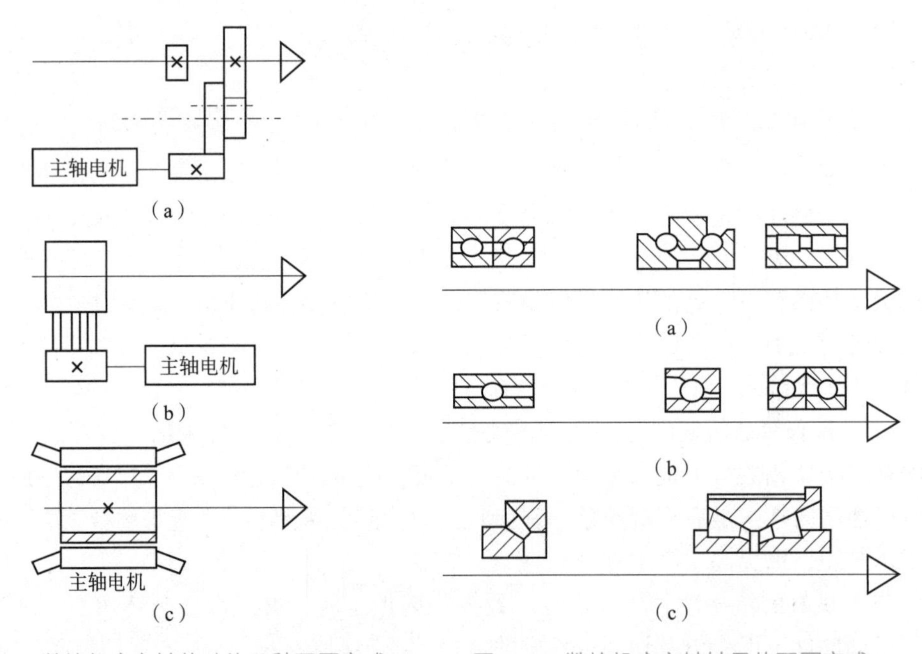

图 6-1　数控机床主轴传动的三种配置方式

图 6-2　数控机床主轴轴承的配置方式

4. 主轴定向装置（主轴准停装置）

主轴定向装置是加工中心换刀过程所要求的特别装置，也称为主轴准停装置。由于刀具装在主轴上，切削时的切削转矩不能完全靠锥孔的摩擦力传递，因此通常在主轴的前端设置一个凸键，当刀具装入主轴时刀柄上的键槽必须与此凸键对准。为保证顺利换刀，主轴必须停止在某一固定方向，主轴定向装置就是为此而设置的。

加工中心主轴定向装置通常采用主轴编码器（通过同步齿形带或齿轮传动与主轴相连接）检测定向，进行初定位。然后由定位销（液动或气动）插入主轴上的销孔或销槽完成精定位，换刀后定位销退出，主轴才可旋转。这种方式比较可靠、准确。

5. 主轴组件的润滑与密封

主轴组件的润滑与密封是机床使用和维护过程中值得重视的两个问题。良好的润滑效果可以降低轴承的工作温度，延长其使用寿命。密封不仅要防止灰尘、切屑和切削液进入，还要防止润滑油的泄漏。

（1）主轴轴承的润滑方式。

数控机床上主轴轴承的润滑方式有油脂润滑、油液循环润滑、油雾润滑、油气润滑等。

1）油脂润滑。

这是目前数控机床的主轴轴承上最常用的润滑方式，特别是在前支承上更常用。如果主轴箱中没有冷却润滑油系统，那么后支承和其他轴承也采用油脂润滑方式。主轴轴承油脂封入量通常为轴承空间容积的 10%，切忌随意填满。油脂过多会加剧主轴发热。采用油脂润滑方式要采取有效的密封措施，以防切削液或润滑油进入轴承中。

2）油液循环润滑。

数控机床的主轴也可以采用油液循环润滑方式。装有 GAMET 轴承的主轴即可使用这种方式。一般主轴轴承的后支承上也常采用这种润滑方式。

3）油雾润滑。

油雾润滑方式是将油液经高压气体雾化后，从喷嘴喷到需润滑部位的润滑方式。由于雾状油液吸热性好，又无须搅拌油液，所以常用于高速主轴轴承的润滑，但缺点是油雾容易吹出，污染环境。

4）油气润滑。

油气润滑方式是针对高速主轴而开发的新型润滑方式。它是用微量油（8~16 min 内流量约 0.03 cm^3 的油）润滑轴承，以抑制轴承发热，其润滑原理如图 6-3 所示。油箱中的油位开关和管路中的压力开关可确保油箱中无油或油的压力不足时，能自动切断主电机电源。

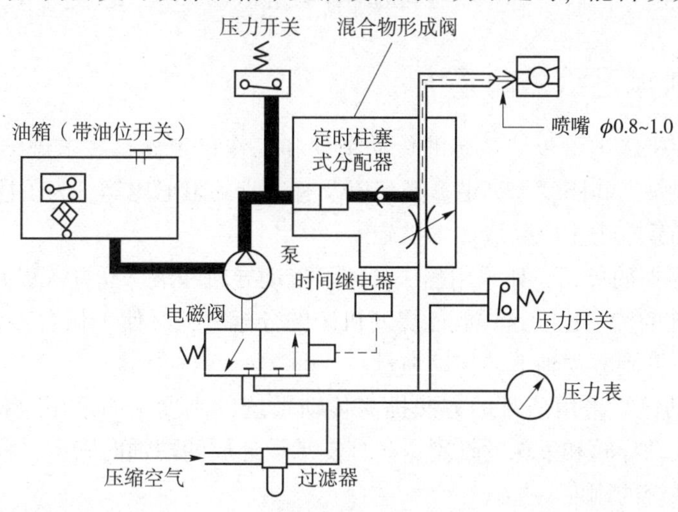

图 6-3　油气润滑原理

另外，用油液润滑角接触轴承时，要注意角接触轴承的泵油效应，必须使油液从小口进入。

（2）主轴的密封。

主轴的密封有接触式密封和非接触式密封两种。接触式密封主要有油毡圈和耐油橡胶密封圈密封两种。非接触式密封的形式如图 6-4 所示。

图 6-4（a）所示结构是利用了轴承盖与轴的间隙密封，其在轴承盖的孔内开槽是为了提高密封效果。这种密封用在工作环境比较清洁的油脂润滑处。图 6-4（b）所示结构是在螺母的外圆上开锯齿形环槽，当油向外流时，靠主轴转动的离心力把油沿斜面甩到端盖的空腔内，油液再流回箱内。图 6-4（c）所示为迷宫式密封结构，在切屑多、灰尘大的工作环境下可获得可靠的密封效果。这种结构适用于油脂或油液润滑的密封。采用非接触式油液密封时，为了防漏必须保证回油能尽快排掉，以保证回油孔的通畅。

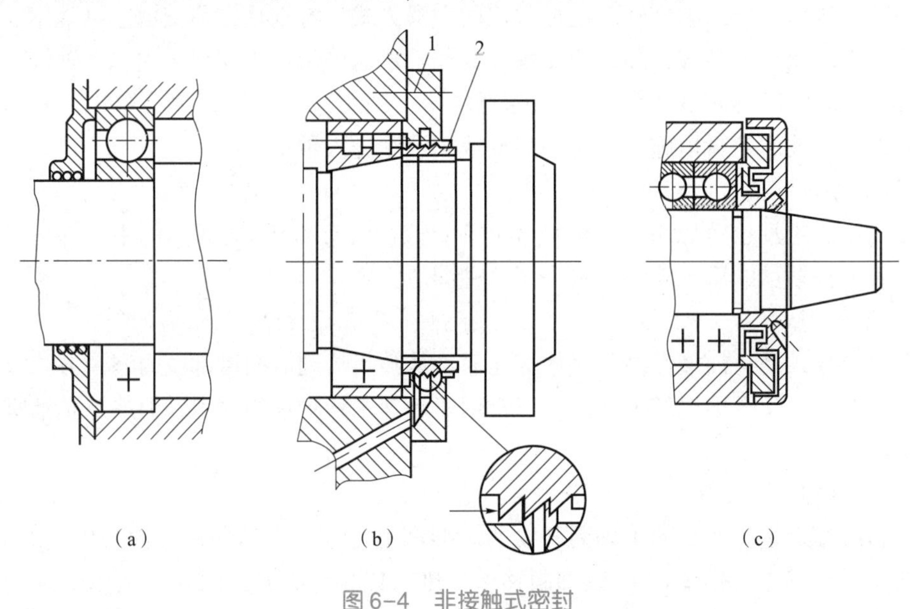

图 6-4 非接触式密封

1—端盖；2—密封圈

三、数控机床进给传动系统

为了确保数控机床进给传动系统的传动精度和工作稳定性，要求进给系统要达到无间隙、低摩擦、低惯量、高刚度、高谐振频率以及有适宜的阻尼比等。为了达到这些要求，在数控机床进给传动系统中主要采取如下措施。

（1）采用低摩擦的传动，如采用静压导轨、滚动导轨和滚珠丝杠等以减少摩擦力。

（2）选用最佳的传动比。这样既能提高机床的分辨率，又使工作台能更快地跟踪指令，同时可减小系统折算到驱动轴上的传动惯量。

（3）缩短传动链，采用预紧的方法提高传动系统的刚度。如采用电机直接驱动丝杠，应用预加负载的滚动导轨和滚珠丝杠副，丝杠支承设计成两端轴向固定，采用预拉伸的结构等办法提高传动系统的刚度。

（4）尽量清除传动间隙，减少反向死区误差。如采用消除间隙的联轴器，采用有消除

间隙措施的传动副等。

数控机床进给传动系统中常用的装置包括滚珠丝杠螺母副、静压蜗杆蜗母条、低摩擦系数的导轨、齿轮齿条等。

1. 滚珠丝杠螺母副

滚珠丝杠螺母副是数控机床理想的运动转换装置。它是由丝杠 1、螺母 2 和位于螺纹滚道之间的滚珠 3 构成，如图 6-5 所示。由于传动时，滚珠与丝杠、滚珠与螺母之间基本上是滚动摩擦，故与普通滚动丝杠螺母副相比有很多优点，如传动效率高（滚珠丝杠螺母副的传动效率 η 为 0.92~0.96）、传动灵敏、不易产生爬行、传动精度和定位精度高、同步性好；滚动摩擦、磨损小，使用寿命长；具有可逆性，不仅可以将旋转运动变成直线运动，也可将直线运动变成旋转运动；在施加适当的预紧力后，可消除轴向间隙，反向时无空行程，还可提高轴向运动精度、刚度和重复定位精度。其缺点是成本高且不能自锁；垂直安装时需要有制动装置。

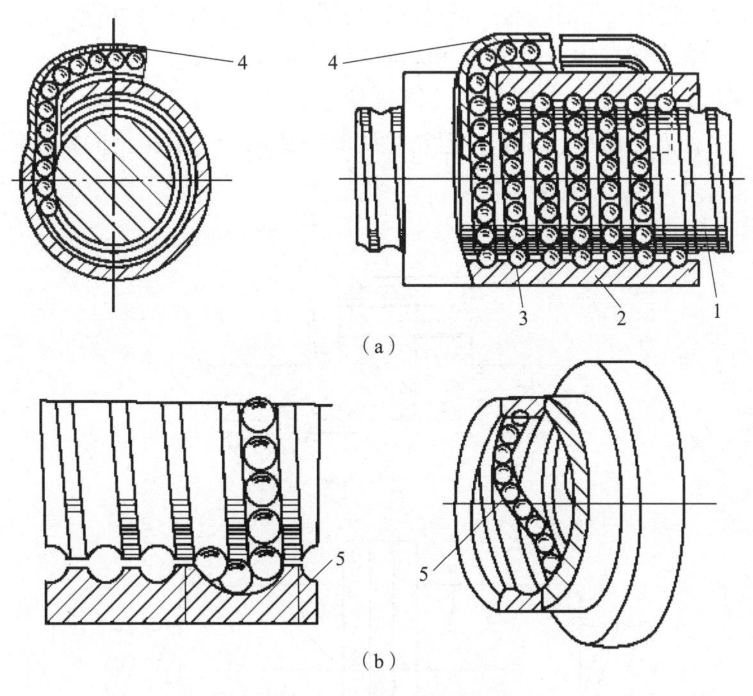

图 6-5　滚珠丝杠螺母副

1—丝杠；2—螺母；3—滚珠；4—弯管；5—反向器

（1）滚珠丝杠螺母副的结构。

滚珠丝杠螺母副的结构有内循环和外循环两种方式，二者在生产中均有应用。图 6-5（a）所示为外循环式。图 6-5（b）所示为内循环式。二者的区别在于滚珠返回的方式不同，一个用弯管 4 引导，返回的滚珠不与丝杠外圆接触；另一个用反向器 5 引导，返回的滚珠经反向器和丝杠外圆之间返回。

从丝杠滚道螺纹型面来看，主要有圆弧型面（图 6-6（a））和双圆弧型面（图 6-6（b））两种，前者容易加工，后者性能较好。

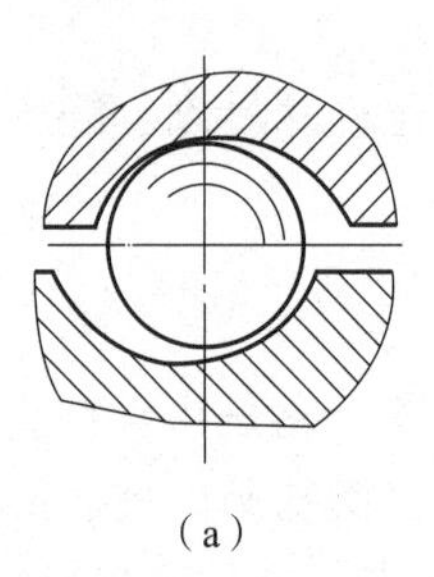
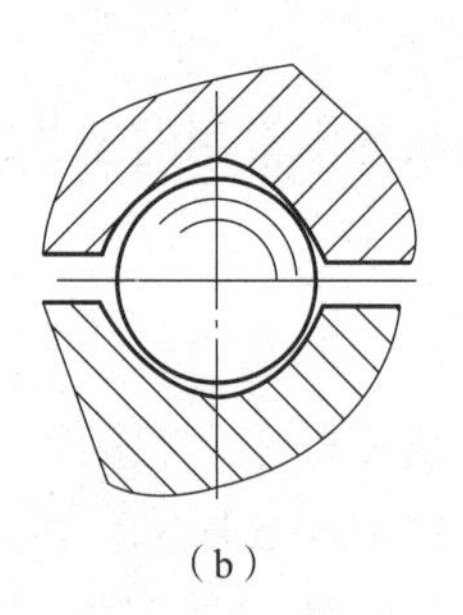

图 6-6　螺纹型面

(2) 滚珠丝杠螺母副间隙调整方法。

滚珠丝杠螺母副
间隙调整方法

滚珠丝杠螺母副的传动间隙是轴向间隙。为了保证反向传动精度和轴向刚度，必须消除轴向间隙。双螺母滚珠丝杠螺母副用预紧的方法消除轴向间隙时，应注意预紧力不宜过大。预紧力过大会使空载力矩增加，从而降低传动效率，缩短使用寿命。此外还要消除丝杠螺母副安装部分和驱动部分的间隙。

常用的双螺母滚珠丝杠螺母副消除间隙的方法有以下几种。

1）垫片调隙式（图 6-7）。调整垫片厚度使左右两螺母产生轴向位移，即可消除间隙和产生预紧力。这种方法简单、可靠，但调整费时，适用于一般精度的机床使用。

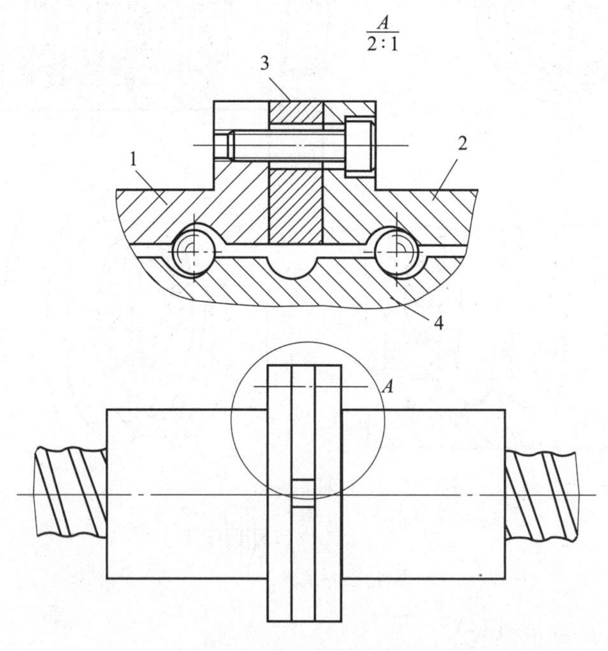

图 6-7　垫片调隙式

1—左螺母；2—右螺母；3—垫片；4—丝杠

2）螺纹调隙式（图 6-8）。左螺母 1 外圆上有螺纹，用两个圆螺母 2、3 将其固定，当转动圆螺母 3 时，即可调整轴向间隙和进行预紧，然后用圆螺母 2 锁紧。这种结构紧凑，工作可靠，滚道磨损时可随时调整，但预紧量不准确。

3）齿差调隙式（图 6-9）。在图 6-9（a）中，螺母 2、螺母 5 的凸缘为圆柱齿轮，两齿

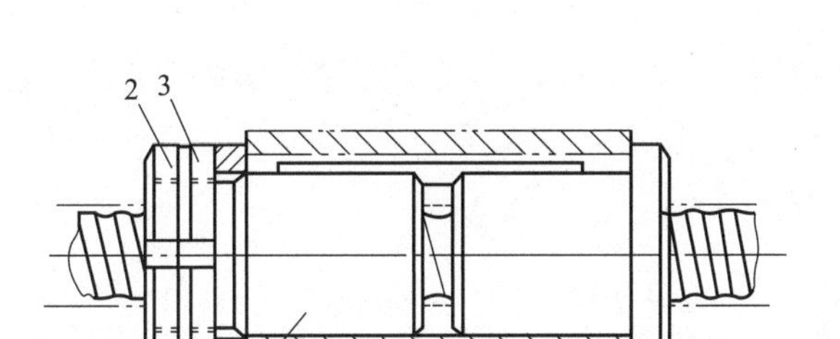

图 6-8　螺纹调隙式

1—左螺母；2，3—圆螺母

轮齿数只相差一齿，分别装在内齿圈 1、内齿圈 4 中。内齿圈用螺钉、定位销紧固在螺母座 3 上。调整时先取下内齿圈，当两个螺母向相同方向都转过一个齿时，其轴向位移量 $s=(1/z_1-1/z_2)t$。例如，两个圆柱齿轮的齿数分别为 $z_1=80$，$z_2=81$，滚珠丝杠的导程为 $t=6$ mm，则 $s=6/6\ 480\approx 0.001$ mm，因此能精确地调整间隙和预紧力，但结构复杂，尺寸较大，适用于高精度传动。图 6-9（b）所示的齿差调隙式结构紧凑且可以预先调整。

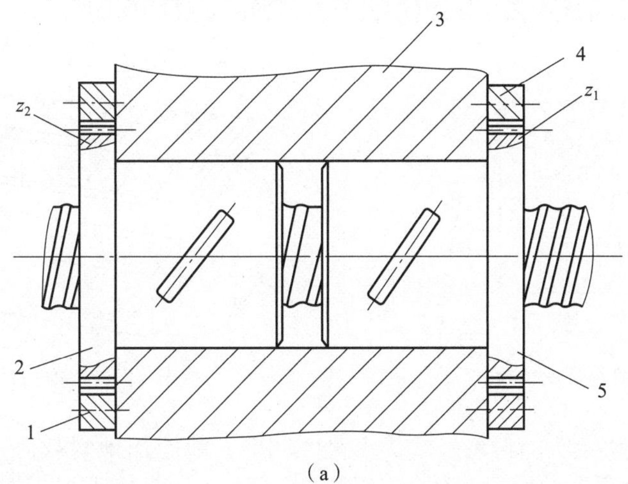

（a）

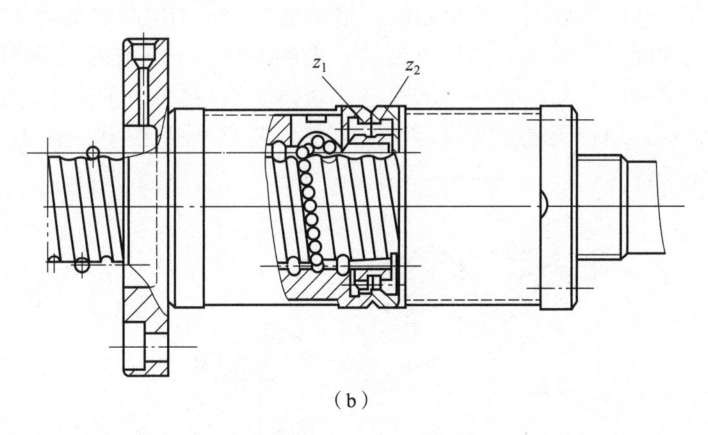

（b）

图 6-9　齿差调隙式

1，4—内齿圈；2，5—螺母；3—螺母座

4）变位螺距预加负荷式（图6-10）。它的螺纹中间为变螺距，结构简单，但负荷量需预先设定且不能改变。

低摩擦系数的导轨

2. 低摩擦系数的导轨

（1）塑料滑动导轨。

塑料滑动导轨即铸铁-塑料或镶钢-塑料滑动导轨。导轨塑料常用聚四氟乙烯导轨软带，采用粘接方法，习惯上称为“贴塑导轨”，用于进给速度为15 m/min以下的中小型数控机床。另一种导轨塑料是树脂型耐磨涂层，以环氧树脂和二硫化钼为基体，加入增塑剂混合成液状或膏状为一组分；加入固化剂混合成另一组分，形成了双组分塑料涂层。

（2）滚动导轨。

滚动导轨摩擦系数低（0.003左右），动态、静态摩擦系数小，几乎不受运动速度变化的影响，定位精度高，灵敏度高，应用也较为普遍。

它有两种类型。第一种是滚动导轨块。滚动体为滚珠或滚柱，滚动体在滚动导轨内做循环运动。这种类型适用于中等载荷的导轨，导轨块装在运动的部件上，每一导轨应至少用两块或更多块，与之相配的导轨多用镶钢淬火导轨，如图6-11所示。

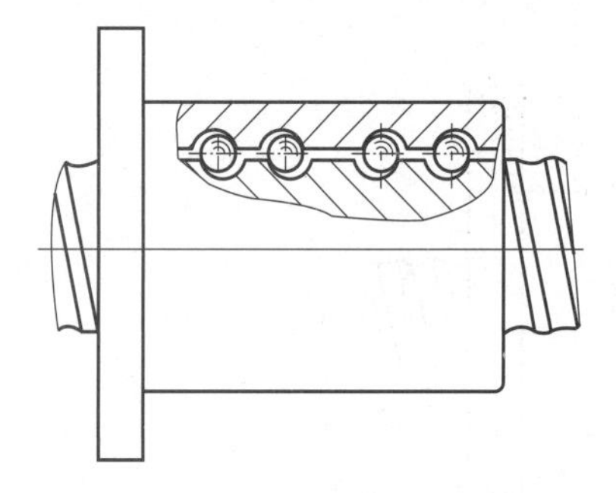

图6-10　变位螺距预加负荷式

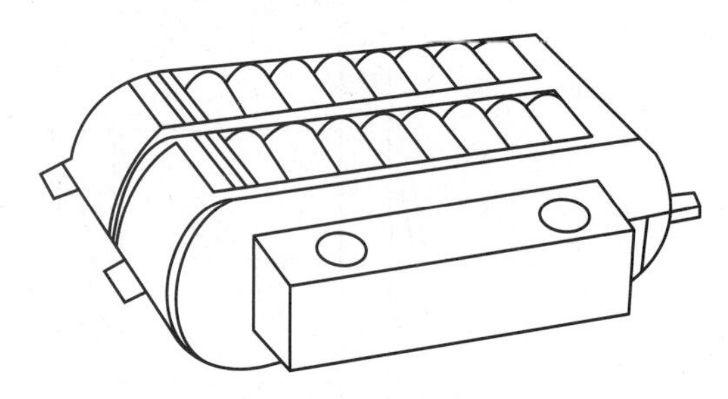

图6-11　滚动导轨块

第二种是直线滚动导轨。它突出的优点是无间隙，能施加预紧载荷。其外形与结构如图6-12和图6-13所示，它由导轨体1、滑块7、滚珠4、保持器3、端盖6等组成。使用时，导轨体固定在不动的部件上，滑块固定在运动部件上；当滑块沿导轨体移动时，滚珠在导轨体和滑块之间的圆弧直槽内滚动，通过端盖内的滚道，从工作载荷区到非工作载荷区，然后再滚动回工作载荷区，不断循环，从而把导轨体和滑块之间的移动变成了滚珠的滚动；为防止灰尘和脏物进入导轨滚道，滑块两端及下部均装有端部密封垫5、侧面密封垫2；滑块上还有润滑油注油杯8。

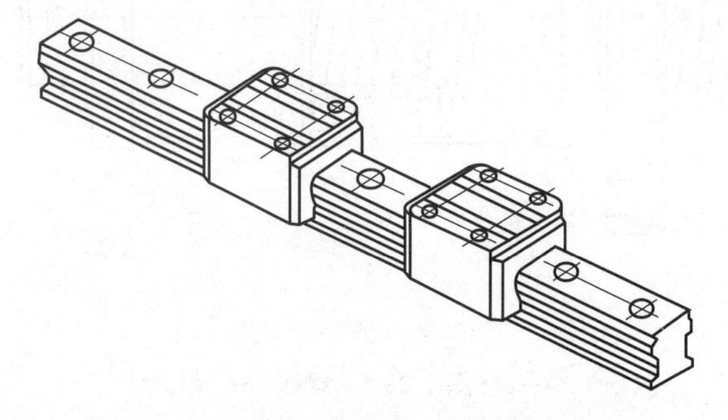

图6-12　直线滚动导轨（单元式）外形

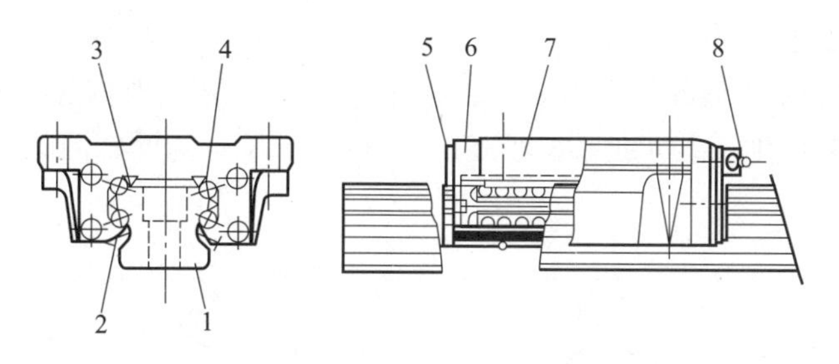

图 6-13　直线滚动导轨（HSR 型）结构

1—导轨体；2—侧面密封垫；3—保持器；4—滚珠；5—端部密封垫；6—端盖；7—滑块；8—润滑油注油杯

（3）静压导轨。

静压导轨通常在两个相对运动的导轨面间通入压力油，使运动部件浮起。在工作过程中，导轨面上油腔中的油压能随外加负载的变化自动调节，保证导轨面间始终处于纯液体摩擦状态。所以静压导轨的摩擦系数极小（约为 0.000 5），功率消耗少。这种导轨不会磨损，因而导轨的精度保持性好，寿命长。它的油膜厚度几乎不受速度的影响，油膜承载能力大、刚性高、吸振性良好。这种导轨的运行很平稳，既无爬行也不会产生振动。但静压导轨结构复杂，并需要有一套过滤效果良好的液压装置，制造成本较高。目前静压导轨一般应用在大型、重型数控机床上。

静压导轨按导轨的形式可分为开式和闭式两种，数控机床上常采用闭式静压导轨。静压导轨按供油方式又可分为恒压（定压）供油和恒流（定量）供油两种。

3. 齿轮齿条

若数控机床的行程很长，一般采用齿轮齿条传动来实现它的进给运动。但齿轮齿条的传动和齿轮传动一样存在齿侧间隙，因此应消除间隙。

当载荷小时齿轮齿条可采用双片薄齿轮错齿调整，分别与齿条的齿槽左右两侧贴紧，来消除间隙。

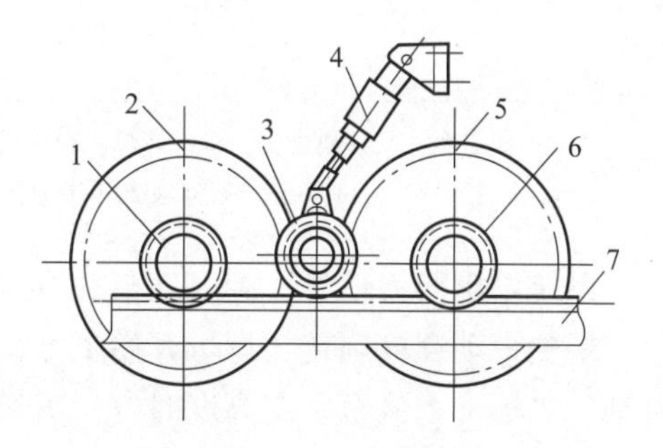

图 6-14　齿轮齿条传动的齿侧间隙消除

1，6—小齿轮；2，5—大齿轮；3—齿轮；4—预紧力装置；7—齿条

当载荷大时可采用径向加载法消除间隙。如图 6-14 所示，两个小齿轮 1 和 6 分别与齿条 7 啮合，并用预紧力装置 4 在齿轮 3 上预加载荷，于是齿轮 3 使相啮合的大齿轮 2 和 5 向外伸开，而与其分别同轴的小齿轮 1 和 6 同时向外伸开，这样就能分别与齿条 7 上齿槽左、右两侧分别贴紧而无间隙。

四、自动换刀装置

自动换刀装置具有根据工艺要求自动更换所需刀具的功能。自动换刀装置应满足换刀时间短、刀具重复定位精度高、刀具储存量大、刀库占地面积小及安全可靠等要求。

1. 自动换刀装置的分类

自动换刀装置根据其组成结构可分为回转刀架式、刀塔式、带刀库的自动换刀装置。

（1）回转刀架式自动换刀装置。

回转刀架是一种简单的自动换刀装置，常用于数控车床。根据加工要求设计成四方刀架、六方刀架，并相应地安装 4 把、6 把刀具。图 6-15 所示为回转刀架（即四方刀架）。其工作原理如下。

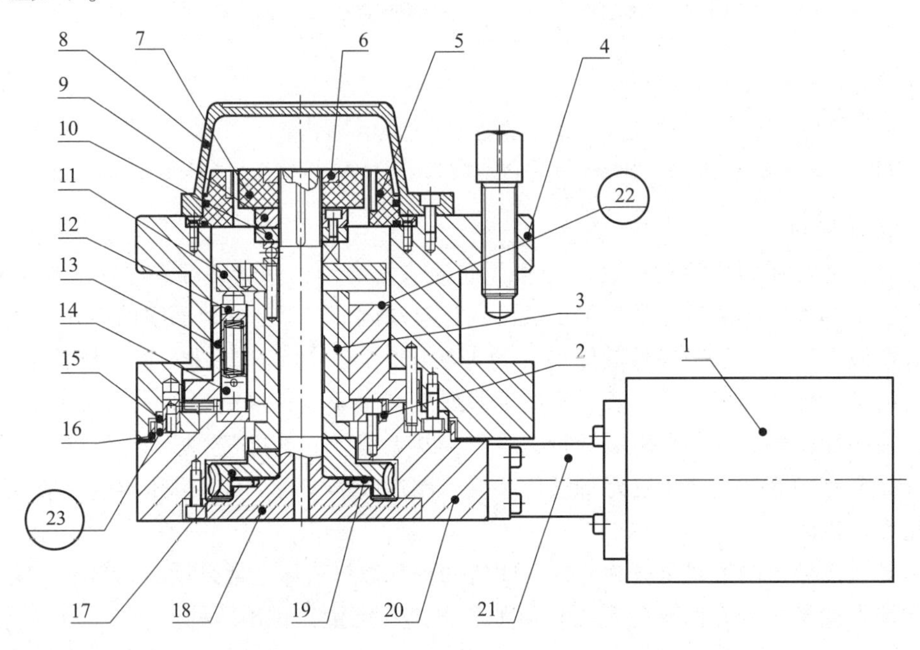

图 6-15　回转刀架

1—电机；2—反靠盘；3—螺杆；4—上刀体；5—罩座；6—小螺母；7—发信盘；8—铝罩；9—大螺母；10—止退圈；11—离合盘；12—离合销；13—螺母；14—反靠销；15—外端齿盘；16—防护罩；17—蜗轮；18—主轴；19—滚针轴承；20—下刀体；21—连接座；22—F 面；23—刀架基面

系统发出换刀信号，控制正转继电器动作，电机正转。联轴器带动蜗杆、蜗杆带动蜗轮、蜗轮带动螺杆旋转，螺母开始上升，同时螺杆带动离合盘转动，离合销在离合盘平面上滑动。当螺母上升至一定高度时（上升高度出厂装配时已调好），三端齿啮合脱开，离合销进入离合盘槽中。此时螺杆带动离合盘、离合销、螺母、上刀体、外端齿及反靠销开始转位。反靠销从反靠盘槽中爬出，即上刀体开始换刀动作。

当上刀体转到所需刀位时，霍尔元件电路发出到位信号，正转继电器松开，反转继电器吸合，电机开始反转，螺杆带动离合盘、离合销、反靠销、上刀体反转。当反靠销在反靠盘平面上移动经过反靠槽时，反靠销被弹簧弹入反靠槽。由于反靠销进入反靠槽，反靠销直角面与反靠槽直角面相互顶住（刀架完成粗定位），阻止了反靠销、离合销、上刀体的转动。此时离合盘在螺杆的带动下继续反转，离合销从离合盘槽中爬出，螺母开始下降，直至三端齿完全啮合，完成精定位，刀架锁紧。此时反转时间到，反转继电器松开，电机停止转动。并向系统发出回答信号，加工程序开始。

（2）刀塔式自动换刀装置。

刀塔式自动换刀装置是倾斜床身数控车床中常见的换刀装置，如图 6-16 所示。其特点是可双向旋转、任意刀位就近选刀、结构紧凑、定位精度高。

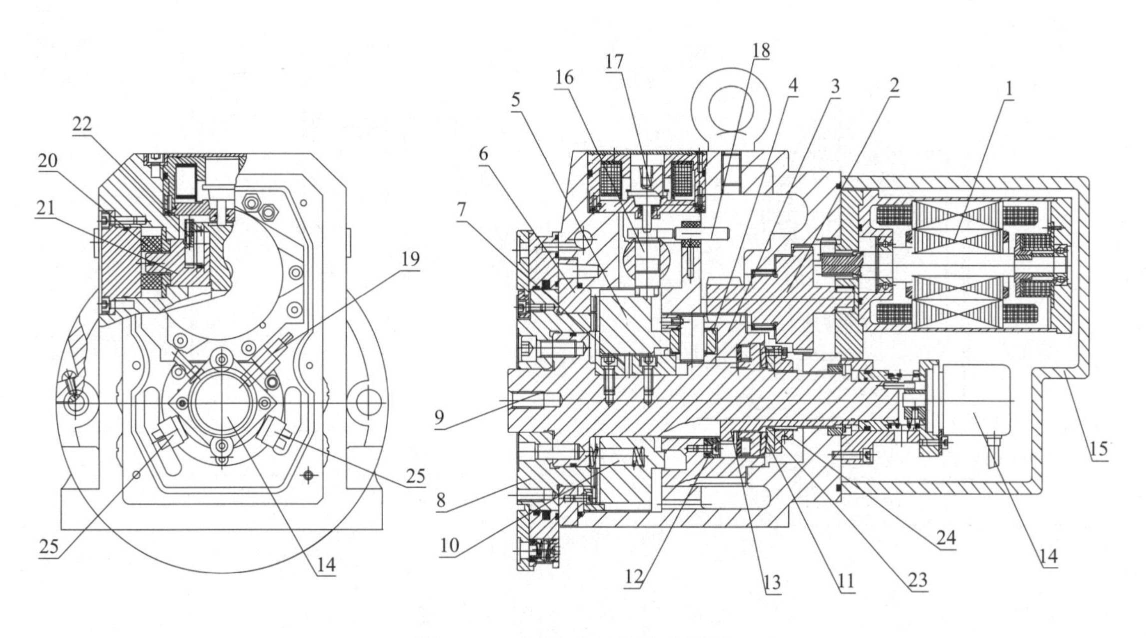

图 6-16　刀塔式自动换刀装置

1—电机；2—二级双联齿轮；3—滚轮座；4—滚轮；5—销盘；6—外端齿盘；7—中轴键；8—内端齿；9—中轴；10—松开弹簧；11—碟形弹簧；12—发信头；13—发信杆；14—编码器；15—后罩；16—定位销；17—电磁吸铁；18—插销接近开关；18—夹紧接近开关；20—减振垫；21—减振杆；22—小销；23—止退圈；24—螺母；25—接线端子

刀塔由电机驱动后，通过齿轮减速、电磁吸铁插销动作、凸轮夹紧完成精定位。工作程序：系统发出换刀指令、刀塔电机内的制动器松开、系统根据要刀位置与当前刀号位置判断电机转动方向、电机转动、凸轮松开、刀塔转位。当刀塔转动到所要刀位前一工位选通信号下沿时，预定位电磁吸铁供电、插销动作、插销接近开关接收插入信号、电机停转、停 50 ms 电机反转、夹紧接近开关接收夹紧信号、电机停转、电机内制动器供电、延时 200 ms 插销电磁铁失电、插销接近开关检测插销是否松开、换刀程序结束。

（3）带刀库的自动换刀装置。

目前大量使用的是带有刀库的自动换刀装置。由于有刀库，所以加工中心只需要有一个夹持刀具进行切削的主轴。当需要用某一刀具进行切削加工时，将该刀具自动地从刀库交换到主轴上，切削完毕后又将用过的刀具自动地从主轴上放回刀库。由于换刀过程是在各部件之间进行的，所以要求参与换刀的各部件的动作必须准确协调。这种换刀装置不仅使主轴刚度得到提高，还提高了加工精度和加工效率，而且刀具的存储容量增多，有利于加工复杂零件。刀库还可以离开加工区，消除了很多不必要的干扰。带刀库的自动换刀装置可以分为无机械手换刀装置和有机械手换刀装置两类。

无机械手换刀装置一般把刀库放在主轴箱可以移动到的位置，如可以将整个刀库或刀库的某一刀位能移动到主轴箱可以到达的位置。刀库中刀具的存放方向一般与主轴箱的装刀方向一致。换刀时通过主轴和刀库的相对运动执行换刀动作，利用主轴取走或放回刀具。图 6-17 所示为加工中心的换刀装置。

图 6-18 所示为如图 6-17（a）所示的卧式加工中心无机械手换刀装置的换刀过程。

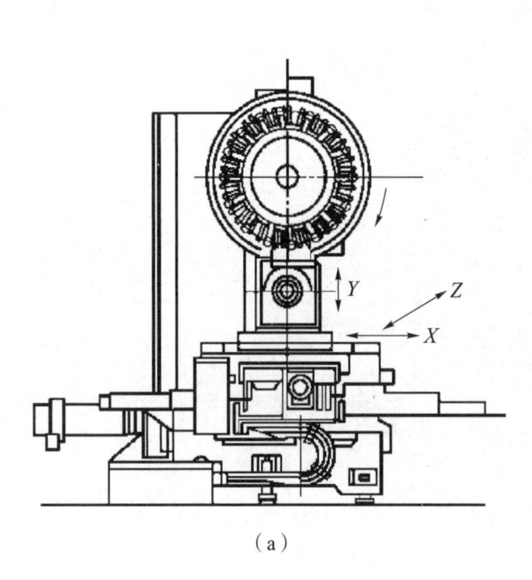

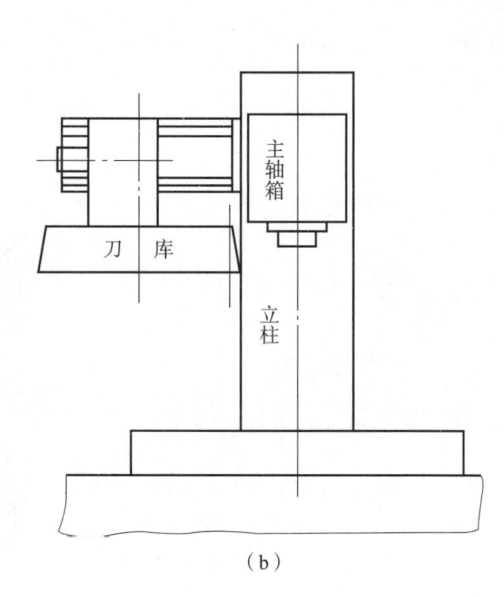

图 6-17　加工中心的换刀装置

（a）卧式加工中心；（b）立式加工中心

图 6-18（a）表示上一工步结束后，主轴准停定位，主轴箱上升。图 6-18（b）表示主轴箱上升到顶部换刀位置，刀具进入刀库的交换位置空位。刀具被刀库上的夹爪固定，主轴上的刀具自动夹紧装置松开。图 6-18（c）表示刀库前移，从主轴孔中把要更换的刀具拔出。

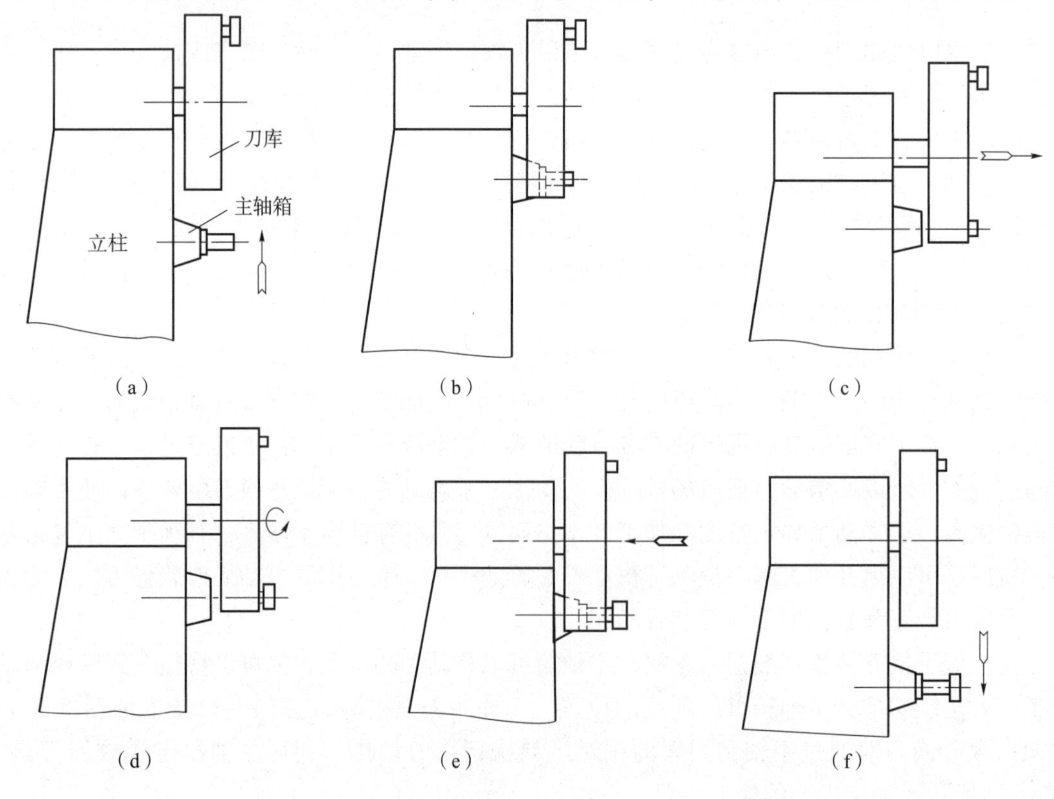

图 6-18　换刀过程

图 6-18（d）表示刀库转位，根据程序把下一工步要用的刀具转换到换刀位置，同时主轴孔清洁装置清洁主轴上的刀具孔。图 6-18（e）表示刀库后退，把需要的刀具插入主轴孔，主轴上的刀具夹紧装置把刀具夹紧。图 6-18（f）表示主轴箱下降到工作位置，开始进行下一步工作。

无机械手换刀装置的优点有结构简单、成本低、换刀可靠性较高；缺点是由于结构所限，刀库的容量不大，且换刀时间较长（一般需要 10~20 s）。因此，无机械手换刀装置多为中小型加工中心所采用。

2. 刀库

数控机床的刀库

刀库用来储存加工刀具及辅助工具。由于多数加工中心的取送刀位置都是在刀库中的某一固定刀位，因此刀库还需要有使刀具运动及定位的机构，以保证换刀的可靠进行。使刀具运动的动力可采用伺服电机或液压马达。刀具定位机构用来保证要更换的每一把刀具和刀套都能准确地停在换刀位置上，其控制部分可以采用简易位置控制器，或类似半闭环进给系统的伺服位置控制。也可以采用电气和机械相结合的销定位方式，一般要求其综合定位精度达到 0. 1~0. 5 mm 即可。

按结构形式刀库可分为圆盘式刀库、链式刀库和箱格式刀库。图 6-19 所示为几种典型的刀库形式。圆盘式刀库结构简单，应用也较多，但由于刀具采用环形排列，所以空间利用率低。因此出现了将刀具在盘中采用双环或多环排列的形式，以增加空间的利用率，但这样刀库的外径扩大，转动惯量也增大，选刀时间也较长，因此圆盘式刀库一般用于刀具容量较

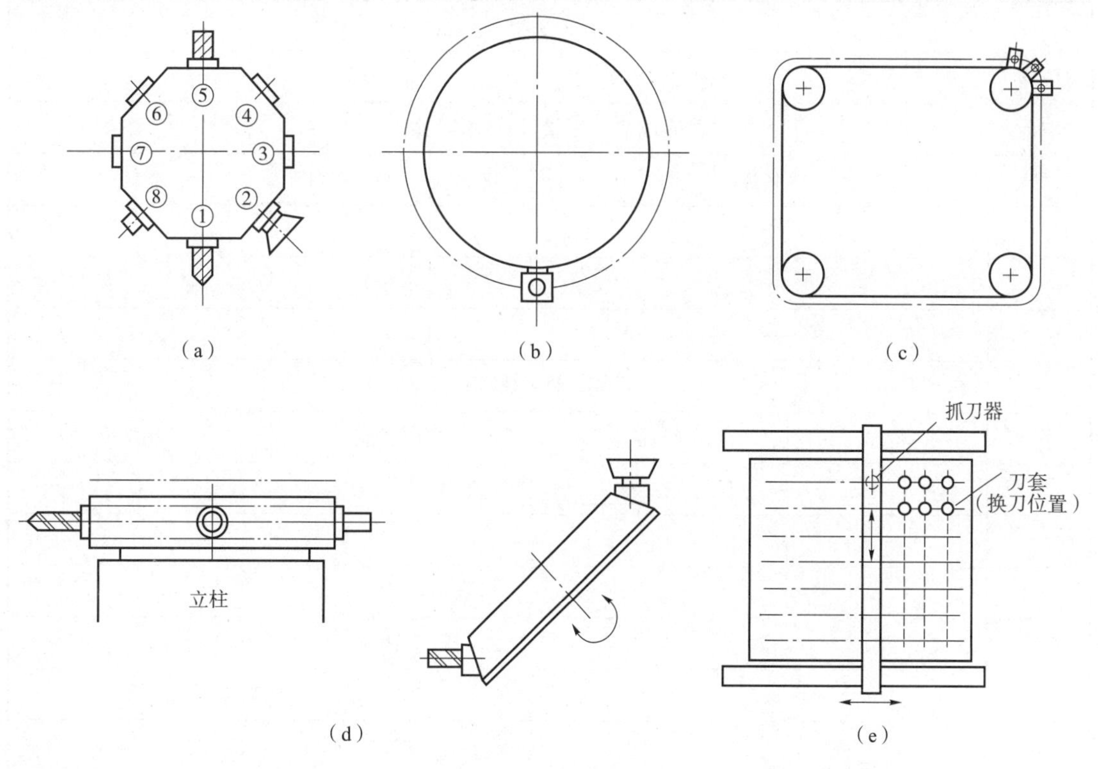

图 6-19　刀库形式

（a）转塔式；（b）圆盘式（侧置式）；（c）链式；（d）圆盘式（顶置式）；（e）箱格式

小的刀库；链式刀库的结构紧凑，刀库容量较大，链的形状可以根据机床的布局配置，也可以将换刀位突出以利于换刀。当必须增加链式刀库的刀具容量时，只需要增加链条的长度，在一定范围内，无须变更刀库的线速度及惯量。一般刀具数量为 30~120 把时都采用链式刀库；箱格式刀库的结构也简单，有线形和箱形两种。线形刀库一般用于无机械手换刀装置，箱形刀库一般容量比较大，往往用于加工单元式加工中心。

另外，按设置部位的不同刀库可分为顶置式、侧置式、悬挂式和落地式等多种类型。按交换刀具还是交换主轴，刀库可分为普通刀库（简称刀库）和主轴箱刀库两种。

刀库选刀方式一般采用“近路移动”原则，即无论采取哪种选刀方式，在根据程序指令把下一工序要用的刀具移动到换刀位置时，都要向小于刀库半圈的方向移动，以节省选刀时间。

【任务实施】

（1）按照随机分组方式，分成各个观摩小组，并选出一名组长，将本组成员的个人信息填入表 6-1 中。

（2）根据观摩及讨论的结果，在表 6-1 中记录各机床机械结构名称、特征、主要功能及适用场合等。

（3）对观摩过程中的疑难问题进行充分讨论，并将问题及解决办法填入表 6-1 中。

表 6-1　机床机械观摩任务工单

<table>
<tr><td colspan="6">任务工单</td></tr>
<tr><td colspan="2">姓名：</td><td colspan="2">班级：</td><td colspan="2">学号：</td></tr>
<tr><td colspan="6">所需材料、工量具、设备列表</td></tr>
<tr><td>序号</td><td colspan="2">名称</td><td>数量</td><td>型号或规格</td><td>备注</td></tr>
<tr><td></td><td colspan="2"></td><td></td><td></td><td></td></tr>
<tr><td></td><td colspan="2"></td><td></td><td></td><td></td></tr>
<tr><td></td><td colspan="2"></td><td></td><td></td><td></td></tr>
<tr><td colspan="6">工作过程（观摩）</td></tr>
<tr><td>序号</td><td>名称</td><td>特征</td><td colspan="2">主要功能</td><td>适用场合</td></tr>
<tr><td></td><td></td><td></td><td colspan="2"></td><td></td></tr>
<tr><td></td><td></td><td></td><td colspan="2"></td><td></td></tr>
<tr><td></td><td></td><td></td><td colspan="2"></td><td></td></tr>
<tr><td></td><td></td><td></td><td colspan="2"></td><td></td></tr>
<tr><td colspan="3">出现问题</td><td colspan="3">解决办法</td></tr>
<tr><td colspan="3"></td><td colspan="3"></td></tr>
</table>

【任务评价】

序号	评价内容	分值	得分
1	能够熟练说出数控机床机械结构	25	
2	能够掌握数控机床进给系统的结构	25	
3	能够掌握数控机床换刀系统的结构	25	
4	具有有效沟通、表达及团结协作的职业精神	25	

任务二　数控机床主轴装调

数控机床
主轴的装调

【学习目标】

（1）了解数控机床主轴结构。

（2）熟练使用数控机床主轴安装使用的工量具。

（3）读懂并掌握数控机床主轴装配图。

（4）掌握数控机床主轴安装工艺。

（5）安全文明操作。

（6）掌握工程工作方法，培养严谨的工作作风，遵守5S管理制度。

【任务描述】

以小组为单位，观摩YL569型数控铣床的主轴结构，并将数控机床主轴装调步骤、装调方法、主要功能、精度等填写到任务工单中。要求观摩过程中务必注意安全，遵守实训场所相关操作规范。

【知识链接】

主轴是数控机床主传动系统的重要组件，用来安装刀具并高速旋转，其装配质量好坏直接影响工件加工质量。本任务以亚龙YL-1506B数控铣床主轴为例，具体介绍主轴装配过程。YL-1506B主轴总装图如图6-20所示。

一、主轴装配前准备

（1）主轴组件中各零件均需要清洗干净，尤其与轴承接触面需蘸酒精擦拭，并检验无污迹。

（2）检查零件定位表面无疤痕、划伤、锈斑，并重点检查接触台阶面与轴承配合外圆面。

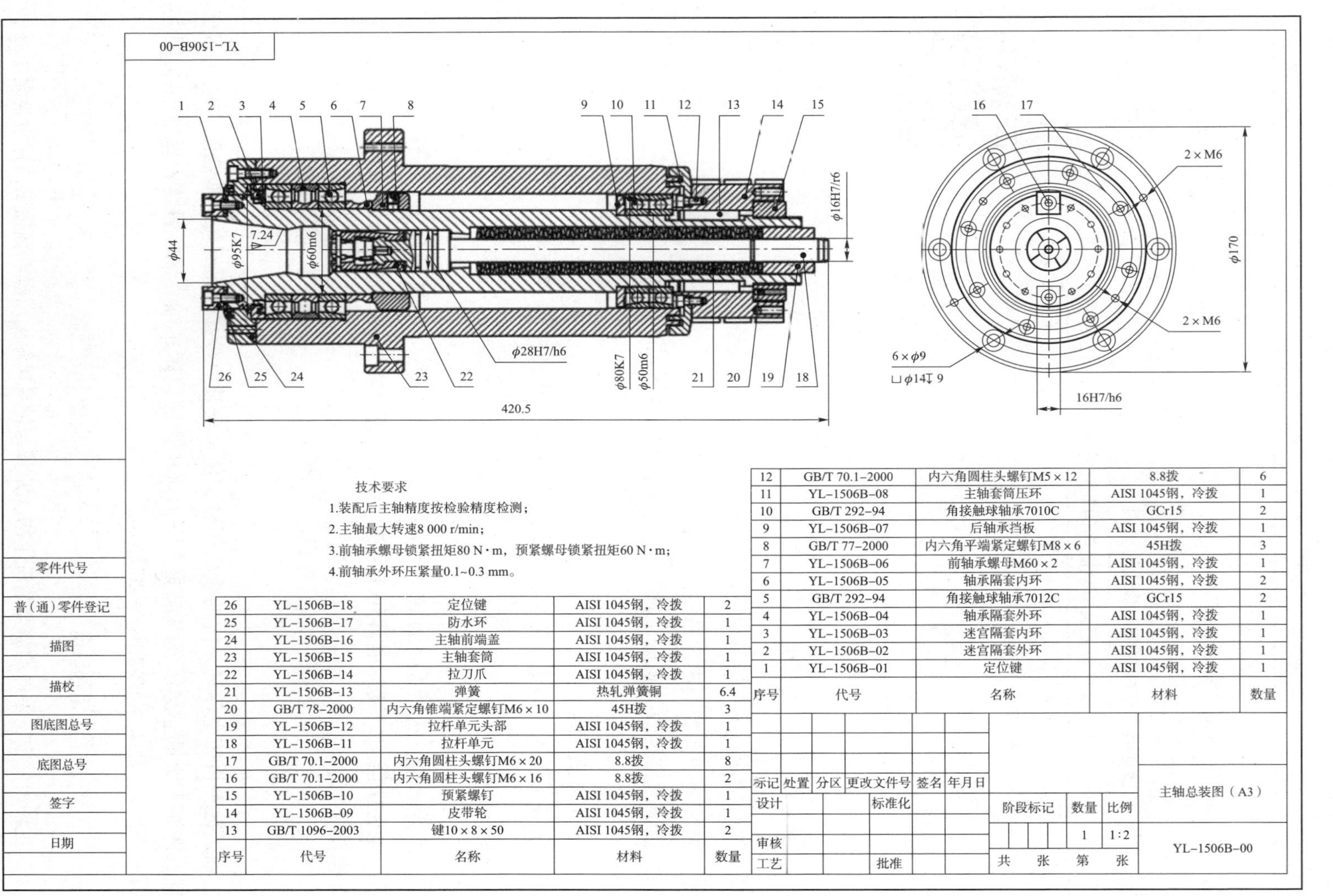

技术要求

1.装配后主轴精度按检验精度检测；

2.主轴最大转速8 000 r/min；

3.前轴承螺母锁紧扭矩80 N·m，预紧螺母锁紧扭矩60 N·m；

4.前轴承外环压紧量0.1~0.3 mm。

序号	代号	名称	材料	数量
26	YL-1506B-18	定位键	AISI 1045钢，冷拨	2
25	YL-1506B-17	防水环	AISI 1045钢，冷拨	1
24	YL-1506B-16	主轴前端盖	AISI 1045钢，冷拨	1
23	YL-1506B-15	主轴套筒	AISI 1045钢，冷拨	1
22	YL-1506B-14	拉刀爪	AISI 1045钢，冷拨	1
21	YL-1506B-13	弹簧	热轧弹簧铜	6.4
20	GB/T 78-2000	内六角锥端紧定螺钉M6×10	45H拨	3
19	YL-1506B-12	拉杆单元头部	AISI 1045钢，冷拨	1
18	YL-1506B-11	拉杆单元	AISI 1045钢，冷拨	1
17	GB/T 70.1-2000	内六角圆柱头螺钉M6×20	8.8拨	8
16	GB/T 70.1-2000	内六角圆柱头螺钉M6×16	8.8拨	2
15	YL-1506B-10	预紧螺钉	AISI 1045钢，冷拨	1
14	YL-1506B-09	皮带轮	AISI 1045钢，冷拨	1
13	GB/T 1096-2003	键10×8×50	AISI 1045钢，冷拨	2
12	GB/T 70.1-2000	内六角圆柱头螺钉M5×12	8.8拨	6
11	YL-1506B-08	主轴套筒压环	AISI 1045钢，冷拨	1
10	GB/T 292-94	角接触球轴承7010C	GCr15	2
9	YL-1506B-07	后轴承挡板	AISI 1045钢，冷拨	1
8	GB/T 77-2000	内六角平端紧定螺钉M8×6	45H拨	3
7	YL-1506B-06	前轴承螺母M60×2	AISI 1045钢，冷拨	1
6	YL-1506B-05	轴承隔套内环	AISI 1045钢，冷拨	2
5	GB/T 292-94	角接触球轴承7012C	GCr15	2
4	YL-1506B-04	轴承隔套外环	AISI 1045钢，冷拨	1
3	YL-1506B-03	迷宫隔套内环	AISI 1045钢，冷拨	1
2	YL-1506B-02	迷宫隔套外环	AISI 1045钢，冷拨	1
1	YL-1506B-01	定位键	AISI 1045钢，冷拨	1

标记	处置	分区	更改文件号	签名	年月日	阶段标记	数量	比例	主轴总装图（A3）
设计			标准化				1	1:2	YL-1506B-00
审核						共　张	第　张		
工艺			批准						

图 6-20　亚龙 YL-1506B 主轴总装图

（3）检查各锐边倒角无毛刺，保证装配时用手触摸光滑顺畅无棱角。

（4）检查紧固螺纹孔的残屑、深度，并用丝锥去除残屑，吹净。

（5）清除干净的零件摆放在无灰尘的干净油纸或布上，清洗过且暂时不用的零件需加上防尘盖。

（6）零件摆放位置应与工作区域保持≥800 mm 以上的距离。

（7）轴承清洗处理。

1）轴承清洗液用两个容器分别盛装，一个为清洗用、一个为清涮轴承用。

2）在初洗轴承过程中，不允许相对转动轴承，可在液体中上下左右晃动。

3）清洗完成后将轴承放在涮洗池中，边刷边转动轴承内外环。

4）清洗过程中不得将轴承放入池底，洗完必须将轴承离池。

5）清洗完轴承在离开液池前甩下轴承上的液珠，并转动轴承后重复此操作。

6）放干净处，用油纸或擦纸遮盖，进行晾干。为缩短晾干时间，可用电吹风吹干。严禁使用空压机风管吹轴承。

二、主轴配合零件间精度检测和零件检查

（1）角接触球轴承（7010C）（7012C）与主轴分别试装。

（2）检验平台测量前用抹布擦拭干净，将配合件放在检验平台上，检测各项精度。

检验前/后轴承（7010C）（7012C）等高精度，要求值≤0.002 mm，内外环逐一检测。检测过程如图 6-21 所示。

（3）皮带轮平衡顶丝（M6×10；GB70）用天平称量后分组，各组质量差≤0.2 g。

（4）反扣盘上紧固螺丝（M5×12；GB70）用天平称量后分组，各组质量差≤0.2 g。

（5）主轴前端盖键螺钉（M5×12；GB70）用天平称量后分组。

（6）皮带轮张紧固定螺钉用天平称量后分组，各组质量差≤0.2 g。

（7）向主轴轴承注入润滑脂。

1）注润滑脂前保证轴承已晾干，清洗过的注射器筒装入润滑脂，然后推压排除空气使注射器留有规定容量，前轴承 3.6 cm^3，后轴承是 2.6 cm^3。

2）对轴承每个滚动体进行均匀注入，并且两面分配。

（8）主轴前轴承装配相关数据测量。

1）用深度尺测量主轴套筒端面到主轴套隔台的数值 *K*1，如图 6-22 所示。

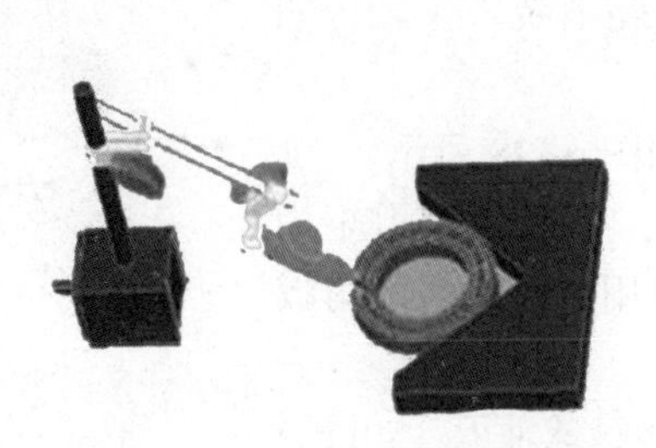

图 6-21　前/后轴承等高精度检测

图 6-22　深度尺测量 *K*1

2）清洗后的轴承，一起叠加放置，分别为角接触球轴承（7012C）、轴承隔套内环、轴

承隔套外环、角接触球轴承（7012C）、迷宫隔环内、外环。用深度尺测量叠加高度数值为 $K2$，如图 6-23 所示。

3）用深度尺测量主轴前端盖凹台深度数值为 H，如图 6-24 所示（在相互垂直的两组位置各测量一次，所得值进行加权计算平均值）。

图 6-23　深度尺测量 $K2$

图 6-24　深度尺测量 H

4）出厂安装按 $K=K2-K1+0.2$ mm 与 H 值的偏差结果修配调整主轴前端盖。

测量各数值时，保证各工件干净，无污渍。等高台在测量前用酒精、擦拭纸擦拭干净。

三、主轴部件装配

（1）主轴前端面朝下竖立在工作台上。

（2）放入迷宫隔环外环，要求迷宫隔环外环环形槽朝上装入主轴。

（3）放入迷宫隔环内环，要求迷宫隔环内环环形槽朝下装入主轴，如图 6-25 所示。

（4）将角接触球轴承（7012C）放置在主轴迷宫环内环上，要求轴承外圈宽端面一侧朝上装入主轴，如图 6-26 所示。

图 6-25　安装迷宫隔环内外环

图 6-26　角接触球轴承安装

（5）将轴承隔套内环装入主轴，再放置轴承隔套外环，将第二个角接触球轴承（7012C）外环宽端面一侧朝下装入主轴，如图 6-27 所示。

（6）将另一个轴承隔套内环装入主轴，如图 6-28 所示。

（7）将前轴承螺母（M60×2）装入主轴，要求锁紧力矩为 80 N · m。使用勾扳手紧固前轴承螺母，再使用 4 mm 内六角扳手将其三颗 M8×6 顶丝紧固，如图 6-29 所示。

（8）用磁性表座吸在主轴上，表头接触角接触球轴承（7012C）外环，旋转测量并调整外圆与主轴同心，允差 ≤0.05 mm，如图 6-30 所示。

图 6-27　轴承隔套内外环与第二个角接触球轴承安装

图 6-28　另一个轴承隔套内环安装

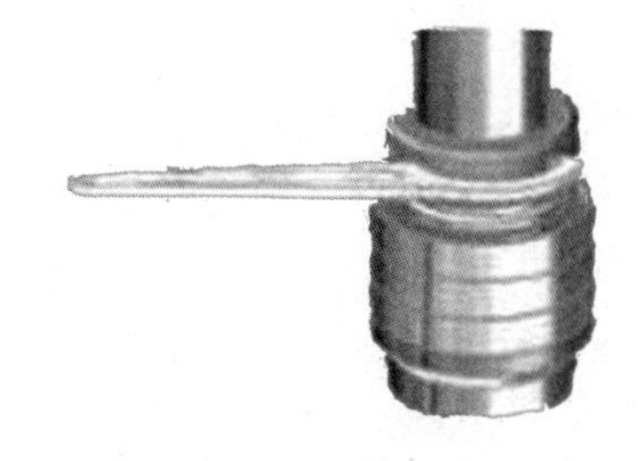

图 6-29　紧固前轴承螺母

（9）用磁性表座吸在后角接触球轴承轴径上，表头接触主轴，检查其回转跳动，允差≤0.04 mm，如图 6-31 所示。

（10）用磁性表座吸在主轴上，磁性表座不动，让表头接触在角接触球轴承外环端面，转动外环，检查其端面跳动，允差≤0.02 mm，如图 6-32 所示。

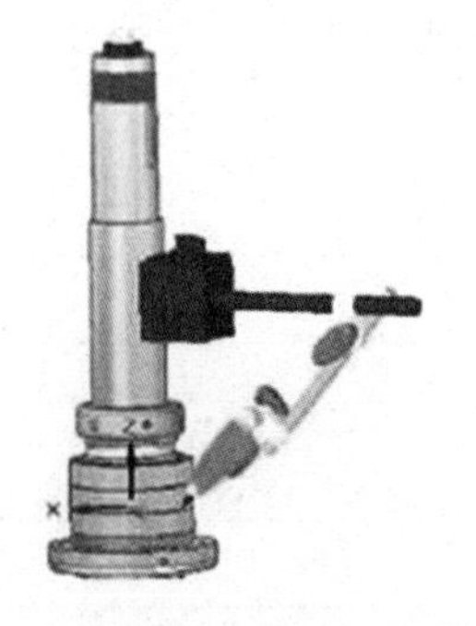

图 6-30　测量并调整外圆与主轴同心

图 6-31　检查后角接触球轴承回转跳动

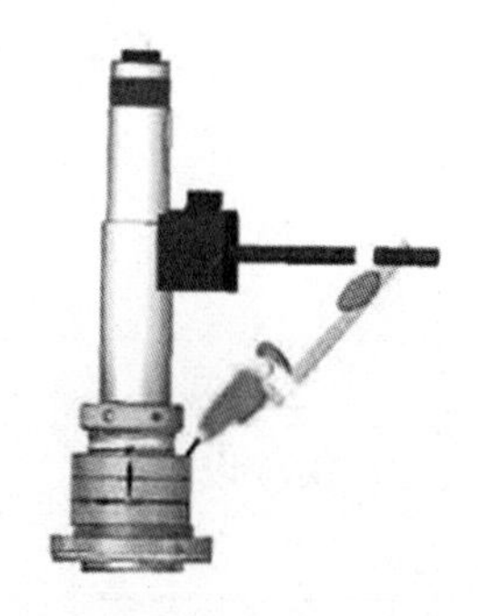

图 6-32　检查角接触球轴承端面跳动

（11）第“9”工序和第“10”工序检验，若跳动超差，可通过调整前轴承螺母（M60×2）上三颗顶丝或轻敲螺母对应方向达到要求止。

（12）装入后轴承挡板（凸面朝上）。

（13）装入角接触球轴承（7010C），组合方式为 DB，将其放置在后轴承挡板上。

（14）将主轴套筒套入主轴。

（15）先使用（M5×12）螺丝组装好主轴套筒压环与皮带轮。

（16）安装键（C10×8×50）。

（17）将组装好的主轴套筒压环与皮带轮装入主轴。

（18）将预紧螺母安装在主轴上，要求锁紧力矩为 60 N·m，使用可调式圆螺母扳手将其安装到位，并调整预紧螺母上的三颗顶丝（M6×10）。

（19）安装主轴前端盖及防水环，并用八颗内六角圆柱头螺钉 M6× 20 的螺丝锁紧。计算所得前轴承外环压紧量 A 在技术要求公差范围内，其中 $A=K2-K1-H$。

（20）安装定位键，并用两颗内六角圆柱头螺钉 M6×16 的螺丝锁紧。

四、检测主轴部件精度

将主轴放置在检测台上，检测主轴跳动，要求跳动≤0.01 mm。

【任务实施】

（1）按照随机分组方式，分成各个观摩小组，并选出一名组长，将本组成员的个人信息填入表 6-2 中。

（2）根据观摩及讨论的结果，在表 6-2 中记录数控机床主轴装调步骤、装调方法、主要功能及精度等。

（3）对观摩过程中的疑难问题进行充分讨论，并将问题及解决办法填入表 6-2 中。

表 6-2　数控机床主轴装调任务工单

任务工单				
姓名：		班级：		学号：
所需材料、工量具、设备列表				
序号	名称	数量	型号或规格	备注
工作过程（观摩）				
序号	装调步骤	装调方法	主要功能	精度
出现问题			解决办法	

【任务评价】

序号	评价内容	分值	得分
1	能够熟练说出数控机床主轴结构	25	
2	能够正确使用数控机床主轴安装使用的工量具	25	
3	能够掌握数控机床主轴安装工艺	25	
4	具有有效沟通、表达及团结协作的职业精神	25	

任务三　数控机床几何精度检测

数控机床精度检测与验收

【学习目标】

（1）了解数控机床几何精度检测内容及相应的工量具。
（2）掌握直线度检测方法。
（3）掌握平行度检测方法。
（4）掌握垂直度检测方法。
（5）掌握主轴跳动检测方法。
（6）掌握平面度检测方法。
（7）掌握工程工作方法，培养严谨的工作作风，遵守5S管理制度。

【任务描述】

以小组为单位，观摩YL569型数控机床的几何精度，并将数控机床各项几何精度检测的项目名称、检测工具、检测方法及允许范围等填写到任务工单中。要求观摩过程中务必注意安全，遵守实训场所相关操作规范。

【知识链接】

一、数控机床几何精度检测内容

数控机床的几何精度综合反映了该机床的各关键零部件及其组装后的几何形状误差，因为在几何精度中有些项目是相互联系、相互影响的，所以机床几何精度的检测必须在机床精调后一次完成，不允许调整一项检测一项。

几何精度检测的项目一般包括直线度、平面度、平行度等。如加工中心几何精度检测通常包括以下内容。

（1）工作台面的平面度。
（2）各坐标轴方向移动的相互垂直度。
（3）X、Y轴坐标方向移动时工作台面的平行度。
（4）主轴的轴向窜动。
（5）主轴孔的径向跳动。
（6）主轴箱沿Z坐标方向移动时主轴轴心线的平行度。
（7）主轴回转轴心线对工作台面的垂直度。
（8）主轴箱在Z坐标方向移动的直线度。

二、数控铣床精度检测实验

1. 实验仪器

（1）数控铣床。

（2）平尺（400 mm，1 000 mm，0 级）两个。

（3）角尺（300 mm×200 mm，0 级）一个。

（4）直验棒（80 mm×500 mm）一个。

（5）BT40 锥度检验棒一个。

（6）百分表两只。

（7）磁力表座两只。

（8）水平仪（200 mm，0. 02/1 000）两个。

（9）等高块三只。

（10）可调量块两只。

2. 实验内容：几何精度验收

1）机床调平。

检测工具：水平仪。

检测方法：如图 6-33 所示，将工作台置于导轨行程的中间位置，将两个水平仪分别沿 X 和 Y 坐标轴置于工作台中央，调整机床垫铁高度，使水平仪水泡处于读数中间位置；分别沿 X 和 Y 坐标轴全行程移动工作台，观察水平仪读数的变化，调整机床垫铁的高度，使工作台沿 X 和 Y 坐标轴全行程移动时水平仪读数的变化范围小于两格，且读数处于中间位置即可。

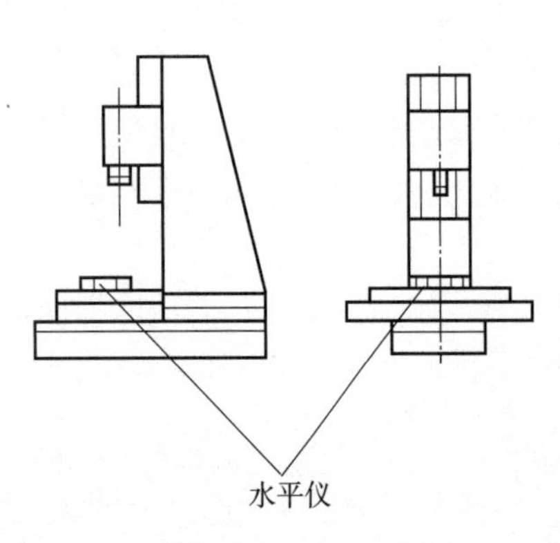

图 6-33　机床调平图解

2）检测工作台面的平面度。

检测工具：百分表、平尺、可调量块、等高块、水平仪。

用平尺检测工作台面的平面度误差的原理：在规定的测量范围内，若所有点被包含在与该平面的总方向平行并相距给定值的两个平面内时，则认为该平面是平的。

检测方法：如图 6-34 所示，首先在检验面上选 A、B、C 三个点作为零位标记，将三个等高量块放在这三点上，这三个量块的上表面就确定了与被检面作比较的基准面。将平尺置于点 A 和点 C 上，并在检验面上点 E 处放一可调量块，使其与平尺的小表面接触。此时，量块 A、B、C、E 的上表面均在同一表面上。再将平尺放在点 B 和点 E 上，即可找到点 D 的偏差。在点 D 处放一可调量块，并将其上表面调到由已经就位的量块上表面所确定的平面上。将平尺分别放在点 A 和点 D 及点 B 和点 C 上，即可找到被检面上点 A 和点 D 及点 B 和点 C 之间的偏差。其余各点之间的偏差可用同样的方法找到。

3）主轴锥孔轴线的径向跳动、主轴端面偏摆、主轴套筒外壁偏摆。

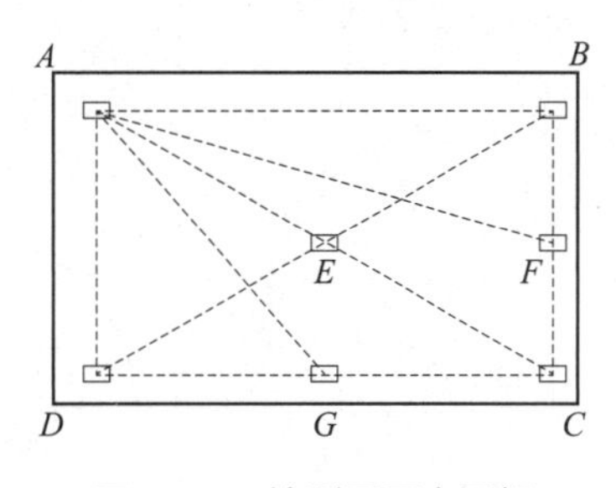

图 6-34　检测平面度图解

检测工具：检验棒、百分表。

检测方法：如图 6-35 所示，将检验棒插在主轴锥孔内，百分表安装在机床固定部件上，百分表测头垂直触及被测表面，旋转主轴，记录百分表的最大读数差值，在 *a*、*b* 处分别测量。标记检验棒与主轴的圆周方向的相对位置，取下检验棒，同向分别旋转检验棒 90°、180°、270°后重新插入主轴锥孔，在每个位置分别检测。取四次检测的平均值为主轴锥孔轴线的径向跳动误差。

检测主轴端面偏摆及主轴套筒外壁偏摆如图 6-36 所示。

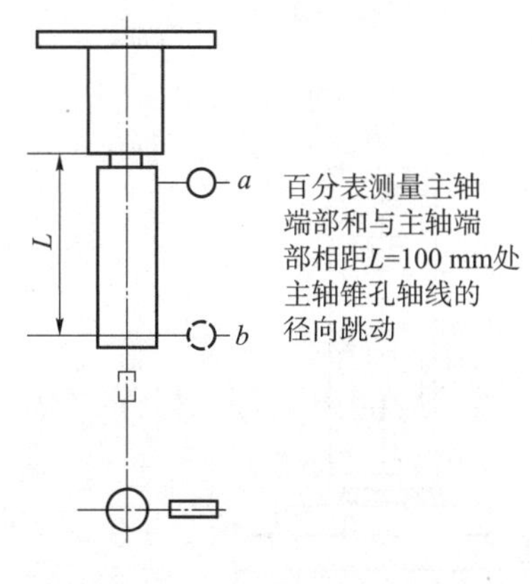

图 6-35　检测主轴锥孔轴线径向跳动图解

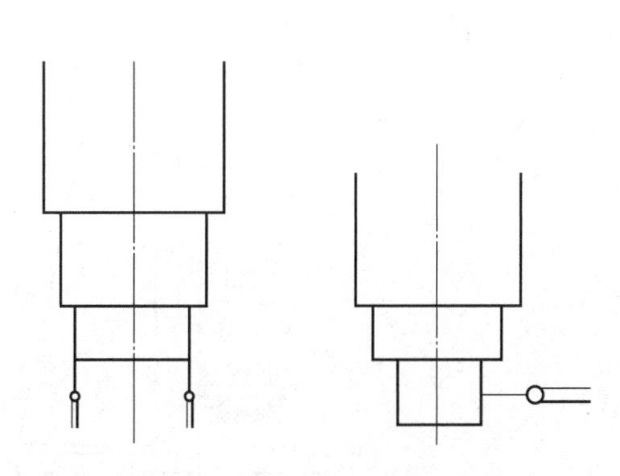
图 6-36　检测主轴端面偏摆及主轴套筒外壁偏摆图解

4）主轴轴线对工作台面的垂直度。

检测工具：平尺、等高块、百分表、表架。

检测方法：如图 6-37 所示，将带有百分表的表架装在主轴上，并将百分表的测头调至平行于主轴轴线位置。被测平面与基准面之间的平行度偏差可以通过百分表测头在被测平面上摆动的检查方法测得。主轴旋转一周，百分表读数的最大差值即垂直度偏差。分别在 *X*-*Z*、*Y*-*Z* 平面内记录百分表在相隔 180°的两个位置上的读数差值。为消除测量误差，可在第一次检验后将验具相对于主轴转过 180°再重复检验一次。

5）主轴箱垂直移动对工作台面的垂直度。

检测工具：等高块、平尺、角尺、百分表。

检测方法：如图 6-38 所示，将等高块沿 *Y* 轴方向放在工作台上，平尺置于等高块上，将角尺置于平尺上（在 *Y*-*Z* 平面内），百分表固定在主轴箱上，百分表测头垂直触及角尺，移动主轴箱，记录百分表读数及方向，其读数最大差值即在 *Y*-*Z* 平面内主轴箱垂直移动对工作台面的垂直度误差；同理，将等高块、平尺、角尺置于 *X*-*Z* 平面内重新测量一次，百分表读数最大差值即在 *X*-*Z* 平面内主轴箱垂直移动对工作台面的垂直度误差。

6）主轴套筒垂直移动对工作台面的垂直度。

检测工具：等高块、平尺、角尺、百分表。

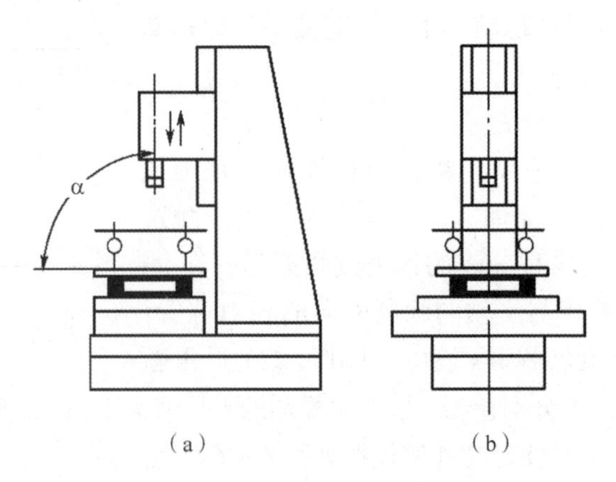

图 6-37　检测主轴轴线对工作台面的垂直度图解

（a）Y-Z 平面；（b）X-Z 平面

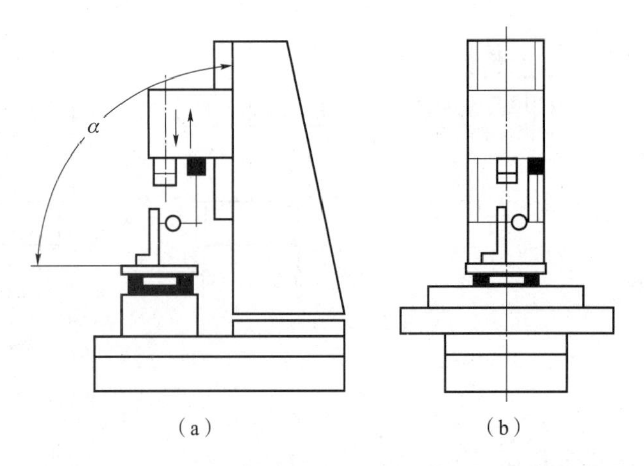

图 6-38　检测主轴箱垂直移动对工作台面的垂直度图解

（a）Y-Z 平面；（b）X-Z 平面

检测方法：如图 6-39 所示，将等高块沿 Y 轴方向放在工作台上，平尺置于等高块上，将角尺置于平尺上，并调整角尺位置使角尺轴线与主轴轴线重合；将百分表固定在主轴上，百分表测头在 Y-Z 平面内垂直触及角尺，移动主轴，记录百分表读数及方向，其读数最大差值即在 Y-Z 平面内主轴套筒垂直移动对工作台面的垂直度误差；同理，百分表测头在 X-Z 平面内垂直触及角尺重新测量一次，百分表读数最大差值即在 X-Z 平面内主轴套筒垂直移动对工作台面的垂直度误差。

7）工作台 X 轴方向或 Y 轴方向移动对工作台面的平行度。

检测工具：等高块、平尺、百分表。

检测方法：如图 6-40 所示，将等高块沿 Y 轴方向放在工作台上，平尺置于等高块上，把百分表测头垂直触及平尺，Y 轴方向移动工作台，记录百分表读数，其读数最大差值即工作台 Y 轴方向移动对工作台面的平行度误差；将等高块沿 X 轴方向放在工作台上，X 轴方向移动工作台，重复测量一次，其读数最大差值即工作台 X 轴方向移动对工作台面的平行度误差。

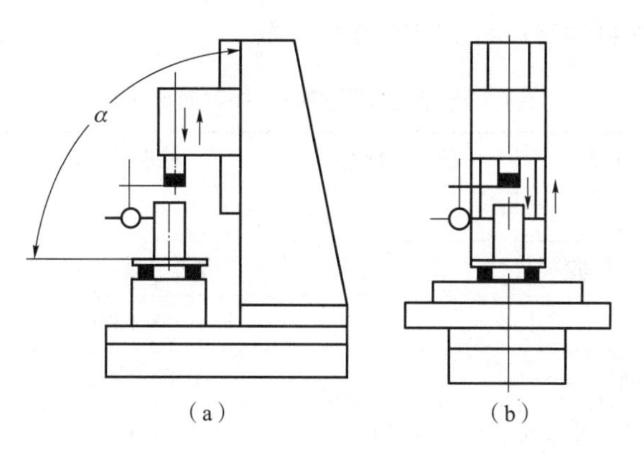

图 6-39　检测主轴套筒垂直移动对工作台面的垂直度图解

(a) Y-Z 平面；(b) X-Z 平面

8）工作台 X 轴方向移动对工作台面基准（T 形槽）的平行度。

检测工具：百分表。

检测方法：如图 6-41 所示，把百分表固定在主轴箱上，使百分表测头垂直触及基准（T 形槽），X 轴方向移动工作台，记录百分表读数，其读数最大差值即工作台 X 轴方向移动对工作台面基准（T 形槽）的平行度误差。

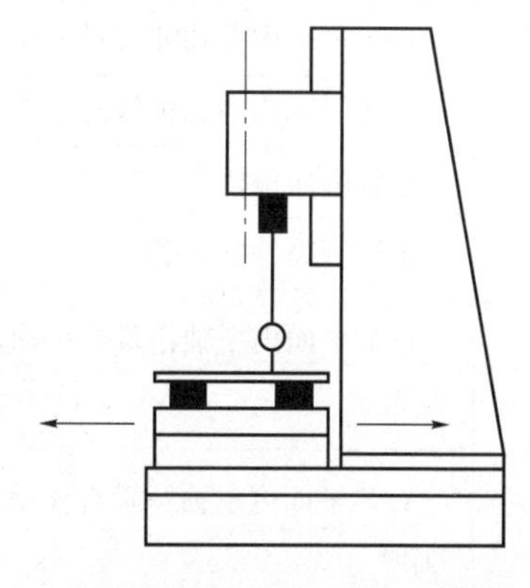

图 6-40　检测工作台 X 轴、Y 轴方向移动对工作台面的平行度图解

9）工作台 X 轴方向移动对 Y 轴方向移动的工作垂直度。

检测工具：角尺、百分表。

检测方法：如图 6-42 所示，工作台处于行程的中间位置，将角尺置于工作台上，把百分表固定在主轴箱上，使百分表测头垂直触及角尺（Y 轴方向），Y 轴方向移动工作台，调整角尺位置，使角尺的一个边与 Y 轴轴线平行，再将百分表测头垂直触及角尺另一边（X 轴方向），X 轴方向移动工作台，记录百分表读数，其读数最大差值即工作台 X 轴方向移动对 Y 轴方向移动的工作垂直度误差。

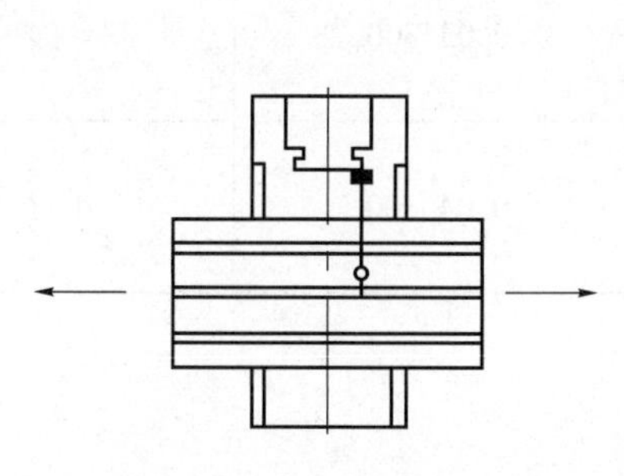

图 6-41　检测工作台 X 轴方向移动对工作台面基准（T 形槽）的平行度图解

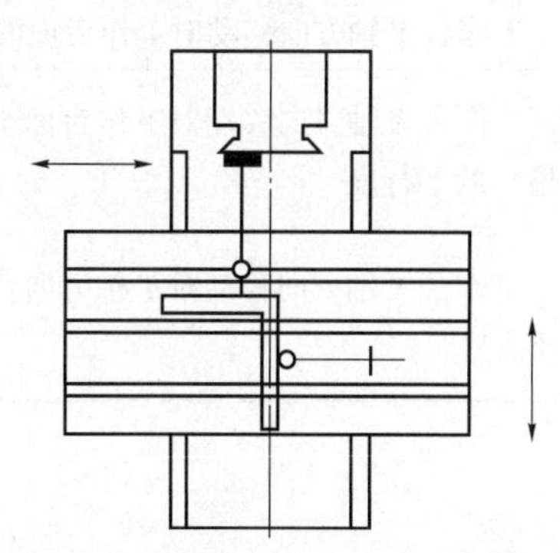

图 6-42　检测工作台 X 轴方向移动对 Y 轴方向移动的工作垂直度图解

将上述各项检测项目的测量结果记入表 6-3 中。

表 6-3 数控铣床精度检测数据记录表

<table>
<tr><td colspan="2">机床型号</td><td>机床编号</td><td>环境温度</td><td>检验人</td><td colspan="2">实验日期</td></tr>
<tr><td colspan="2"></td><td></td><td></td><td></td><td colspan="2"></td></tr>
<tr><td>序号</td><td colspan="3">检验项目</td><td>允差范围/mm</td><td>检验工具</td><td>实测/mm</td></tr>
<tr><td>1</td><td colspan="3">机床调平</td><td>0. 06/1 000</td><td></td><td></td></tr>
<tr><td>2</td><td colspan="3">工作台面的平面度</td><td>0. 08/全长</td><td></td><td></td></tr>
<tr><td rowspan="2">3</td><td colspan="3">靠近主轴端部主轴锥孔轴线的径向跳动</td><td>0. 01</td><td></td><td></td></tr>
<tr><td colspan="3">距主轴端部 L(L=100 mm) 处主轴锥孔轴线的径向跳动</td><td>0. 02</td><td></td><td></td></tr>
<tr><td rowspan="4">4</td><td colspan="3">Y-Z 平面内主轴轴线对工作台面的垂直度</td><td rowspan="2">0. 05/300 (α≤90°)</td><td></td><td></td></tr>
<tr><td colspan="3">X-Z 平面内主轴轴线对工作台面的垂直度</td><td></td><td></td></tr>
<tr><td colspan="3">主轴端面偏摆</td><td>0. 01</td><td></td><td></td></tr>
<tr><td colspan="3">主轴套筒外壁偏摆</td><td>0. 01</td><td></td><td></td></tr>
<tr><td rowspan="2">5</td><td colspan="3">Y-Z 平面内主轴箱垂直移动对工作台面的垂直度</td><td rowspan="2">0. 05/300 (α≤90°)</td><td></td><td></td></tr>
<tr><td colspan="3">X-Z 平面内主轴箱垂直移动对工作台面的垂直度</td><td></td><td></td></tr>
<tr><td rowspan="2">6</td><td colspan="3">Y-Z 平面内主轴套筒移动对工作台面的垂直度</td><td rowspan="2">0. 05/300 (α≤90°)</td><td></td><td></td></tr>
<tr><td colspan="3">X-Z 平面内主轴套筒移动对工作台面的垂直度</td><td></td><td></td></tr>
<tr><td rowspan="2">7</td><td colspan="3">工作台 X 轴方向移动对工作台面的平行度</td><td>0. 056 (α≤90°)</td><td></td><td></td></tr>
<tr><td colspan="3">工作台 Y 轴方向移动对工作台面的平行度</td><td>0. 04 (α≤90°)</td><td></td><td></td></tr>
<tr><td>8</td><td colspan="3">工作台 X 轴方向移动对工作台面基准(T 形槽)的平行度</td><td>0. 03/300</td><td></td><td></td></tr>
<tr><td>9</td><td colspan="3">工作台 X 轴方向移动对 Y 轴方向移动的工作垂直度</td><td>0. 04/300</td><td></td><td></td></tr>
</table>

【任务实施】

(1) 按照随机分组方式,分成各个观摩小组,并选出一名组长,将本组成员的个人信息填入表 6-4 中。

（2）根据观摩及讨论的结果，在表6-4中记录数控机床各项几何精度检测的项目名称、检测工具、检测方法及允许范围等。

（3）对观摩过程中的疑难问题进行充分讨论，并将问题及解决办法填入表6-4中。

表6-4　数控机床几何精度检测任务工单

<table>
<tr><td colspan="6">任务工单</td></tr>
<tr><td colspan="2">姓名：</td><td colspan="2">班级：</td><td colspan="2">学号：</td></tr>
<tr><td colspan="6">所需材料、工量具、设备列表</td></tr>
<tr><td>序号</td><td colspan="2">名称</td><td>数量</td><td>型号或规格</td><td>备注</td></tr>
<tr><td></td><td colspan="2"></td><td></td><td></td><td></td></tr>
<tr><td></td><td colspan="2"></td><td></td><td></td><td></td></tr>
<tr><td></td><td colspan="2"></td><td></td><td></td><td></td></tr>
<tr><td colspan="6">工作过程（观摩）</td></tr>
<tr><td>序号</td><td>项目名称</td><td>检测工具</td><td colspan="2">检测方法</td><td>允许范围</td></tr>
<tr><td></td><td></td><td></td><td colspan="2"></td><td></td></tr>
<tr><td></td><td></td><td></td><td colspan="2"></td><td></td></tr>
<tr><td></td><td></td><td></td><td colspan="2"></td><td></td></tr>
<tr><td></td><td></td><td></td><td colspan="2"></td><td></td></tr>
<tr><td colspan="3">出现问题</td><td colspan="3">解决办法</td></tr>
<tr><td colspan="3"></td><td colspan="3"></td></tr>
</table>

【任务评价】

序号	评价内容	分值	得分
1	能够熟练说出机床机械结构的名称；了解数控机床几何精度检测相应的工量具	30	
2	能够掌握数控机床几何精度检测内容	30	
3	能够熟练运用相应的工量具进行数控机床的几何精度检测	30	
4	有效培养沟通表达及团结协作的职业精神	10	

复习思考题

（1）数控机床机械结构主要包括哪些部分？

（2）数控机床主传动系统有哪些要求？

（3）数控机床进给系统常用的机构有哪些？

（4）数控铣床主轴安装过程中有哪些注意事项？

（5）数控机床几何精度检验的意义？

（6）数控机床主轴锥孔轴线的径向跳动如何测量？

参考文献

[1] 杨中力，胡宗政，左维. 数控机床故障诊断与维修 [M]. 大连：大连理工大学出版社，2019.
[2] 周兰，陈少艾. FANUC 0i-D/0i Mate-D 数控系统连接调试与 PMC 编程 [M]. 北京：机械工业出版社，2022.
[3] 蒋洪平，王蓓，刘彩霞. 数控设备故障诊断与维修 [M]. 北京：北京理工大学出版社，2018.
[4] 于涛，武洪恩. 数控技术与数控机床 [M]. 北京：清华大学出版社，2019.

Fault Diagnosis and Maintenance of CNC Machine Tools

主编　韩志国　李文强　王同庆
主审　李云梅

北京理工大学出版社
BEIJING INSTITUTE OF TECHNOLOGY PRESS

Foreword

In order to better serve China's "Belt and Road" Initiative and Luban Workshop and export high-quality technical talents for the country, teachers of the Luban Workshop project team, upholding the concept of modern education, have compiled this book in combination with China's vocational skill standards and relying on advanced training equipment. Based on the technical skills and theoretical knowledge requirements of the national vocational qualification standards for CNC machine tool installation, commissioning and maintenance workers, combined with the most advanced practical training equipment in the field of domestic vocational education at present, this book flexibly applies the systematic scientific education theory of working process, and fully reflects the knowledge and skills that students need to master in each module and specific task. This book has made scientific and reasonable innovations and breakthroughs in contents and structure. It is ability-based, actual task-oriented, and compiled according to the principle of project-based teaching mode. This book combines theory with practice, and students can learn while practicing according to the needs of actual tasks, so as to truly apply knowledge and skills.

The module and task contents designed in this book meet all the requirements and regulations of national vocational standards for fitters on elementary theoretical knowledge and related skills. This book is written in both Chinese and English, which is in line with international textbooks. It can not only meet the learning needs of domestic students majoring in CNC equipment application and maintenance, but also satisify the requirments of overseas training on relevant skills and self-study by foreign students.

Han Zhiguo, Li Wenqiang, and Wang Tongqing serve as the chief editors of this book, Yang Guoxing, Zhou Jing, Shang Dandan and Wang Shiwei (enterprise participant) serve as deputy editors, Zhou Shuqing, Li Hong, and Zhang Chuan participated in the book editing. Specific tasks allocation: Han Zhiguo edited module Ⅰ and Ⅴ; Li Wenqiang edited module Ⅵ, Wang Tongqing edited module Ⅲ; Yang Guoxing and Zhou Shuqing edited module Ⅱ; Zhou Jing edited module Ⅳ; Wang Shiwei provided some related project materials and cases for the tasks. Li Hong and Zhang Chuan sorted out all the materials; Li Wenqiang compiled and edited the entire book finally. President Li Yunmei of Tianjin Light Industry Vocational Technical College served as the chief reviewer of this book.

Due to the limited knowledge of authors, some mistakes and omissions in this book are inevitable. Your suggestions for improvement will be gratefully received.

Editors

Contents

Contents

Module I Overview of CNC Machine Tool Fault Diagnosis and Maintenance

With the development of electronic technology and automation technology, computer numerical control(CNC) technology is more and more widely used. CNC equipment is advanced machining equipment with a high degree of automation and complex structure, which is essential for enterprises. In order to give full play to the efficiency of CNC equipment, correct operation and careful maintenance are necessary. Correct operation can prevent abnormal wear of CNC machine tools and avoid sudden faults; careful daily maintenance can keep the equipment in good technical condition, delay the deterioration process, and timely identify and eliminate hidden faults, thus ensuring safe operation. Fault diagnosis is the first step for CNC machine tool maintenance. It can not only quickly find out the cause of faults and eliminate them, but also prevent their occurrence and expansion. The essence of fault diagnosis is to judge the type of fault through information input from the outside or the system itself on the premise of a fault occurred in the diagnosed object, determine the fault location (component), then estimate the possible time, severity and cause of the fault, and even provide evaluation, decision-making and maintenance suggestions.

The main contents of modern fault diagnosis should include three parts: real-time monitoring technology, fault analysis (diagnosis) technology and fault repair methods. From information acquisition to fault location and then to fault troubleshooting, they develop as separate technical fields and also develop coordinately as technologies for fault diagnosis.

Task I Fault Diagnosis and Maintenance Tasks of CNC Machine Tools

Learning Objectives

(1) To understand the fault classification for CNC machine tools.

(2) To understand the fault diagnosis and maintenance tasks of CNC machine tools.

(3) To master the knowledge and skills required for specific fault diagnosis and maintenance tasks of CNC machine tools.

(4) To be able to analyze and solve problems independently.

(5) To master the engineering working methods, cultivate a rigorous work style, and abide by the 5S management system.

Task Description

Read and learn the fault diagnosis and maintenance tasks of CNC machine tools in groups, and fill the fault classification, fault phenomena, fault diagnosis methods and etc, of CNC machine tools into the task work order. Make sure to be careful and pay attention to the relevant operation specifications of the practical training site during learning.

Knowledge Link

I. Classification of faults

1. Classification by faults locations

(1) Faults of the main unit. The main unit of the CNC machine tool usually includes the mechanical, lubrication, cooling, chip removal, hydraulic, pneumatic and protection parts. Common faults of the main unit mainly include:

1) mechanical transmission fault caused by improper installation, commissioning and operation of mechanical components;

2) faults caused by interference and excessive friction of moving parts such as guide rails and spindles;

3) faults caused by damage and poor connection of mechanical parts, etc.

The main unit faults mainly include loud transmission noise, poor machining accuracy, great operation resistance, non-operation of mechanical components, damage to mechanical components, etc. Poor lubrication, pipeline blockage and poor sealing of hydraulic and pneumatic systems are the common causes of such faults. The regular maintenance of CNC machine tools, the control and elimination of "three leakages" (leakage of water, air and oil) are important measures to reduce some faults in the main unit.

(2) Electrical control system faults are generally divided into "weak current" faults and "strong current" faults according to the types of components used.

The "weak current" part mainly refers to the electronic components and integrated circuits in the control system. The weak current part of the CNC machine tool includes CNC, PMC, MDI/C RT, servo drive unit, I/O unit, etc.

"Weak current" faults are divided into hardware faults and software faults. The former refers to the faults of integrated circuit chips, discrete electronic components, connectors and external connection components in the above parts; the latter refers to faults such as operation errors and data loss under normal hardware condition. Common examples include machining program errors, change

or loss of system programs and parameters, computer calculation errors, etc.

The "strong current" part refers to the relays, contactors, switches, fuses, power transformers, motors, electromagnets, travel switches and other electrical components in the main circuit or high-voltage and high-power circuits of the control system and their control circuits. Although it is convenient to maintain and diagnose faults of the strong current part, the probability of fault occurrence is higher than that of the "weak current" part due to its high voltage and large current working state. Therefore, maintenance personnel must pay much attention to this aspect.

2. Classification by nature of faults

(1) Determinate faults. These refer to the hardware damage in the control system main unit or faults that will inevitably occur on CNC machine tools as long as certain conditions are met. This kind of fault phenomenon is the most common on CNC machine tools, but it also brings convenience to maintenance because of its certain regularity.

Determinate faults are unrecoverable. Once a fault occurs, the machine tool will not automatically return to normal if it is not repaired. However, as long as the root cause of the fault is found, the machine tool can return to normal immediately after maintenance. Correct operation and careful maintenance are important measures to eliminate or avoid faults.

(2) Random faults. These refer to faults that occur occasionally in the operating process of CNC machine tools. Such faults have hidden causes and are difficult to find out their regularity, so they are often called "soft faults". The cause analysis and fault diagnosis of random faults are difficult. Generally speaking, the occurrence of faults is often related to many factors such as component installation quality, parameter setting, component quality, imperfect software design, and influence of operating environment.

Random faults are recoverable. After the fault occurs, the machine tool can usually return to normal by restarting and other measures, but the same fault may occur during operation.

To reduce and avoid such faults, measures such as strengthening the maintenance and inspection of the CNC system, ensuring the sealing of the electrical cabinet, reliable installation and connection, and correct grounding and shielding can be taken.

3. Classification by faults indication forms

(1) Faults with alarm display. The fault display of CNC machine tools can be in the form of indicator alarm and display alarm.

1) Indicator alarm refers to alarm through state indicators (generally composed of LEDs or small indicators) on each unit of the control system. It is still possible to roughly analyze and judge the location and nature of the fault according to the state indicators of the CNC system even when the display fails. Therefore, these state indicators should be carefully checked during maintenance and troubleshooting.

2) Display alarm refers to the alarm number and information shown on the CNC display device. Since the CNC system generally has a strong self-diagnosis function, if the diagnostic software and display circuit of the system work normally, once the system fails, the fault information can be

displayed on the display in the form of an alarm number and text. The CNC system can display dozens or thousands of alarms, which is important information for fault diagnosis.

The display alarm can be divided into CNC alarm and PMC alarm. The former is a fault display set by the CNC machine tool manufacturer, and the possible cause of the fault can be determined according to the *Maintenance Manual of CNC Machine Tool* provided by the manufacturer. The latter is the PMC alarm information text set by the CNC machine tool manufacturer, which belongs to the fault display on the machine tool side. The possible cause of the fault can be determined according to the relevant contents in the *Maintenance Manual of CNC Machine Tool* provided by the manufacturer.

(2) Faults without alarm display. When such faults occur, neither the machine tool nor the system will give an alarm display. It is usually difficult to analyze and diagnose these faults which can only be confirmed through careful analysis and judgment. Especially for some early CNC systems, there are more faults without alarm display due to the weak diagnosis function of the system itself or no PMC alarm information text.

For faults without alarm display, it is usually necessary to analyze the specific situation according to the changes before and after the fault occurs. The principle analysis method and PMC program analysis method are the main methods for solving these faults.

4. Classification by causes of faults

(1) Faults of the CNC machine tool itself. Such faults are caused by the CNC machine tool itself with nothing to do with external service environment conditions. Most of the faults related to CNC machine tools belong to this type.

(2) External faults of the CNC machine tool. Such faults are caused by external factors. They include excessively low, high, or fluctuating power supply voltage; incorrect phase sequence of the power supply or unbalanced three-phase input voltage; excessively high ambient temperature; harmful gas, moisture, and dust input; and external vibration and interference, etc.

In addition, human factors are also one of the external causes of CNC machine tool faults. According to relevant data, the external faults of CNC machine tools caused by a person who operates the machine tool for the first time or lacks skills account for more than one third of the total faults of CNC machine tools in the first year of use.

In addition to the above common fault classification methods, there are many other different classification methods. For example, faults can be divided into destructive faults and non-destructive faults according to whether they are destructive at the time of occurrence. According to the occurrence of faults and the specific functional parts that need maintenance, faults can be divided into CNC device faults, feed servo system faults, spindle drive system faults, automatic tool change system faults, etc. This classification method is commonly used in maintenance.

II. Fault diagnosis and maintenance tasks for CNC machine tools

Modern CNC machine tools are highly integrated with machinery, electricity, hydraulics and light, and the electromechanical interface is highly complex. They are typical technology-intensive

mechatronics products, which require workers of CNC machine tool installation, commissioning and maintenance to have a certain theoretical foundation and maintenance and adjustment skills in mechanical and microelectronic aspects. Therefore, high requirements are put forward for them in the following aspects.

1. Assembly of mechanical functional components and CNC machine tools

(1) Be able to read the final assembly drawing or component assembly drawing of CNC machine tools.

(2) Be able to draw the connector assembly drawing.

(3) Be able to prepare tools and tooling according to the assembly and commissioning requirements of CNC machine tools.

(4) Be able to complete the assembly of more than two mechanical functional components or the final assembly of more than one model of CNC machine tools.

(5) Be able to detect and adjust the geometric accuracy and working accuracy of CNC machine tools after final assembly.

(6) Be able to read the three-coordinate measurement report, laser inspection report, and analyze and adjust general errors.

2. Adjustment of mechanical functional components and CNC machine tools

(1) Be able to read the electrical schematic diagram and electrical wiring diagram of CNC machine tools.

(2) Be able to complete the data input after initialization of the CNC system when the machine tool is powered on for test run.

(3) Be able to adjust the functions of the system operation panel and CNC machine tool operation panel.

(4) Be able to carry out CNC machine tool test run (such as idling).

(5) Be able to adjust the CNC machine tool performance by modifying common parameters.

(6) Be able to operate more than two models of CNC systems.

(7) Be able to carry out machining programming of more than two models of CNC systems.

(8) Be able to prepare tools and fixtures according to the machining process requirements of parts.

(9) Be able to complete the machining of test run workpieces.

(10) Be able to use general measuring instruments to detect the machined workpieces, analyze and adjust errors.

3. Maintenance of mechanical functional components and CNC machine tools

(1) Be able to read the final assembly drawing or component assembly drawing of CNC machine tools.

(2) Be able to read the electrical schematic diagram and electrical wiring diagram of CNC machine tools.

(3) Be able to read the hydraulic and pneumatic schematic diagrams of CNC machine tools.

(4) Be able to disassemble and assemble the whole CNC machine tool (such as disassembly and assembly of spindle box and bed, saddle and bed, machining center spindle box and bed, workbench and bed, etc.).

(5) Be able to judge common mechanical, electrical and hydraulic (pneumatic) faults through the diagnostic function of CNC machine tools.

(6) Be able to eliminate mechanical faults of CNC machine tools.

(7) Be able to eliminate strong current faults of CNC machine tools.

4. Electrical assembly of CNC machine tools

(1) Be able to read the electrical assembly drawing, electrical schematic diagram and electrical wiring diagram of CNC machine tools.

(2) Be able to read the final assembly drawing of CNC machine tools.

(3) Be able to read the hydraulic and pneumatic schematic diagrams of CNC machine tools.

(4) Be able to read mechanical drawings related to electricity (CNC tool rest, tool magazine and manipulator, etc.).

(5) Be able to install all circuits of more than two models of CNC machine tools according to the requirements of electrical schematic diagrams, including distribution board, electrical cabinet, console, spindle frequency converter, cable connection between various components of machine tools, etc.

5. Electrical adjustment of CNC machine tools

(1) Be able to read PMC ladder diagrams(LD) and correct errors.

(2) Be able to adjust the CNC machine tool with system parameters, PMC parameters and frequency converter parameters.

(3) Be able to transmit the machine tool parameters and PMC programs (such as LDs) to the CNC controller through the machine tool communication port when the CNC machine tool is powered on for test run.

(4) Be able to carry out commissioning of various functions of CNC machine tools.

(5) Be able to carry out machining programming through the CNC systems.

(6) Be able to carry out test run and adjustment of CNC machine tools.

(7) Be able to select common tools for machining test run workpieces.

(8) Be able to adjust CNC machine tool horizontally.

(9) Be able to detect the geometric accuracy of CNC machine tools.

(10) Be able to read the three-coordinate measurement report and carry out general analysis (such as perpendicularity, parallelism, coaxiality and position degree).

(11) Be able to use general measuring instruments to detect shaft and disk workpieces and carry out error analyses.

6. Electrical maintenance of CNC machine tools

(1) Be able to read the electrical assembly drawing, electrical schematic diagram and electrical wiring diagram of CNC machine tools.

(2) Be able to read the final assembly drawing of CNC machine tools.

(3) Be able to read hydraulic and pneumatic schematic diagrams.

(4) Be able to read mechanical drawings related to electricity (CNC tool rest, tool magazine and manipulator, etc.).

(5) Be able to check fault points through instruments and meters.

(6) Be able to diagnose common mechanical, electrical and hydraulic (pneumatic) faults of CNC machine tools through the diagnosis function of CNC systems and PLC LDs.

(7) Be able to complete the maintenance of common strong and weak current faults of CNC machine tools with more than two specifications.

Task Implementation

(1) Divide students randomly into groups and appoint the group leaders. Fill the personal infornation of the group members in Table 1-1.

(2) According to the observation and discussion results, record the fault classification, fault phenomena and fault diagnosis methods for fault diagnosis and maintenance of CNC machine tools in Table 1-1.

(3) Fully discuss the problems found during the observation, and fill the problems and solutions in Table 1-1.

Table 1-1 Task Work Order for Fault Diagnosis and Maintenance of CNC Machine Tools

Task Work Order				
Name:		Class:	Student ID:	
Operating process(Observation)				
S/N	Fault classification	Fault phenomenon	Fault diagnosis method	Remark
Problems			Solutions	

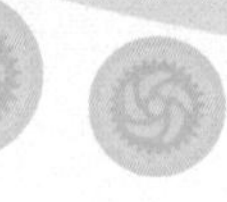

Task Evaluation

S/N	Evaluation contents	Total point	Score
1	Be proficient in the fault classification for CNC machine tools	25	
2	Be able to correctly draw and proficiently explain the skills that CNC machine tool maintainers should have in all aspects	25	
3	Be able to use the fault diagnosis and maintenance tasks of CNC machine tools for fault analysis of CNC machine tools	25	
4	Have a professional spirit of effective communication, expression and solidarity	25	

Task II General Steps for Fault Diagnosis and Maintenance of CNC Machine Tools

Learning Objectives

(1) To get familiar with the precautions before maintenance of CNC machine tools.

(2) To master the general steps for fault diagnosis and maintenance of CNC machine tools.

(3) To master the principles to be followed for fault diagnosis and maintenance of CNC machine tools.

(4) To be able to analyze and solve problems independently.

(5) To master the engineering working methods, cultivate a rigorous work style, and abide by the 5S management system.

Task Description

Read and learn this task in groups, and fill the precautions, general steps, principles, applications and etc. for fault diagnosis and maintenance of CNC machine tools into the task work order. Make sure to be careful and pay attention to the relevant operation specifications of the practical training site during learning.

Knowledge Link

I. Precautions before maintenance of CNC machine tools

Fault diagnosis begins with mechanical equipment fault diagnosis, which mainly refers to the

state monitoring and fault diagnosis of manufacturing equipment and manufacturing process. Manufacturing equipment mainly refers to machine tools, fixtures, measuring instruments and tools; manufacturing process refers to the process itself and process parameters. The state monitoring and fault diagnosis of mechanical equipment during operation includes two aspects: one is to monitor the operating state of the equipment; the other is to analyze and diagnose the faults of the equipment after abnormal conditions are found. Before using the CNC machine tool, carefully read the operation manual of the machine tool and other relevant materials, and pay attention to the following points:

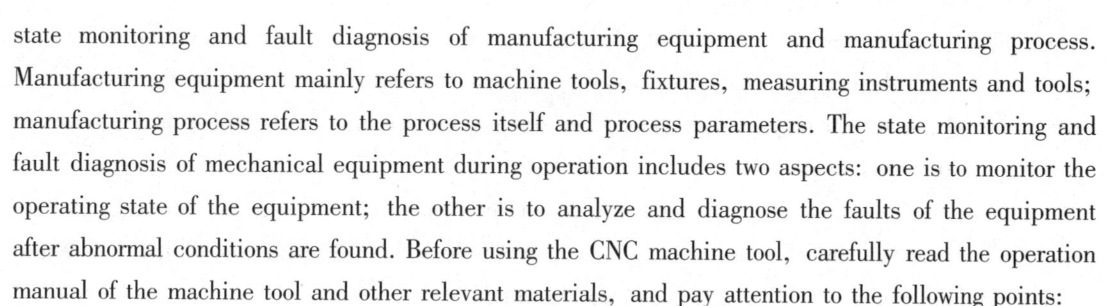

(1) The operation and maintenance personnel of the machine tool must be professionals or technically trained personnel who have mastered the corresponding professional knowledge of the machine tool, and must operate the machine tool according to safe operation procedures and regulations;

(2) Non-professionals are not allowed to open the electrical cabinet door. Before opening the electrical cabinet door, it must be confirmed that the main power switch of the machine tool has been turned off. Only professional maintenance personnel are allowed to open the electrical cabinet door for power-on maintenance;

(3) Except for some parameters that can be used and changed by the user, other system parameters, spindle parameters, servo parameters, etc., cannot be modified by the user without permission, otherwise it will cause damage to equipment and workpieces as well as personal injury;

(4) After the parameters are modified, during the first machining, the machine tool shall be locked without tools and workpieces, and a single program segment shall be used for the trial operation of the machine tool. The machine tool shall not be used until it is confirmed to be normal;

(5) The PMC program of the machine tool is designed by the manufacturer according to the needs of the machine tool and does not need to be modified. Incorrect modification and operation of the machine tool may cause damage to the machine tool or even injury to the operator;

(6) It is recommended that the machine tool run continuously for not more than 24 hours. If the continuous operation time is too long, the service life of the electrical system and some mechanical components will be affected, thus affecting the accuracy of the machine tool;

(7) All connectors of the machine tool are not allowed to be pulled out or plugged in with electricity, otherwise, serious consequences will be caused.

II. General steps for fault diagnosis of CNC machine tools

No matter which fault period the CNC machine tool is in, the general steps for fault diagnosis of the CNC machine tool are the same. When the CNC machine tool fails, generally do not turn off the power supply unless there is an emergency of danger, damage to the CNC machine tool or personal safety. Keep the original state of the machine tool unchanged as much as possible and record some signals and phenomena, including detailed records of fault phenomena, operation mode and contents when fault occurs, fault No. and contents displayed by fault indicators. Fault diagnosis is generally carried out according to the following steps.

(1) Learn about the fault in detail. For example, when the CNC machine tool chatters, vibrates

or overshoots, it is necessary to figure out whether the fault occurs in all axes or a certain axis; if it is a certain axis, whether it occurs in the whole process or at a certain position; whether it occurs as soon as the axis moves or only in high speed, certain speed of feeding, acceleration or deceleration. In order to further understand the fault, it is necessary to carry out a preliminary inspection of the CNC machine tool, and focus on checking the display contents on the fluorescent screen, the fault indicator and state indicator in the control cabinet, etc. When the fault condition allows, it is better to start the machine to check the fault condition in detail.

(2) Carry out analysis according to the fault situation, narrow down the scope, and determine the direction and means to locate the fault source. After obtaining a comprehensive understanding of the fault phenomenon, the next step is to analyze the possible location of the fault according to the fault phenomenon. Some faults are less connected with other parts, so it is easy to determine the search direction. However, some faults may have many causes, and it is difficult to determine the search direction of the fault source by a simple method. Therefore, it is necessary to carefully consult the relevant data of the CNC machine tool, find out various factors related to the fault, determine several search directions, and search one by one.

(3) Locate the fault source step by step. Fault source location is generally carried out step by step, from easy to difficult and from the outside to the inside. The difficulty lies in the complexity of technology and the difficulty of disassembly and assembly, of which the former refers to the difficulty of judging whether there is a fault. The first step of fault diagnosis is to check those parts that are easily accessible or only need simple disassembly, and then check the parts that aren't easily accessible and need a lot of disassembly work.

III. Principles for fault diagnosis and maintenance of CNC machine tools

1. Solution before operation

After a fault occurs, the maintenance personnel should first understand the whole process of the fault occurrence from the machine tool operator, consult the technical instructions and relevant technical drawings of the machine tool, and consider the solution to the fault before starting maintenance.

2. Check before power-on

After the solution is determined, check, test and analyze the machine tool first to determine whether the fault is destructive or non-destructive before powering on the machine tool.

If it is a destructive fault, the danger must be eliminated first; if it is a non-destructive fault, the machine tool can be powered on, and then further dynamic observation, inspection and testing shall be carried out on the running machine tool to find out the location of the fault.

3. Software before hardware

Software malfunction of the CNC system may also lead to failures of the CNC machine tool. Malfunction may include loss of software parameter, improper use and incorrect operation of

software.

Therefore, after the machine tool is powered on, check whether the software functions normally first to avoid greater fault.

4. Exterior before interior

In case of a fault of the CNC machine tool, firstly, the maintenance personnel should check whether the fault is attributed to any mechanical component like the travel switch and the button switch.

If the mechanical components are confirmed fine, check whether the hydraulic components function is normally, for example, whether the connections of hydraulic components are loose.

In the end, check whether the electrical contact components such as printed circuit board sockets and electric control cabinet sockets are loose. Poor signal contact of these components caused by mechanical vibration, oil stain, dust, temperature and humidity changes will result in signal distortion, leading to fault of the CNC machine tool.

5. Mechanical before electrical

The CNC machine tool is a piece of highly automated, complex and advanced machining equipment composed of various mechanical, hydraulic and electrical components.

The diagnosis of the CNC machine tool should be carried out in a certain order. Experience shows that most faults are caused by the malfunction of mechanical components like the travel switch. Besides, mechanical faults are easily detectable while electrical faults are not.

Therefore, it will be more than double effective to check and rule out possible mechanical faults for maintenance.

6. Common faults before special faults

Common faults have extensive influence and are the principal contradiction, while special faults only affect local areas and are secondary contradiction.

For example, if all coordinate axes of the CNC machine tool fail feeding, or the power grid or main power supply fails, it indicates a common fault. Only when common faults are eliminated can special faults be solved.

7. Simple before complex

The CNC machine tool may have a combination of multiple faults of various complexities. Minor faults are easy to repair while the difficult ones may be challenging. Maintenance personnel should eliminate the small and easy faults first. The maintenance process may be inspiring and explain the complex faults, or turn the complex faults into minor ones. Anyway, it will be conducive to the elimination of complex faults.

8. General before special

A certain fault of the CNC machine tool may be attributed to many reasons. Priority should be given to possible common factors leading to faults in maintenance, and then move on to uncommon special factors.

Task Implementation

(1) Divide students randomly into groups and appoint the group leaders. Fill the persnonal information of the group members in Table 1-2.

(2) Based on the results of observation and discussion, record the precautions, general steps, principles, applications and etc. for fault diagnosis and maintenance of the CNC machine tool in Table 1-2.

(3) Fully discuss the problems found during the observation, and fill the problems and solutions in Table 1-2.

Table 1-2 Task Work Order for General Steps for Fault Diagnosis and Maintenance of CNC Machine Tools

<table>
<tr><td colspan="5">Task Work Order</td></tr>
<tr><td colspan="2">Name:</td><td colspan="2">Class:</td><td>Student ID:</td></tr>
<tr><td colspan="5">Operating process(Observation)</td></tr>
<tr><td>S/N</td><td>Precaution</td><td>General step</td><td>Principle</td><td>Application</td></tr>
<tr><td></td><td></td><td></td><td></td><td></td></tr>
<tr><td></td><td></td><td></td><td></td><td></td></tr>
<tr><td></td><td></td><td></td><td></td><td></td></tr>
<tr><td></td><td></td><td></td><td></td><td></td></tr>
<tr><td colspan="3">Problems</td><td colspan="2">Solutions</td></tr>
<tr><td colspan="3"></td><td colspan="2"></td></tr>
</table>

Task Evaluation

S/N	Evaluation contents	Total point	Score
1	Be able to clearly explain the precautions for fault diagnosis and maintenance of CNC machine tools	25	
2	Clearly explain the general steps and principles for fault diagnosis and maintenance of CNC machine tools	25	
3	Be able to proceed the general steps for CNC machine tool fault diagnosis and maintenance and carry out CNC machine tool fault diagnosis and maintenance	25	
4	Have a professional spirit of effective communication, expression and solidarity	25	

Task III Introduction to Commonly Used Tools and Measuring Instruments for Fault Diagnosis and Maintenance of CNC Machine Tools

Learning Objectives

(1) To know the commonly used tools and measuring instruments for fault diagnosis and maintenance of CNC machine tools.

(2) To master the usage of commonly used tools and measuring instruments for fault diagnosis and maintenance of CNC machine tools.

(3) To be able to analyze and solve problems independently.

(4) To master the engineering working methods, cultivate a rigorous work style, and abide by the 5S management system.

Task Description

Learn to use the tools and instruments for fault diagnosis and maintenance of CNC machine tools in groups, and fill the classification, names, main functions, applications and etc. of the commonly used tools and measuring instruments for fault diagnosis and maintenance of CNC machine tools into the task work order. Make sure to be careful and pay attention to the relevant operation specifications of the practical training site during learning.

Knowledge Link

I. Tools for disassembly and assembly

(1) Single-end hook wrench. There are fixed hook wrenches and adjustable hook wrenches, as shown in Fig. 1 – 1. Both are used to pull the round nuts with straight grooves or holes in the circumferential direction.

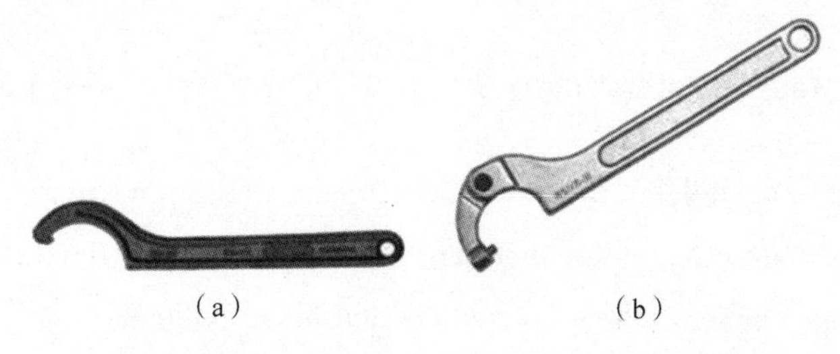

(a)　　(b)

Fig. 1–1 Single-End Hook Wrenches

(a) Fixed hook wrench; (b) Adjustable hook wrench

(2) Round nut wrenches with a groove or a hole on the end face(Fig.1-2). There are socket wrenches and double-pin fork wrenches.

(a) (b)

Fig. 1-2 Round Nut Wrenches

(a) Round nut wrench with groove on end face; (b) Round nut wrench with hole on end face

(3) Snap ring pliers(Fig.1-3). There are pliers for shaft and pliers for hole.

(4) Elastic hammers. There are wooden hammers and copper hammers(Fig. 1-4).

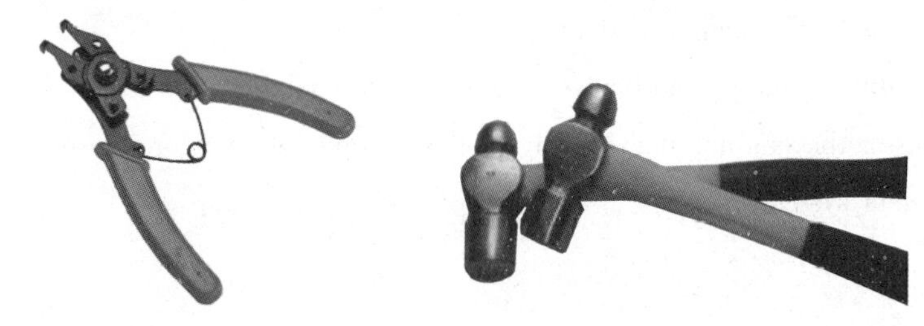

Fig. 1-3 A Snap Ring Plier

Fig. 1-4 Copper Hammers

(5) Tapered flat key pullers(Fig.1-5). There are impact pullers and resistance pullers, as shown in Fig. 1-5.

(6) Pin pullers(Fig.1-6). Tool to pull internal threaded small shaft and tapered pin (commonly known as pin puller).

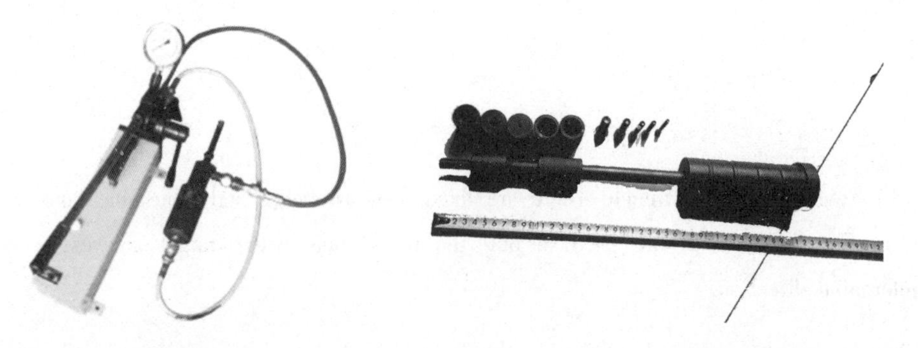

Fig. 1-5 A Tapered Fat Key Puller

Fig. 1-6 A Pin Puller

(7) Pullers(Fig.1-7). Pullers are often used to remove rolling bearings, pulley couplings and other parts installed on shafts. Generally, there are screw pullers and hydraulic pullers. As for screw pullers, there are two-jaw pullers, three-jaw pullers and hinge pullers.

(8) Pin punchs(Fig.1-8).

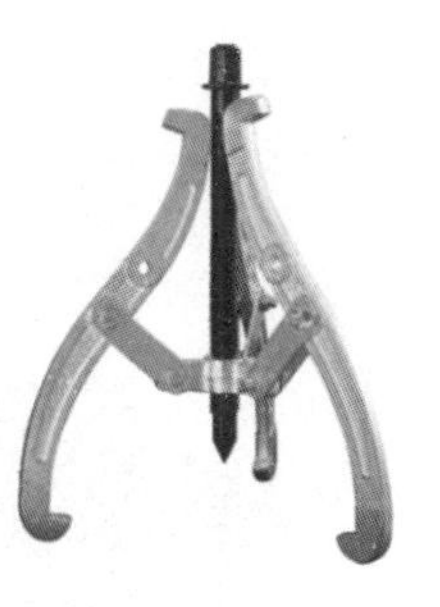

Fig. 1-7 A Three-Jaw Puller

Fig. 1-8 Pin Punches

II. Tools for mechanical maintenance

(1) Rulers(Fig.1-9). There are leveling rulers, tool-edge rulers and 90° angle squares.

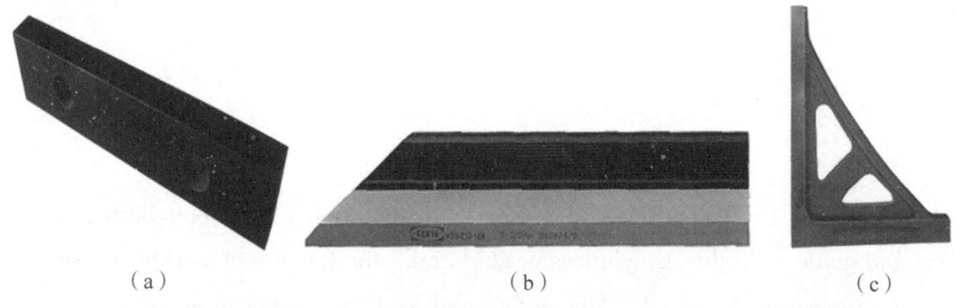

(a) (b) (c)

Fig. 1-9 Rulers

(a) Leveling ruler; (b) Tool-edge ruler; (3) 90° angle square

(2) Shim plates (Fig. 1 - 10). They are shim plates with angle planes of 90° and 55° respectively, and shim plates for level gauge.

(3) Inspection bars(Fig.1-11). There are dial inspection bar with standard shank, cylindrical inspection bar and special inspection bars.

Fig. 1-10 A Shim Plate

Fig. 1-11 A Dial Inspection Bar with Standard Shank

(4) Lever micrometers(Fig.1-12). With a measuring accuracy up to 0.001 mm, it can be used for measurement of parts with strict geometric accuracy requirement.

(5) Universal bevel protractors(Fig. 1 - 13). It is used to measure the internal and external angles of a workpiece. According to the vernier reading, there are 2 ft① protractors and 5 ft protractors; while divided by the blade shape, there are circular protractors and sectorial protractors,

① 1 ft=0. 034 8 m.

as shown in Fig. 1−13.

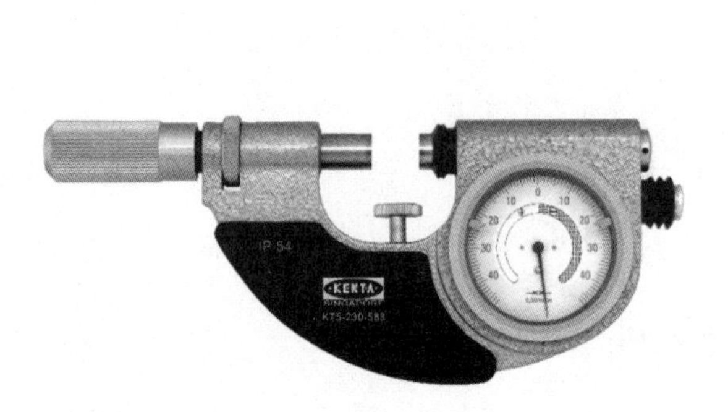

Fig. 1−12 A Lever Micrometer

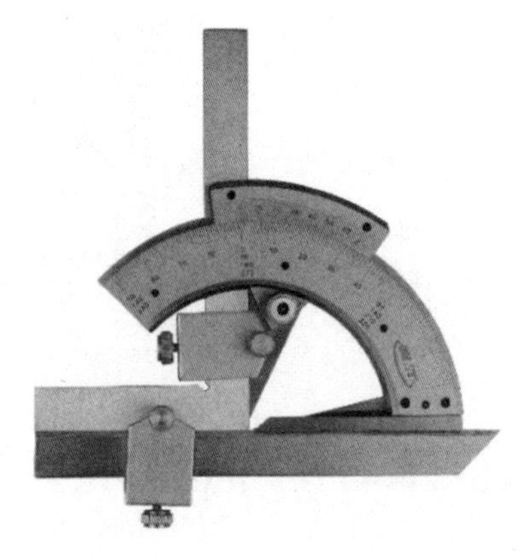

Fig. 1−13 A Universal Bevel Protractor

III. Tools for maintenance of CNC machine tools

1. Dial gauges

A dial gauges(Fig. 1 − 14) is used to measure the parallelism between parts, the parallelism between axis and guide rail, the straightness of guide rail, the flatness of workbench surface, as well as the end face circle runout, radial circle runout and axial movement of spindle.

2. Lever dial gauges

A lever dial gauges(Fig.1−15) is used to measure workpieces subject to space constraints, such as inner hole runout and keyway. Be sure to keep the movement direction under measurement perpendicular to the probe center so as to avoid measurement error.

Fig. 1−14 A Dial Gauge

Fig. 1−15 A Lever Dial Gauge

3. Dial indicators and lever dial indicators

The dial indicator and lever dial indicators(Fig.1−16) function by the same principles as those of the dial gauge and lever dial gauge. The only difference lies in the division values. They are often used for repairing precision machine tools.

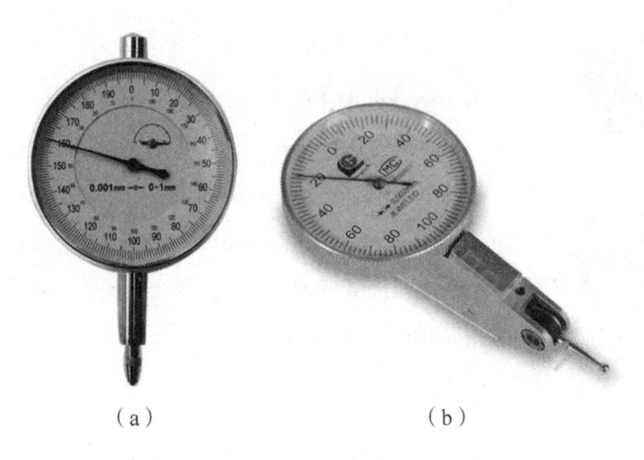

(a)   (b)

Fig. 1-16 Dial Indicators

(a) A dial indicator; (b) A lever dial indicator

4. Comparators

There are torsion spring comparators and lever gear comparators. The former is especially suitable for runout measuring with strict accuracy requirements. Fig. 1-17 shows a comparator.

5. Level gauges

The level gauge is one of the most commonly used measuring instruments in machine tool manufacturing and maintenance. It is used to measure the straightness of guide rail in vertical plane, the flatness of workbench surface and the perpendicularity and parallelism between parts. According to the function principles, there are leveling level gauges and electronic level gauges. Leveling level gauges are further divided into bar level gauges (Fig. 1-18), frame level gauge, and combined diagram level gauge.

Fig. 1-17 A Comparator

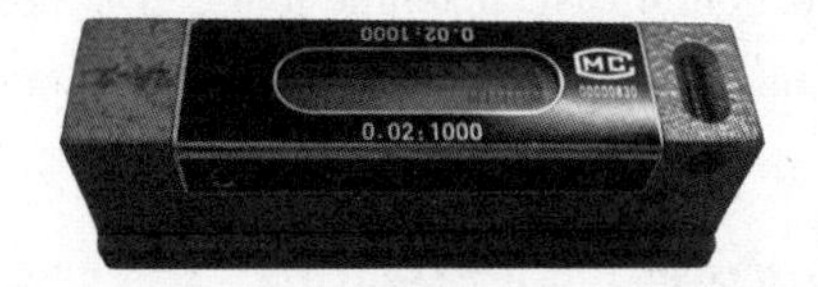

Fig. 1-18 A Bar Level Gauge

6. Optical flatness gauges

The optical flatness gauges (Fig. 1-19) is commonly used to check the straightness of lathe bed guide rail in the horizontal and vertical planes and the flatness of inspection plate during mechanical maintenance. It is one of the advanced instruments for guide rail straightness measurement at present.

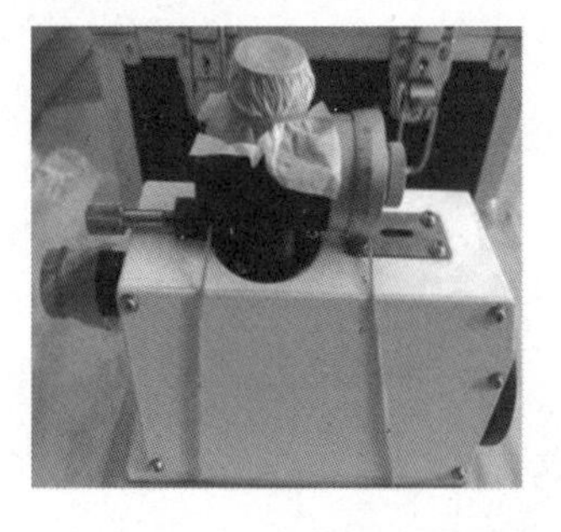

Fig. 1-19 Optical Flatness Gauge

7. Theodolites

A theodolite(Fig. 1-20) is one of the high-precision instruments commonly used in machine tool accuracy inspection and maintenance. It is often used for accurate measurement on the division accuracy of CNC milling machine's and machining center's horizontal turntables and universal turntables. It is usually used in a combined optical system with collimator.

8. Tachometer

A tachometer(Fig.1-21) is often used to measure the speed of servo motors and is one of the important means to check the servo speed governing system. Centrifugal tachometers and digital tachometers are both commonly used tachometers Fig. 1-21 shows a centrifugal tachometer.

Fig. 1-20 A Theodolite

Fig. 1-21 A Centrifugal Tachometer

IV. Measuring instruments for maintenance of CNC machine tools

For fault detection of CNC machine tools, it is necessary to use certain instruments, which may directly indicate the state of fault points from the perspective of quantitative analysis and play a decisive role.

1. Vibration measuring instruments

A vibrometer is used to measure the operation conditions of the spindle, motor, and even the complete unit of CNC machine tools. Sensors of various types may be adopted depending on the parameters to be measured, vibration frequency and dynamic range, sensor installation conditions, bearing forms (rolling or sliding) of machine tools, and other factors. Commonly used sensors include eddy current displacement sensors, magnetoelectric speed sensors, and piezoelectric acceleration sensors.

For vibration measuring judgement, generally, the best accessibility and availability will be the absolute judgment criterion. It is formulated targeted on typical objects, such as international general

standards ISO 20816 series for mechnical vibration.

2. Infrared thermometers

For infrared temperature measurement, the measuring of object surface temperature will be achieved by measuring the radiant power of the object according to the infrared radiation principle. The infrared detector and corresponding optical system receive invisible infrared radiation energy from the object under measurement, and convert it into other easily detectable energy forms for display and recording.

Photoelectric detectors and heat-sensitive detectors may both be used as infrared thermometers and these two differ for their responses to infrared radiation. Infrared thermometers are used to test the CNC machine tool components that are prone to heating, such as power modules, wire contacts and spindle bearings. HCW series of Kunming Institute of Physics, China; HCW-1 and HCW-2 of Xibei Optical Instrument Factory, China, and CYCLOPS and SOLD models of LAND Company of the United States, etc., are all prevailing products in the market.

Infrared thermal TV, optical scanning thermal imager and focal plane thermal imager are also used for temperature measurement based on infrared principle. The infrared diagnosis is mainly performed through the temperature-based judgment, comparison with similar objects, file analysis, relative temperature difference method and thermal image anomaly method.

3. Laser interferometers

A laser interferometer carries out high-precision (position and geometry) accuracy correction for the machine tool, three-testing machine, and various positioning devices. It also performs measurement of various parameters, such as linear position accuracy, repeated positioning accuracy, angle, straightness, perpendicularity, parallelism, and flatness. There are also optional functions such as automatic pitch error compensation (applicable to most CNC systems), measurement and assessment of machine tool dynamic characteristics, calibration of gyration coordinate indexing accuracy, and trigger pulse input and output functions.

The laser interferometer is used to check the accuracy of machine tools and measure the length, angle, straightness, right angle, etc. It features high accuracy, high efficiency, convenient use, length measuring range of more than ten meters or even dozens of meters, and the accuracy reaching to micron level.

Task Implementation

(1) Divide students randomly into groups and appoint the group leaders. Fill the personal information of the group members in Table 1-3.

(2) According to the results of observation and discussion, record the classification, names, main functions and applications of commonly used tools and measuring instruments for fault diagnosis and

maintenance of CNC machine tools in Table 1−3.

(3) Fully discuss the problems found during the observation, and fill the problems and solutions in Table 1−3.

Table 1−3 Task Work Order Commonly Used Tools and Measuring Instruments for Maintenance of CNC Machine Tool

Task Work Order				
Name:		Class:	Student ID:	
Operating process(Observation)				
S/N	Classification	Name	Main function	Application
Problems			Solutions	

Task Evaluation

S/N	Evaluation contents	Total point	Score
1	Be able to state the names of tools and measuring instruments	25	
2	Be able to describe the correct usage of tools and measuring instruments	25	
3	Be able to use the tools and measuring instruments skillfully	25	
4	Have a professional spirit of effective communication, expression and solidarity	25	

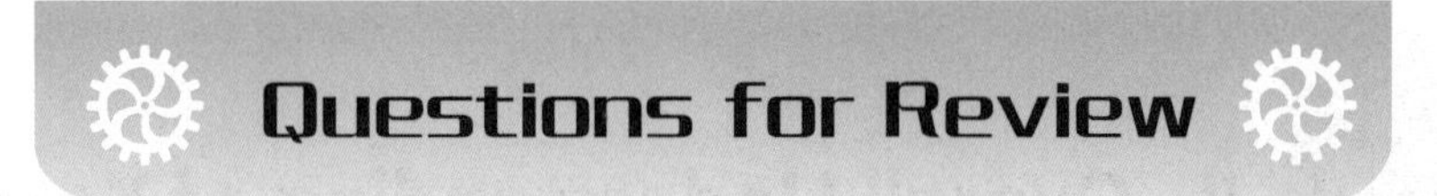

Questions for Review

(1) What is the classification for CNC machine tool faults?

(2) What are the main tasks for fault diagnosis and maintenance of CNC machine tools?

(3) What are the principles to be followed for fault diagnosis and maintenance of CNC machine tools?

(4) What are the general steps for fault diagnosis and maintenance of CNC machine tools?

(5) What are the tools and measuring instruments commonly used for fault diagnosis and maintenance of CNC machine tools?

(6) What are the basic methods for fault diagnosis and maintenance of CNC machine tools?

(7) What is the development direction of CNC machine tool fault diagnosis and maintenance technology?

Module II Overall Understanding of YaLong YL-569 0i MF CNC Equipment

Practical training equipment is an indispensable carrier for practical teaching. While using such equipment, both teachers and students should fully understand the technical parameters of the equipment and strictly abide by relevant instructions. This module mainly introduces the basic composition, technical parameters, available practical training projects, instructions for use and power-on process of YaLong YL-569 0i MF CNC milling machine.

Task I Introduction to YaLong YL-569 0i MF CNC Equipment

Learning Objectives

(1) To understand the basic composition and technical parameters of YaLong YL-569 0i MF (abbr. as YL-569 0i MF) CNC milling machines.

(2) To master the specific functional modules of YL-569 0i MF CNC milling machines.

(3) To master the practical training projects of YL-569 0i MF CNC milling machines.

(4) To be able to analyze and solve problems independently.

(5) To master the engineering working methods, cultivate a rigorous work style, and abide by the 5S management system.

Task Description

Observe YL-569 0i MF CNC equipment in groups, and fill the configuration names, main components, specifications, quantity and etc. of YL-569 0i MF into the task work order. Make sure to be careful and pay attention to the relevant operation specifications of the practical training site during observation.

Knowledge Link

I. Equipment overview

YL-569 CNC machine tool is suitable for CNC machine tool assembly, commissioning and

maintenance skill competitions in vocational colleges and the teaching and practical training of CNC assembly, commissioning and maintenance majors, CNC machining majors and mechatronics majors. It is developed with production-oriented and learning-oriented functions according to the teaching characteristics of CNC maintenance majors in vocational colleges in combination with the actual needs of enterprises and the requirements of job skills. The equipment is of a modular structure. Through different combinations, it offers practical training projects such as electrical assembly and commissioning of CNC machine tools, system commissioning, mechanical geometric accuracy inspection and maintenance of functional CNC machine tool components, so as to cultivate talents meeting the enterprises' requirements. It is also suitable for CNC machine tool assembly, commissioning and maintenance skill competitions and vocational skill appraisal of CNC assembly and commissioning workers. Some YL-569 CNC milling machines are shown in Fig. 2-1.

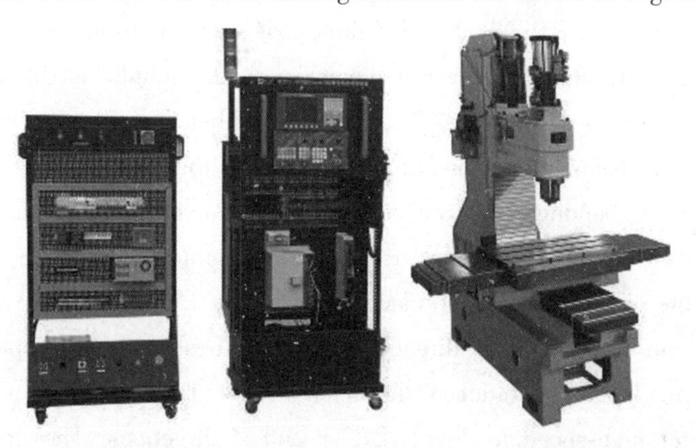

Fig. 2-1 Some YL-569 CNC Milling Machines

II. Equipment functions

YL-569 CNC machine tools consist of electrical control unit, electrical installation practice unit, mechanical unit, tool magazine unit, tools, gauges and etc.

The electrical control unit mainly consists of the CNC system, feeding drive, spindle unit, PMC unit, cooling control circuit, interface unit, protection circuit, and power supply circuit. This unit, embedded with an intelligent appraisal system, has a complete electrical control of the CNC machine tool. It can be directly connected with functional components of the CNC machine tool for practice training of electrical commissioning and maintenance. Its vertical structure conforms to the real CNC electrical installation environment and its component layout is consistent with the actual machine tool. It also complies with *Electrical safety of machinery—Electrical equipment of machines—Part 1: General requirements*(GB/T 5226.1—2019) and is more suitable for skill training based on the actual job requirements.

The intelligent fault maintenance system configured allows training of CNC machine tool maintenance through fault generation, fault analysis, fault diagnosis, line inspection, fault point determination and other process. With computer software, the system supports student login, automatic scoring, score statistics, etc., as well as the convenient training result appraisal functions.

It also performs online knowledge assessment of CNC technology through network, which greatly reducing teachers' workload in fault setting, scoring and statistics. It is a set of CNC technology education equipment integrating practice, examination, and assessment.

The electrical installation practice unit is mainly designed for the repeated training for CNC machine tool electrical installation skills. It is composed of PMC interface unit, cooling control circuit, interface unit, protection circuit and power supply circuit, etc. The electrical mounting board may be a reusable mesh plate structure, which is connected with the electrical control unit through the interface conversion unit to allow commissioning of functional CNC machine tool components.

The power supply is provided with leakage switch and phase loss protection circuit, which will act automatically in case of leakage, short circuit and phase loss to protect the equipment.

The machine tool module is an *X*, *Y* and *Z* three-axis servo-controlled vertical machine tool with compact and reasonable structure and overall dimensions. The spindle is driven by a servo motor. The main basic parts such as the base, slide carriage, workbench, pillar and spindle box are made of high-rigidity cast iron. Reinforcing ribs are arranged inside the base. The slide carriage is a box to ensure the high rigidity, bending resistance and damping capacity of the basic parts. Resin sand molded basic parts are subjected to aging treatment for stable accuracy in long-term use, thus guaranteeing the stable performance of the machine tool.

The guide rails in *X*, *Y* and *Z* directions are all linear, which cooperate with automatic lubrication to allow high-speed operation of the machine tool. High-precision and high-strength ball screws are adopted for high-speed feeding in *X*, *Y* and *Z* directions. The drive motor is directly connected with the lead screw through an elastic coupling, and the feeding servo motor transmits power to the high-precision ball screw directly without backlash, thus ensuring the positioning accuracy of the machine tool. High-speed, high-precision and high-rigidity spindle units of well-known brands are adopted for excellent axial and radial bearing capacities and a rotating speed of up to 10,000 r/min. The spindle uses central purging. After the tool is released, the inner taper of the spindle is quickly purged with high-pressure gas at the center to ensure the clamping accuracy of the tool.

The guide rails and lead screws in *X*, *Y* and *Z* directions are provided with sealing protection to ensure the cleanness, thus further ensuring the transmission and motion accuracy of the machine tool. Advanced centralized automatic lubrication device is adopted for automatic intermittent lubrication at specified time point and amount, thus ensuring the stable and reliable operation.

Hat-type tool magazine is widely used. This tool changing device is advantageous for simple structure, low cost and good tool changing reliability.

The tools and measuring devices for machine tool are mainly used for the practical training of testing projects specified in the national standard.

III. Equipment training practice

(1) Learning about the electrical composition of CNC machine tools.

(2) Learning about the electrical control and PMC of CNC machine tools.

(3) Learning about the spindle and feeding shaft control.

(4) Practical training on installation of the CNC machine tool control circuit.

(5) Reading and plotting of electrical schematic diagrams and assembly drawings.

(6) Parameters setting for the spindle, feeding shaft, system and servo drive.

(7) Data backup.

(8) Basic function commissioning of spindle, feeding shaft, cooling, lubrication and other modules.

(9) Fault diagnosis and maintenance of CNC machine tools.

(10) Geometric accuracy inspection of CNC machine tools.

(11) Practical training on setting for stroke limit parameters.

(12) Practical training on homing parameter setting.

(13) Practical training on pitch compensation setting.

(14) Practical training on tool magazine installation and commissioning.

(15) Practical training on upgrading, reconstruction and maintenance of CNC machine tools.

(16) Milling of CNC parts.

(17) Programming and operation of CNC machine tools.

IV. Equipment configuration

See Table 2−1 for the standard oconfiguration of equipment, Table 2−2 for tools and measuring instruments, and Table 2−3 for materials.

Table 2−1 Standard Configuration of Equipment

S/N	Name	Main components and their specifications	Quantity
1	Equipment electrical platform	569 CNC machine tool training equipment	1 pc.
2	CNC system	FANUC 0i MF	1 pc.
3	Drive unit	FANUC AC servo system	1 set
4	Handwheel unit	Manual pulse generator	1 pc.
5	Tool magazine unit	Hat-type magazine with 12 pieces of tools	1 pc.
6	Lubrication unit	Electric lubricating pump	1 pc.
7	Mechanical unit of CNC machine tool	YL−557A	1 pc.

Table 2−2 Tools and measuring instruments

S/N	Name	Model or specification	Quantity
1	Wire stripper	HS−700D	1 pc.
2	Diagonal plier	DL2206	1 pc.
3	Crimping plier	HS−30J	1 pc.
4	Crimping plier	HS−06WF	1 pc.
5	Needle-nose plier	DL2106	1 pc.
6	Scissor	Civil type	1 pc.

Continued

S/N	Name	Model or specification	Quantity
7	Multimeter	MY60	1 pc.
8	Phillips screwdriver	3 mm×75 mm	1 pc.
9	Phillips screwdriver	5 mm×150 mm	1 pc.
10	Straight screwdriver	3 mm×75 mm	1 pc.
11	Straight screwdriver	5 mm×150 mm	1 pc.
12	Electroprobe	Neon tube type	1 pc.
13	Allen wrench	7-piece set	1 set
14	Square	Grade 0 marble	1 pc.
15	Cotton cloth	—	1 pc.
16	Lubricating grease	—	1 pc.
17	Lever dial gauge	0-0.8 mm/0.01 mm	1 pc.
18	Magnetic gauge stand	CZ-6A	1 pc.
19	Rubber hammer	Button head	1 pc.
20	Bar level gauge	200 mm	2 pcs.
21	Copper rod	ϕ25 mm×240 mm	1 pc.
22	Tool kit	430 mm×230 mm×200 mm	1 pc.
23	Marker pen	0.8-3 mm	1 pc.
24	Dial gauge	0-10 mm/0.01 mm	1 pc.
25	Spindle inspection rod	Matching with spindle taper	1 pc.

Table 2-3 Materials

S/N	Name	Model or specification	Quantity
1	Multi-core flexible cord	RV1.5 mm black	1 reel
2	Multi-core flexible cord	RV0.75 mm black	1 reel
3	Multi-core flexible cord	RV0.75 mm red	1 reel
4	Multi-core flexible cord	RV0.75 mm blue	1 reel
5	Multi-core flexible cord	RV0.75 mm white	1 reel
6	Grounding wire	RV1.5 mm yellow-green wire	10 m
7	Insulated terminal	QE1008 on 0.75 cable	1 pack
8	Cold-pressed terminal	SV2-4 on 2.5 cable	1 pack
9	Cold-pressed terminal	SV1.25-4 on 0.75 cable	1 pack
10	Cable tie	150 black	100 pcs.
11	Cable marker	ϕ3.5 mm(blank)	3 m
12	Cable marker	ϕ5.5 mm(blank)	3 m

V. Technical parameters

1. Parameters of electrical control unit

(1) Power supply: three-phase five-wire, AC 380 V±10%, 50 Hz;

(2) Dimensions of CNC console: length (mm) ×width (mm) ×height (mm) = 800×600×1, 800;

(3) Leakage protection: leakage action current ≤30 mA;

(4) Automatic phase loss protection and overload protection.

2. Parameters of electrical installation practice unit

(1) Power supply: three-phase five-wire, AC 380 V±10%, 50 Hz;

(2) Dimensions of CNC console: length × width × height = 800 mm×600 mm×1 600 mm.

3. Machine tool parameters

Workbench specification: 700 mm×420 mm;

X-coordinate stroke: 550 mm;

Y-coordinate stroke: 400 mm;

Z-coordinate stroke: 450 mm;

Distance from spindle center to mounting surface of pillar guide rail slider: about 453 mm;

Distance from spindle end face to workbench surface: 110-560 mm;

Fast moving speed in *X*, *Y* and *Z* directions: 0-48, 000 mm/min;

Maximum speed of spindle: 10, 000 r/min;

Spindle taper: BT40;

Spindle power: 5.5 kW;

Tool magazine type: hat type;

Tool magazine capacity: 12 pcs;

T-slot of workbench: 3 mm×14 mm×110 mm;

Bearing capacity of workbench: 300 kg;

Positioning accuracy of *X*, *Y* and *Z* coordinates (national standard): 0.008 mm;

Repeated positioning accuracy of *X*, *Y* and *Z* coordinates (national standard): 0.006 mm;

Gas source pressure: 0.6-1 MPa;

Overall dimensions of optical device: about 1, 600 mm×1, 600 mm×2, 100 mm;

Protection: simple semi-protection.

The accuracy of the machine tool fully complies with national standards, and the electrical control and operation fully comply with the requirements of standard machine tools.

The CNC milling machine maintenance training system shall be implemented in strict accordance with *General conditions for machining centres—Part 1: Geometric tests for machines with horizontal spindle and with accessory heads (horizontal Z-axis)* (GB/T 18400. 1—2010), *Test Conditions of Machining Centres—Part 4: Accuracy and Repeatability of Positioning of Linear and Rotary Axes* (GB/T 18400.4—2010), and *Test Conditions for Machining Centres—Part 6: Accuracy*

of Feeds, Speeds and Interpolations(GB/T 18400.6—2001). It is suitable for part machining with high accuracy requirements, complex shapes, multiple processes, long cycle periods and diversified varieties.

Normal service conditions are shown below.

Ambient temperature: 0-40 ℃;

Humidity: ≤85%.

Task Implementation

(1) Divide students randomly into groups and appoint the group leaders. Fill the personal information of the group members in Table 2-4.

(2) According to the results of the observation and discussion, record the names of configuration, main components, specifications and quantity of YL-569 0i MF CNC equipment in Table 2-4.

(3) Fully discuss the problems found during the observation, and fill the problems and solutions in Table 2-4.

Table 2-4 Task Work Order for Observation of YL-569 0i MF CNC Equipment

<table>
<tr><td colspan="6">Task Work Order</td></tr>
<tr><td colspan="2">Name:</td><td colspan="2">Class:</td><td colspan="2">Student ID:</td></tr>
<tr><td colspan="6">List of materials, tools, measuring instruments, and equipment required</td></tr>
<tr><td>S/N</td><td colspan="2">Name</td><td>Quantity</td><td>Model or specification</td><td>Remark</td></tr>
<tr><td></td><td colspan="2"></td><td></td><td></td><td></td></tr>
<tr><td></td><td colspan="2"></td><td></td><td></td><td></td></tr>
<tr><td></td><td colspan="2"></td><td></td><td></td><td></td></tr>
<tr><td colspan="6">Operating process(Observation)</td></tr>
<tr><td>S/N</td><td>Configuration name</td><td>Main component</td><td colspan="2">Specification</td><td>Quantity</td></tr>
<tr><td></td><td></td><td></td><td colspan="2"></td><td></td></tr>
<tr><td></td><td></td><td></td><td colspan="2"></td><td></td></tr>
<tr><td></td><td></td><td></td><td colspan="2"></td><td></td></tr>
<tr><td></td><td></td><td></td><td colspan="2"></td><td></td></tr>
<tr><td colspan="3">Problems</td><td colspan="3">Solutions</td></tr>
<tr><td colspan="3"></td><td colspan="3"></td></tr>
</table>

Task Evaluation

S/N	Evaluation contents	Total point	Score
1	Be able to describe the standard configurations of the equipment clearly	25	
2	Be able to state the machine tool parameters of the equipment correctly	25	
3	Be able to describe YL-569 0i MF CNC equipment briefly	25	
4	Have a professional spirit of effective communication, expression and solidarity	25	

Task II Instructions for Equipment Use

Learning Objectives

(1) To understand the composition of YL-569 CNC milling machines.

(2) To master the precautions in the use of YL-569 CNC milling machines.

(3) To develop good habits of safe and civilized production.

(4) To master the engineering working methods, cultivate a rigorous work style, and abide by the 5S management system.

Task Description

Observe YL - 569 CNC milling machines in groups, and fill the precautions skills to be acquired, main functions, applications, and etc.in the use of YL-569 CNC milling machines in the task work order. Make sure to be careful and pay attention to the relevant operation specifications of the practical training site during observation.

Knowledge Link

Precautions in equipment use

Practical training equipment is an important link in educational reform. It plays an irreplaceable role in teaching experiment and practice, skill training and appraisal. It enables the students to develop their practical ability, thinking ability and innovation ability with the combination of theory

with practice and the association between teaching and production. Proper use and maintenance of the equipment are crucial, which will not only facilitate operation and learning, but will also prolong the service life and application cycle of the equipment, thus giving full play to the tangible asset in cultivating talents as intangible assets. Attention shall be paid to the following matters during the use of equipment.

(1) Before using the equipment, read the product's technical specifications, operation instructions, and experimental guidebook carefully. Operate and test the equipment according to the technical specifications and procedures proposed by the manufacturer, with special attention paid to text and graphic symbols. Turn off the power supply, water supply, and gas supply in sequence after the use of the equipment.

(2) Lay emphasis on environmental protection of equipment, reduce exposure to strong sunlight, water immersion, and invasion of corrosives, make sure that the insulation resistance, withstand voltage coefficient, grounding device and indoor temperature, humidity and purity of equipment comply with relevant regulations, and learn to operate under safe conditions.

(3) It is suggest to operate the equipment within the scope of conventional technical parameters and avoid operation with ultimate technical parameters. Equipment operation beyond the scope of technical requirements is strictly forbidden. That is to say, routine tests are allowed, test with ultimate parameters are limited, and destructive tests are banned.

(4) During testing and training, check the circuits built before activating them.

(5) Protect the equipment from deformation and damage caused by heavy objects, strong forces, and mechanical impacts, as well as excessive load beyond the equipment's bearing capacity and impact resistance.

(6) Handle and place the unit boards, unit modules, and dashboards carefully and steadily to avoid damage caused by dragging, dropping, or smashing.

(7) In case of electric leakage, phase loss, short circuit, dashboard and light malfunction, electric sparks, mechanical noise, or peculiar smell, smoke, etc., use the emergency stop button and cut off power supply for inspection and equipment maintenance. Do not operate or use the equipment with faults.

(8) Try to minimize the damage caused by electrical disasters, magnetic interference, and vibration to the equipment beyond the allowable range.

(9) Pay attention to abnormal resistance including friction and impact during mechanical movement, and apply lubrication as necessary.

(10) The equipment that is not used for a long time shall be inspected and maintained regularly before its operation.

Task Implementation

(1) Divide students randomly into groups and appoint the group leaders. Fill the personal information of the group members in Table 2-5.

(2) According to the observation and discussion results, record the precautions, and common applications of electrical parts of YL−569 milling machines in Table 2−5.

(3) Fully discuss the problems found during the observation, and fill the problems and solutions in Table 2−5.

Table 2−5 Task Work Order for Instructions for Equipment Use

<table>
<tr><td colspan="6">Task Work Order</td></tr>
<tr><td colspan="2">Name:</td><td colspan="2">Class:</td><td colspan="2">Student ID:</td></tr>
<tr><td colspan="6">List of materials, tools, measuring instruments, and equipment required</td></tr>
<tr><td>S/N</td><td colspan="2">Name</td><td>Quantity</td><td>Model or specification</td><td>Remark</td></tr>
<tr><td></td><td colspan="2"></td><td></td><td></td><td></td></tr>
<tr><td></td><td colspan="2"></td><td></td><td></td><td></td></tr>
<tr><td></td><td colspan="2"></td><td></td><td></td><td></td></tr>
<tr><td colspan="6">Operating process</td></tr>
<tr><td>S/N</td><td>Precaution</td><td>Skill to be acquired</td><td colspan="2">Main function</td><td>Application</td></tr>
<tr><td></td><td></td><td></td><td colspan="2"></td><td></td></tr>
<tr><td></td><td></td><td></td><td colspan="2"></td><td></td></tr>
<tr><td></td><td></td><td></td><td colspan="2"></td><td></td></tr>
<tr><td></td><td></td><td></td><td colspan="2"></td><td></td></tr>
<tr><td colspan="3">Problems</td><td colspan="3">Solutions</td></tr>
<tr><td colspan="3"></td><td colspan="3"></td></tr>
</table>

Task Evaluation

S/N	Evaluation contents	Total point	Score
1	Be familiar with and tell the composition of YL−569 CNC milling machines	25	
2	Master the precautions in the use of YL−569 CNC milling machines	25	
3	Be able to develop good habits of safe and civilized production	25	
4	Have a professional spirit of effective communication, expression and solidarity	25	

Task III Power-on Operation Process

Learning Objectives

(1) To understand the installation requirements for YL-569 CNC milling machines.

(2) To understand the power supply requirements for YL-569 CNC milling machines.

(3) To understand the placement requirements for the electrical cabinet of YL-569 CNC milling machines.

(4) To develop good habits of safe and civilized production.

(5) To master the engineering working methods, cultivate a rigorous work style, and abide by the 5S management system.

Task Description

Observe YL-569 CNC milling machine in groups, and fill the machine tool installation requirements, main power supply requirements, electrical cabinet placement requirenents, power-on operation process, and etc. of the machine tool in the task work order. Make sure to be careful and pay attention to the relevant operation specifications of the practical training site during observation.

Knowledge Link

I. Machine tool installation requirements

In order to ensure the accuracy of the machine tool, the machine tool shall meet the following conditions during installation:

(1) It shall be installed somewhere below 1,000 m above sea level;

(2) The ambient temperature should be within 5-40 ℃;

(3) When the ambient temperature reaches 40 ℃, the humidity shall not be larger than 50%;

(4) The light in the working environment shall not be less than 500 lx;

(5) Installation environment: It shall be free of acid, corrosion, oil mist, and dust as far as possible;

(6) Installation position: Avoid direct sunlight or excessive vibration;

(7) Machining shouldn't be done before the level of the machine tool is properly adjusted.

II. Requirements for main power supply of electrical cabinet

(1) Voltage: three-phase five-wire、AC 380 V.

(2) Voltage fluctuation: max. ±10%.

(3) Power frequency: (50±1) Hz.

(4) It is strictly prohibited to lead out the main power supply from the switchboard with an interference source (for example, the switchboard supplies power for an electric welding machine and an EDM machine tool); otherwise, abnormal operation may result.

(5) The machine tool shall be separately grounded for protection (it is forbidden to use a neutral line to connect the machine tool instead of having the machine tool grounded for protection). If the common grounding is used, the other equipment (such as electric welding machines and EDM machine tools) sharing the common grounding shall not generate large interference values.

III. Placement requirements of electrical cabinet

(1) Make sure that scrap iron, coolant, or oil will not splash onto the electrical cabinet;

(2) The electrical cabinet shall be placed on the right side of the machine tool;

(3) When connecting the aviation plug connector, attention should be paid to ensure that it is connected in place.

Task Implementation

(1) Divide students randomly into groups and appoint the group leaders. Fill the personal information of the group members in Table 2−6.

(2) According to the observation and discussion results, record the machine tool installation requirements, main power supply requirements, electrical cabinet placement requirements, power-on operation process, and etc. in Table 2−6.

(3) Fully discuss the problems found during the observation, and fill the problems and solutions in Table 2−6.

Table 2−6 Task Work Order for Power-on Operation Process

<table>
<tr><td colspan="5">Task Work Order</td></tr>
<tr><td colspan="2">Name:</td><td colspan="2">Class:</td><td>Student ID:</td></tr>
<tr><td colspan="5">List of materials, tools, measuring instruments, and equipment required</td></tr>
<tr><td>S/N</td><td>Name</td><td>Quantity</td><td>Model or specification</td><td>Remark</td></tr>
<tr><td></td><td></td><td></td><td></td><td></td></tr>
<tr><td></td><td></td><td></td><td></td><td></td></tr>
<tr><td></td><td></td><td></td><td></td><td></td></tr>
<tr><td colspan="5">Operating process(Observation)</td></tr>
<tr><td>S/N</td><td>Machine tool installation requirement</td><td>Main power supply requirement</td><td>Electrical cabinet placement requirement</td><td>Power-on operation process</td></tr>
<tr><td></td><td></td><td></td><td></td><td></td></tr>
<tr><td></td><td></td><td></td><td></td><td></td></tr>
<tr><td></td><td></td><td></td><td></td><td></td></tr>
<tr><td></td><td></td><td></td><td></td><td></td></tr>
<tr><td colspan="3">Problems</td><td colspan="2">Solutions</td></tr>
<tr><td colspan="3"></td><td colspan="2"></td></tr>
</table>

Task Evaluation

S/N	Evaluation contents	Total point	Score
1	Be able to understand the installation requirements of YL-569 CNC milling machines	25	
2	Be able to understand the power supply requirements of YL-569 CNC milling machines	25	
3	Be able to understand the placement requirements for the electrical cabinet of YL-569 CNC milling machines	25	
4	Have a professional spirit of effective communication, expression and solidarity	25	

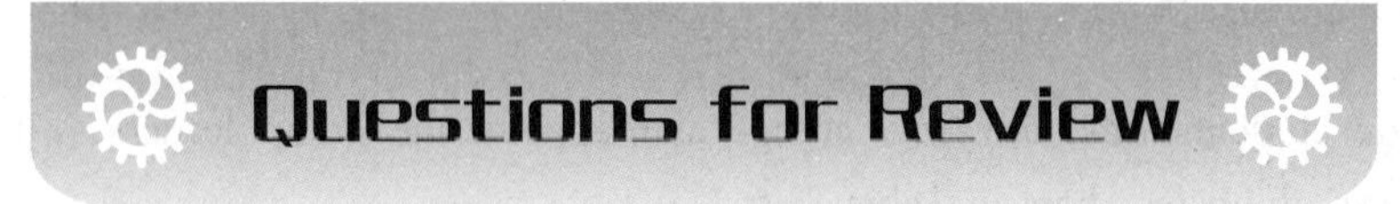

Questions for Review

(1) What is the basic composition of YL-569 CNC milling machines?

(2) What are the main technical parameters of YL-569 CNC milling machines?

(3) What are the practical training projects that can be carried out with YL-569 CNC milling machines?

(4) What are the precautions in the use of YL-569 CNC milling machines?

(5) What are the installation requirements of YL-569 CNC milling machines?

Module III Electrical Installation and Commissioning of CNC Machine Tools

CNC machine tools are highly automated equipment, integrating the latest technologies such as machinery, automation, computers, measurement and microelectronics. The electrical system of the CNC machine tool is the basis for realizing automation of the machine tool. This module mainly introduces the common electrical components of CNC machine tools, the making and reading of electrical schematic diagrams, the introduction to typical machine tool circuits, and the practical training contents related to the design and implementation of the cooling pump start/stop circuit and the tool magazine tool changing circuit.

Task I Understanding of Typical Electrical Components

Learning Objectives

(1) To understand the classification of commonly used electrical components in CNC machine tools.

(2) To master the general structure and working principle of commonly used electrical components in CNC machine tools.

(3) To master the performance and selection methods of commonly used electrical components in CNC machine tools.

(4) To be familiar with the specifications, models and meanings of commonly used electrical components.

(5) To be able to analyze and solve problems independently.

(6) To master the engineering working methods, cultivate a rigorous work style, and abide by the 5S management system.

Task Description

Observe the electrical control unit of the YL-569 CNC milling machine in groups, and fill names, structural characteristics, main functions, applications and etc. of the electrical components

of each machine tool in the work task order. Make sure to be careful and pay attention to the relevant operation specifications of the practical training site during observation.

Knowledge Link

I. Basic knowledge of electrical components

1. Definition of electrical components

Electrical devices or components that can automatically or manually switch on and off the circuit according to operation signals or external signals (mechanical force, electrodynamic force and other physical quantities), so as to continuously or intermittently change circuit parameters or states and realize switching, control, protection, detection, conversion and adjustment of circuits or electrical equipment.

2. Classification of electrical components

(1) Classification by voltage.

High-voltage electrical components: It refers to all kinds of electrical components with an operating voltage higher than AC 1, 200 V or DC 1, 500 V, e.g., high-voltage circuit breakers, disconnectors, reactors, voltage transformers, current transformers, and lightning rods.

Low-voltage electrical components: It refers to all kinds of electrical components with an operating voltage lower than AC 1, 200 V or DC 1, 500 V, e.g., contactors, starters, automatic switches, fuses, relays, and master switches.

(2) Classification by control functions.

Master devices: They are electrical devices used to give signals and commands, e.g., buttons, master controllers, and change-over switches.

Distribution devices: They are electrical devices used for the transmission and distribution of electric energy, e.g., switches, low-voltage circuit breakers, and isolators.

Actuators: They are used to complete a certain action or transmit power, e.g., contactors, solenoid valves, and electromagnets.

Control electrical devices: They are used to control the on-off of a circuit, e.g., relays.

Protection devices: They are used to protect the safety of power supplies, circuits, and electrical components so that they will not operate under short circuits or overload conditions and can be protected from damage, e.g., fuses, thermal relays, overcurrent/ undercurrent/ overvoltage/ undervoltage relays, and leakage protectors.

(3) Classification by operation modes.

Manual devices: They are electrical devices operated by manual operation, e.g., knife switches, disconnectors, and button switches.

Automatic devices: They are electrical components that automatically operates according to the change of a signal or a certain physical quantity, e.g., high and low voltage circuit breakers,

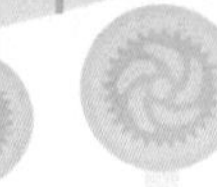

contactors, and relays.

(4) Classification by action principle.

Electromagnetic devices: They work according to the principle of an electromagnet, e. g., contactors and relays.

Non-electromagnetic devices: They are electrical device that operates by external force (human or mechanical) or a change in some non-electric quantity, e. g., buttons, travel switches, speed relays, and thermal relays.

3. Development direction of low-voltage electrical components

At present, it is developing in the direction of small size, light mass, safety and reliability, and convenient use. New electronic control devices are being vigorously developed, such as proximity switches, photoelectric switches, electronic time relays, solid-state relays and contactors, to meet the requirement for rapid electronization of control systems.

II. Master devices

Master devices are electrical devices that sends commands or signals in the automatic control system. They are mainly used to control contactors, relays, or other electrical coils so as to switch on or off the circuit and thus achieve the purpose of controlling production machinery, e. g., control buttons, travel switches, proximity switches, universal change-over switches, master controllers, and other master switches (such as pedal switches, knob switches, and emergency switches), etc.

1. Button switches (buttons)

A button switch is usually used for connecting or disconnecting small current control circuits in the event of a short circuit. A button switch consists of the button cap, return spring, bridge contacts and etc. It is usually made with normally open contacts and a compound structure, as shown in Fig. 3-1. The indicator button can be equipped with a signal lamp to display signals, the emergency button is equipped with a mushroom-shaped button cap for emergency operations, and the knob is operated by twisting and rotating by hand. Some button switches are shown in Fig. 3-2.

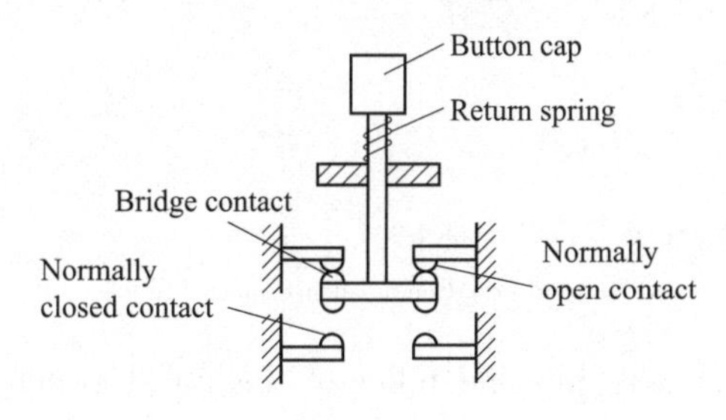

Fig. 3-1 Schematic Diagram of a Button Switch Structure

Fig. 3-2 Some Button Switches

The rated voltage of the button switch is AC 380 V、DC 220 V, and the rated current is 5 A. Button caps come in a variety of colors, generally red for the stop button and green for the start

button. Buttons are mainly selected according to the required number of contacts, application occasions and colors. The graphic symbols of button switches are shown in Fig. 3-3.

SB SB SB

(a) (b) (c)

Fig. 3-3 Graphic Symbols of Button Switches

(a) A start button; (b) A stop button; (c) A combination button

2. Travel switches

Travel switches(Fig. 3-4), also known as limit switches or position switches, which are master switches that uses the collision of some moving parts of production machinery with the switch operating mechanism to make the contacts act and send out control signals. It is mainly used to control the movement direction, travel and position protection for production machinery.

The graphic symbols of travel switches are shown in Fig. 3-5.

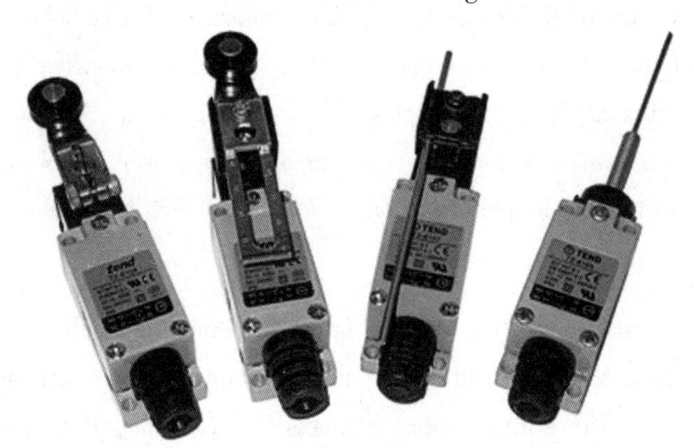

Fig. 3-4 Some Travel Switches

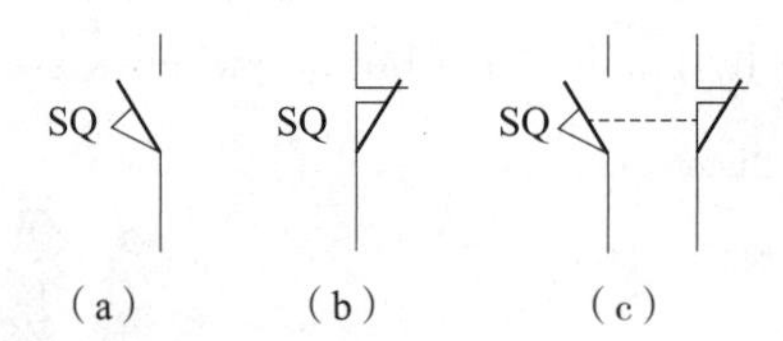

Fig. 3-5 Graphic Symbols of Travel Switches

(a) Normally open contact; (b) Normally closed contact; (c) Combination contact

When the stopper of the moving part of machinery hits the roller of the travel switch, the transmission lever rotates together with the rotating shaft to make the cam push the striker. When the striker is pressed to a certain position, it pushes the microswitch to act quickly to open its normally closed contact and close its normally open contact. After the stopper on the roller moves away, the reset spring restores all parts of the travel switch to their original positions. Structure and working principle of a travel switch is shown in Fig. 3-6. The use of travel switches are shown in Fig. 3-7.

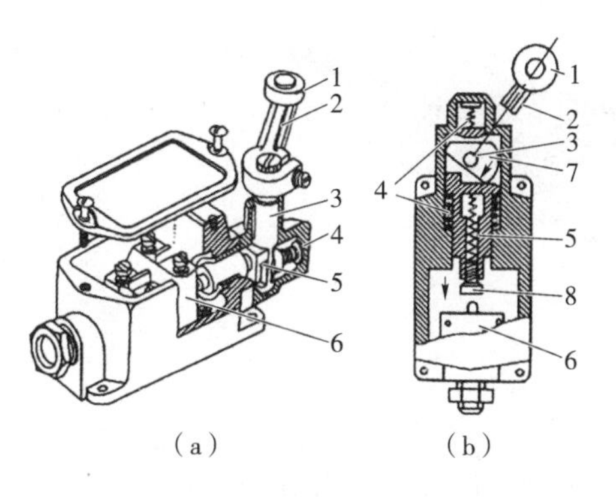

Fig. 3-6 Structure and Working Principle of a Travel Switch

(a) Structure; (b) Working principle

1—Roller; 2—Lever; 3—Rotating shaft; 4—Return spring; 5—Striker; 6—Micrswitch; 7—Camshaft; 8—Adjusting screw

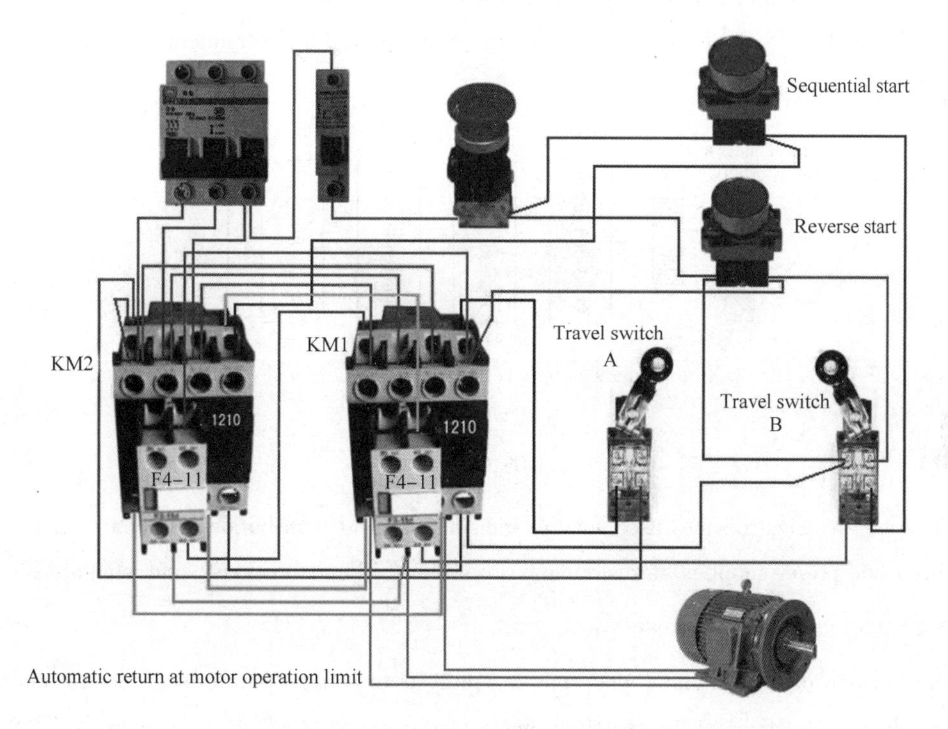

Fig. 3-7 Schematic Diagram for Use of Travel Switches

3. Proximity switches

Proximity switches are also called non-contact travel switches. When an object approaches it to a certain distance, it will send out action signals. Unlike a mechanical travel switch, it does not require mechanical force to be applied, but obtains signals through the change of medium energy between its inductive head and the measured object. A proximity switch is shown in Fig. 3-8.

Proximity switches are more and more widely used because of their advantages such as stable and reliable operation, long service life, high repeated positioning accuracy, high operating frequency, and fast action. The graphic symbols of proximity switches are shown in Fig. 3-9.

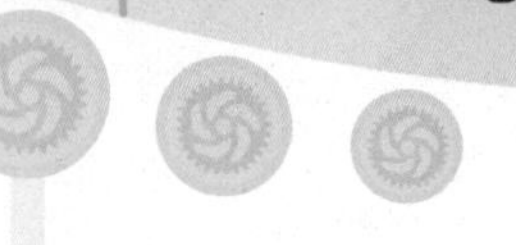

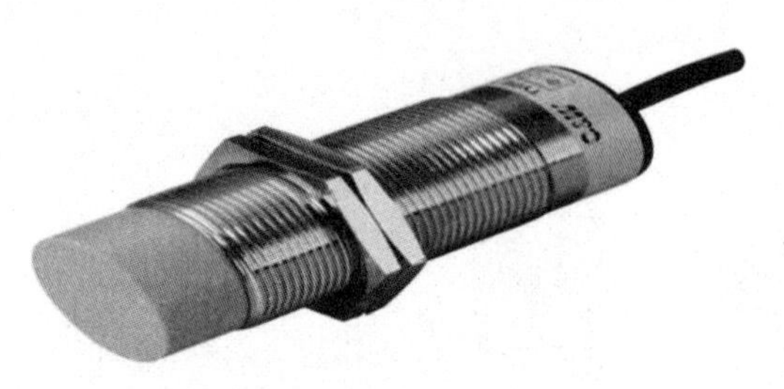

Fig. 3-8 A Proximity Switch

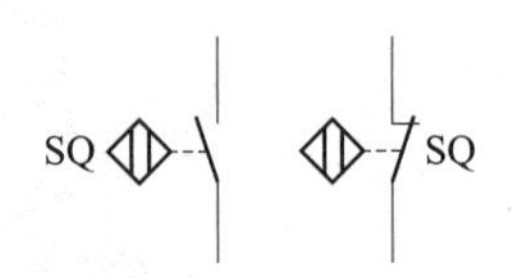

Fig. 3-9 Graphic Symbols of Proximity Switches

Proximity switches can be classified into high-frequency oscillation type, capacitance type, inductance bridge type, permanent magnet type and Hall effect type according to their working principles, among which the high-frequency oscillation is the most commonly used type. The circuit of high-frequency oscillation proximity switch is mainly composed of three parts: oscillation circuit, amplifying circuit and output circuit. The working principle of a high-frequency ossillation proximity switch is shown in Fig. 3-10.

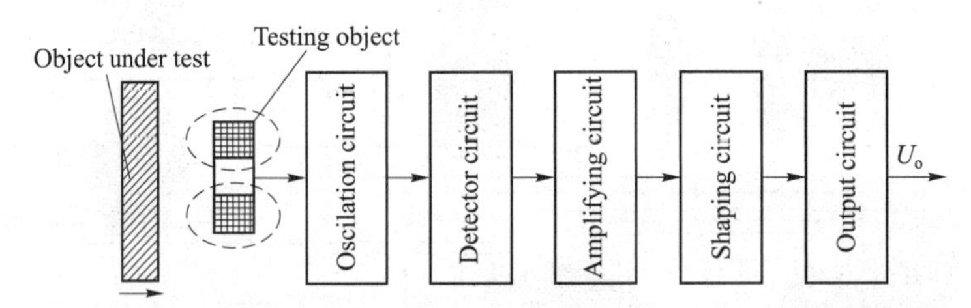

Fig. 3-10 Working Principle of a High-frequency Oscillation Proximity Switch

III. Power distribution components

They are electrical devices used for the transmission and distribution of electric energy, e.g., switching mode power supplies, transformers, low-voltage circuit breakers, and isolators.

1. Switching mode power supplies

The function of the switching mode power supply is to convert one form of electric energy into another. The specifications of the switching mode power supply used in the devices for THWSKW-2A CNC machine tool maintenance comprehensive training are input AC 220 V, output DC 24 V, 5 A, as shown in Fig. 3-11.

2. Transformers

A transformer is a device that transforms AC voltage, current and impedance. When AC current flows through the primary coil, AC magnetic flux will be generated in the iron core (or magnetic core), so as to induce voltage (or current) in the secondary coil. A transformer consists of an iron core (or magnetic core) and a coil. The coil has two or more windings, of which the winding connected to the power supply is called the primary coil and the other windings are called the secondary coils. In a generator, no matter whether the coil moves through a magnetic field or a

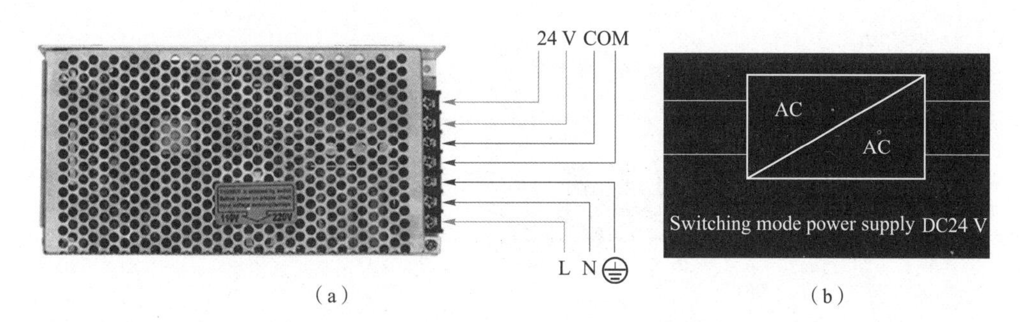

Fig. 3-11 A Switching Mode Power Supply

(a) Switching mode power supply; (b) Graphic symbol of switching mode power supply

magnetic field moves through a fixed coil, an electromotive force can be induced in the coil. In both cases, the magnetic flux value remains unchanged, but the magnetic flux value of the chain intersecting with the coil changes. This is due to the principle of mutual induction. A transformer is a device that uses the electromagnetic mutual induction effect to transform voltage, current, and impedance.

The transformer is used for isolation and transforming voltage, current and impedance. Fig. 3-12 shows the appearance and graphic symbol of a transformer.

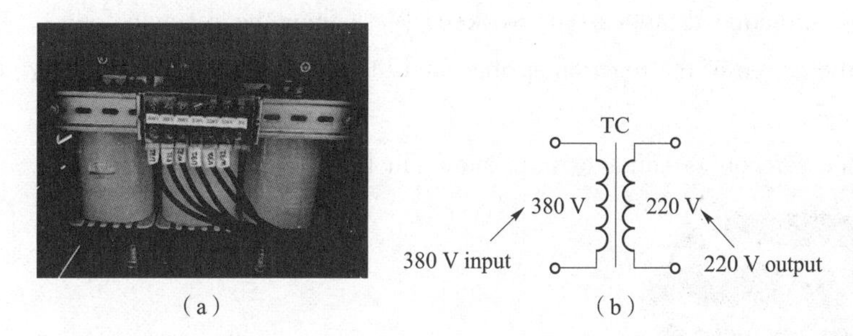

Fig. 3-12 A Transformer

(a) Appearance; (b) Graphic symbol

3. Low-voltage circuit breakers

Low-voltage circuit breakers, also known as automatic air switch, are used to distribute electric energy, start asynchronous motors infrequently and protect power lines and motors. It can automatically cut off the circuit in case of serious overload, short circuits or undervoltage. It is a kind of protective device widely used in low-voltage distribution lines. Fig.3-13 shows the appearance and graphic symbol of a low-voltage circuit breaker.

IV. Actuator

It is used to complete a certain action or transmit power, e.g., contactors, solenoid valves, and electromagnets.

The contactor is an automatic switching device used to frequently turn on or off the main circuit or large-capacity control circuit. It is mainly used in electric power, power distribution and consumption.

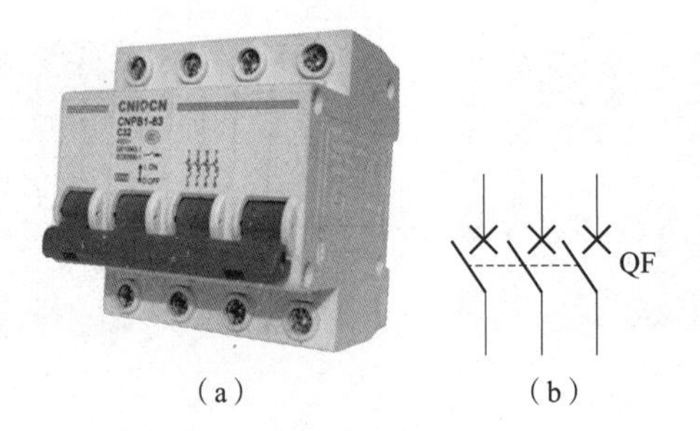

Fig. 3-13 A low-Voltage Circuit Breaker

(a) Low-voltage circuit breaker; (b) Graphic symbol of low-voltage circuit breaker

Contactors are usually divided into AC contactors and DC contactors. Fig. 3-14 shows the appearance of an contactor.

When the electromagnetic coil is energized, the iron core is magnetized to generate magnetic flux. Thus, an electromagnetic force is generated at the armature gap to pick up the armature, which causes the main contact to close and connect the main circuit. At the same time, the armature also drives the auxiliary contact to act. The normally closed auxiliary contact is opened first, and then the normally open auxiliary contact is closed. When the solenoid coil is de-energized or the voltage drops significantly, the attraction disappears or weakens (less than the reaction force), the armature is released under the action of the reaction spring, and the main contact and auxiliary contact return to their original states.

Some graphic symbols of contactors are shown in Fig. 3-15.

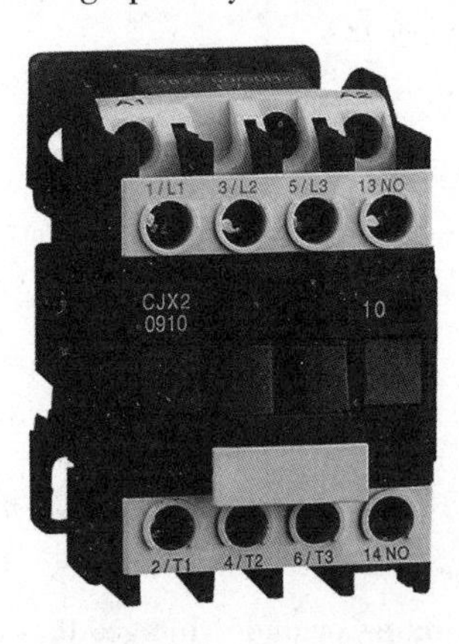

Fig. 3-14 A Contactor

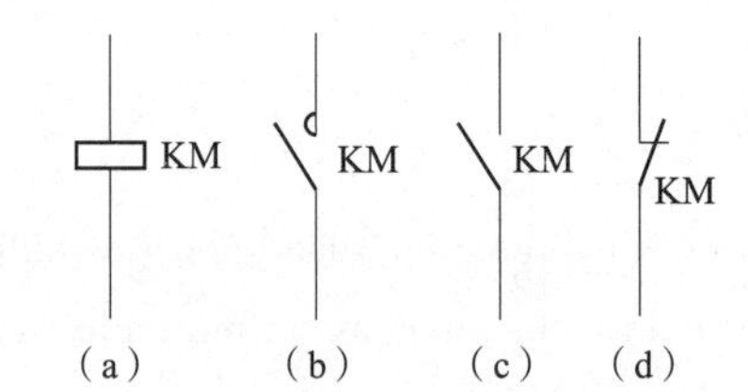

Fig. 3-15 Some Graphic Symbols of Contactors

(a) Coil; (b) Main contact; (c) Auxiliary normally open contact; (d) Auxiliary normally closed contact

V. Control electrical components

Control electrical components are used to control the on-off of a circuit, including switching devices and those relays used for control. Switching devices are widely used in power distribution systems and electric drive control systems for isolation of power supplies, as well as protection

and control of electrical equipment. Such devices can be divided into two categories: manual switches and automatic switches, including knife switches, combination switches, and automatic switches.

1. Electromagnetic relays

Electromagnetic relays refer to relays mainly rely on electromagnetic force or a certain physical quantity. They are control components that convert electrical or non-electrical signals into electromagnetic force or make the output state change stepwise by using changes in various physical quantities, so as to drive other components or devices in the same circuit or another circuit to act through the relay's contacts or break variables.

Main differences between electromagnetic relays and contactors are as below.

(1) The relay can respond to the changes of various input quantities, while the contactor only acts under the action of certain voltage signals;

(2) The relay is used to switch the low-current control circuits and protection circuits, while the contactor is used to control high-current circuits;

(3) The relay has no arc extinguishing device or main/auxiliary contact.

2. Intermediate relays

An intermediate relay is a certain kind of voltage relay, which is usually used to transmit signals and at the same time, control multiple circuits. It can also be directly used to control small-capacity motors or other electrical actuating components. Its main application lies in the expansion of the contact capacity (contact in parallel) or amount of contacts with the help of intermediate relays when the amount of contacts or contact capacity of other relays is insufficient, playing an intermediate conversion role in the entire process.

Graphic symbols of some intermediate relays are shown in Fig. 3-16.

3. Current relays

A relay that operates according to the input (coil) current. The coil is connected in series in a circuit, with thick wires, a small number of turns and small impedance.

Graphic symbols of some current relays are shown in Fig. 3-17.

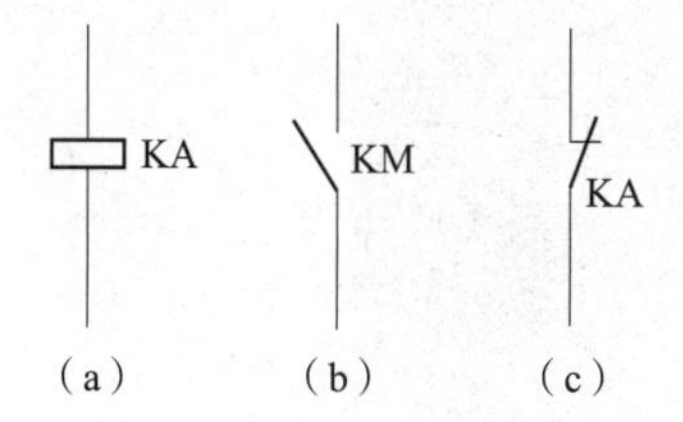

Fig. 3-16 Graphic Symbols of Some Intermediate Relays

(a) Coil; (b) Normally open contact; (c) Normally closed contact

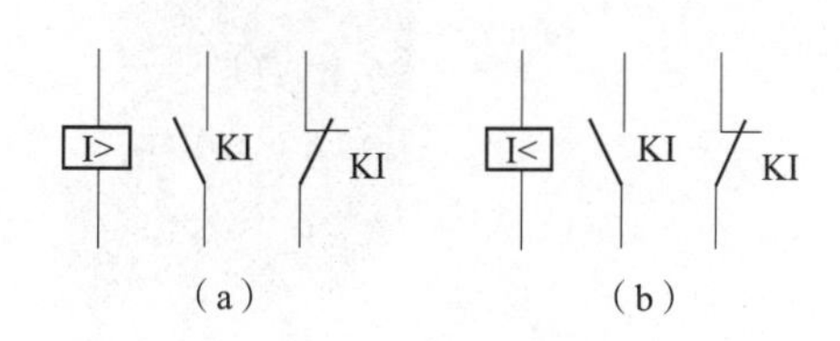

Fig. 3-17 Graphic Symbols of Some Current Relays

(a) Overcurrent relay; (b) Undercurrent relay

Main technical indicators of current relay:

Action current I_q: the current value required to start the action of the current relay;

Return current I_f: the current value when the current relay returns to its original state after action;

Return coefficient K_f: the ratio of return value to action value, $K_f = I_f / I_q$.

4. Voltage relays

They are relays that operate according to the input (coil) voltage.

The coil of voltage relay is connected in parallel with other devices in the circuit, with many turns and thin wires.

Overvoltage relay: the setting range of action voltage is (105%–120%) U_n.

Undervoltage relay: the adjustment range of pick-up voltage is (30%–50%) U_n.

The adjustment range of release voltage is (7%–20%) U_n.

Graphic symbols of voltage relays are shown in Fig. 3–18.

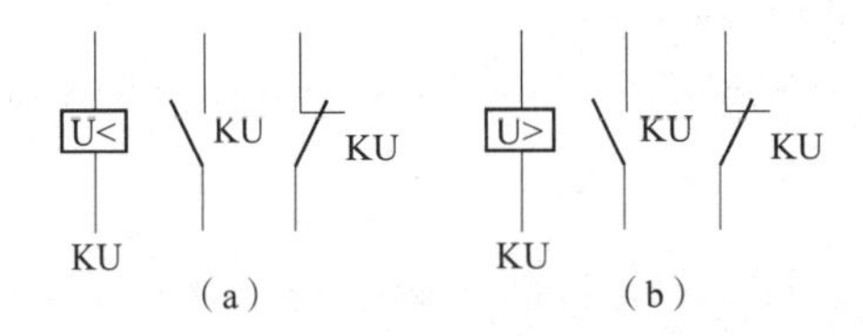

Fig. 3–18 Graphic Symbols of Voltage Relays

(a) Undervoltage relay; (b) overvoltage relay

5. Time relays

Time relays are control devices that can output signals only after a certain delay upon receiving the signal, so as to realize delayed contact connection or disconnection. The appearances of time relays are shown in Fig. 3–19.

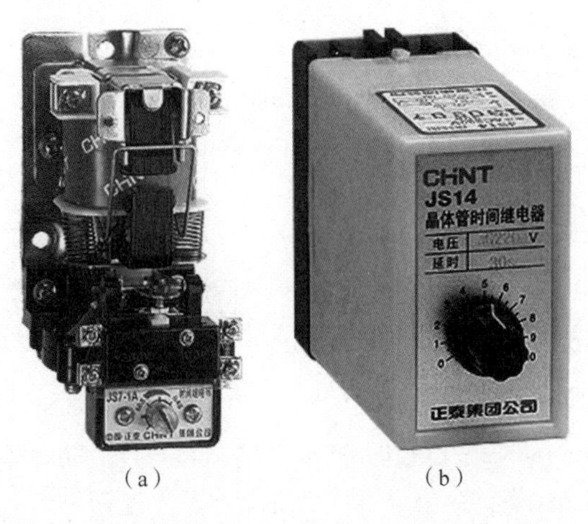

Fig. 3–19 Appearances of Time Relays

(a) An air damping type time relay (JS7 series); (b) A transistor type time relay (JS14 series)

Graphic symbols of some time relays are shown in Fig. 3-20.

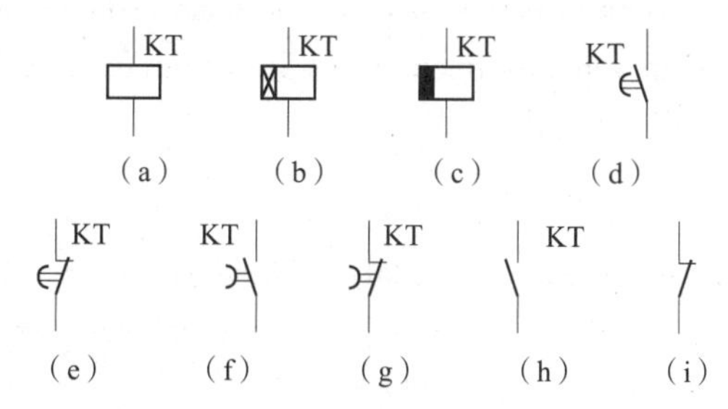

Fig. 3-20 Graphic Symbols of some Time Relays

(a) General symbol of coil; (b) Power-on delay coil; (c) Power-off delay coil; (d) Time-delayed closing normally open contact; (e) Time-delayed opening normally closed contact; (f) Time-delayed opening normally open contact; (g) Time-delayed closing normally closed contact; (h) Instantaneous normally open contact; (i) Instantaneous normally closed contact

6. Thermal relays

A thermal relay is a certain kind of protective device that works on the principle of the current heating effect. In the actual operation of motors, short-term overload is allowed. However, long-term overload, undervoltage operation, or open-phase operation may cause the current of the motor to exceed its rated value, causing the motor to heat up. If the winding temperature rise exceeds the rated value, it will damage the insulation of the winding, shorten the service life of the motor, and even burn out the motor winding in severe cases. Therefore, overload protection measures must be taken. The most common method is to use a thermal relay for overload protection. The appearances of thermal relays are shown in Fig. 3-21.

Fig. 3-21 Appearances of Thermal Relays

A thermal relay is mainly composed of thermal elements, bimetallic strips, contacts and action mechanisms. The schematic diagram of bimetallic thermal velay is shown in Fig. 3-22.

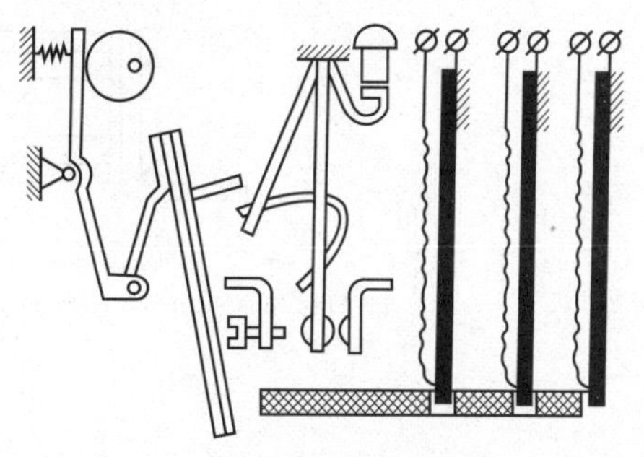

Fig. 3-22 Schematic Diagram of Bimetallic Thermal Relay

In Fig. 3-22, a group of hopping mechanisms is composed of leaf springs and bow springs. Cam can be used to adjust the action current. The compensation bimetallic strip compensates for the influence of ambient temperature change. When the ambient

temperature changes, the main bimetallic strip and the compensation bimetallic strip made of the same material will bend in the same direction, so that the push distance between the guide plate and the compensation bimetallic strip remains unchanged. In addition, the thermal relay can be reset automatically or manually by adjusting the screw.

7. Speed relays

It is used for speed detection and is commonly used in reverse braking of three-phase AC asynchronous motors. When the speed is close to 0, the negative-phase-sequence power supply will be automatically cut off.

VI. Protective electrical components

Fuses(also know as safety wire) in low-voltage distribution lines provide protection mainly in the event of short circuits and overload. A fuse has the advantages of simple structure, small size, light mass, reliable operation, and low price, so it is widely used in strong current and weak current systems. A fuse is mainly composed of a fuse link and an insulating tube or base (also known as a fuse shell) for placing the fuse link. When the fuse is connected in series with other devices in the circuit, the load current flows through the fuse link, and the resistive loss of the fuse link makes it heat up and the temperature rises.

When the circuit works normally, the heating temperature is lower than its melting temperature, so the fuse will not be burned out for a long time.

When the circuit is overloaded or short, the current is greater than the normal allowable heating current for the fuse link, so that the temperature of the fuse link rises sharply and exceeds its melting point, causing the fuse to be burned out, and thus breaking the circuit and protecting the circuit and equipment.

There are roughly two methods for arc extinguishing of fuses. One is to place the fuse link in a sealed insulating tube. The insulating tube is made of high-strength materials, and this material can decompose a large amount of gas at the high temperature of the arc, so that a high pressure is generated in the tube to compress the arc and increase the potential gradient of the arc, thus achieving the purpose of arc extinguishing. The other method is to place the fuse link in a fuse tube with insulating sand filler (such as quartz sand) . When the fuse link disconnects the circuit and generates an arc, the quartz sand can absorb the arc energy, and the metallic vapor can be dissipated into the gaps of the sand particles, so that the fuse link cools down quickly, thus achieving the purpose of arc extinguishing, as shown in Fig. 3-23.

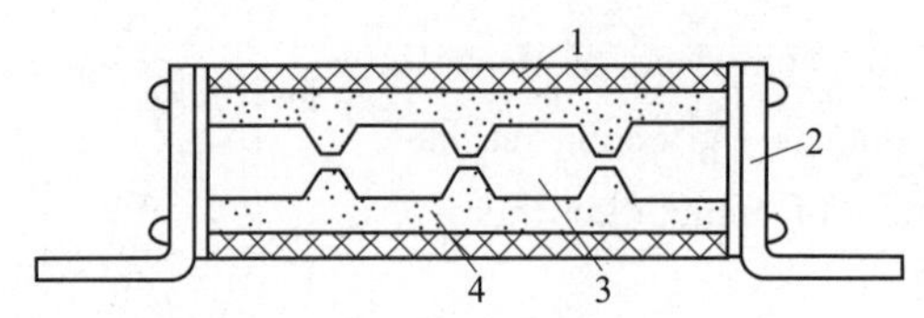

Fig. 3-23 Arc Extinguishing of Fuses

1—Fuse tube; 2—End cover and terminal board; 3—Fuse piece; 4—Quartz sand

Task Implementation

(1) Divide students randomly into groups and appoint the group leaders. Fill the personal

information of the group members in Table 3-1.

(2) According to the observation and discussion results, record the names, structural characteristics, main functions and applications of electrical components of YL-569 CNC machine tool in Table 3-1.

(3) Fully discuss the problems found during the observation, and fill the problems and solutions in Table 3-1.

Table 3-1 Task Work Order for Machine Tool Electrical Components Observation

Task Work Order					
Name:	Class:		Student ID:		
List of materials, tools, measuring instruments, and equipment required					
S/N	Name		Quantity	Model or specification	Remark
Operating process(Observation)					
S/N	Name	Structural characteristic	Main function	Application	
Problems			Solutions		

Task Evaluation

S/N	Evaluation contents	Total point	Score
1	Be familiar with and know the names of machine tool electrical components	25	
2	Be able to correctly draw graphic symbols of common electrical components	25	
3	Master the general structure and working principle of commonly used electrical components in CNC machine tools	25	
4	Have a professional spirit of effective communication, expression and solidarity	25	

Task II Reading of Electrical Schematic Diagrams

Learning Objectives

(1) To understand the classification of electrical control system diagrams.

(2) To understand the preparation principles of electrical system diagrams.

(3) To master the drawing and reading of electrical schematic diagrams.

(4) To master the engineering working methods, cultivate a rigorous work style, and abide by the 5S management system.

Task Description

Observe the electrical schematic diagram of CNC machine tools in groups, and fill the name of each typical circuit, operating process, main functions, applications and etc. in the task work order. Make sure to be careful and pay attention to the relevant operation specifications of the practical training site during observation.

Knowledge Link

I. Electrical control system diagrams

According to the mechanical movement requirements for electrical control systems, electrical graphic symbols and text symbols uniformly specified by the state are adopted. The diagram showing all basic compositions and connection relationships of circuits, equipment or complete sets of devices in detail according to the working sequence of electrical equipment and devices is the electrical control system diagram.

Common electrical control system diagrams include electrical schematic diagrams, plug-in definition diagrams, and plug-in connection diagrams. Electrical schematic diagram is the most commonly used in the analysis of electrical control principle of machine tools.

An electrical schematic diagrams, a plug-in definition diagrams and a plug-in connection diagrams are shown in Fig. 3-24.

1. Reading steps of electrical schematic diagrams

(1) Preparations: Understand the requirements of the production process and technology for circuits; Be familiar with the positions and purposes of various electrical equipment and control devices; Comprehend the significance of graphic and text symbols in diagrams.

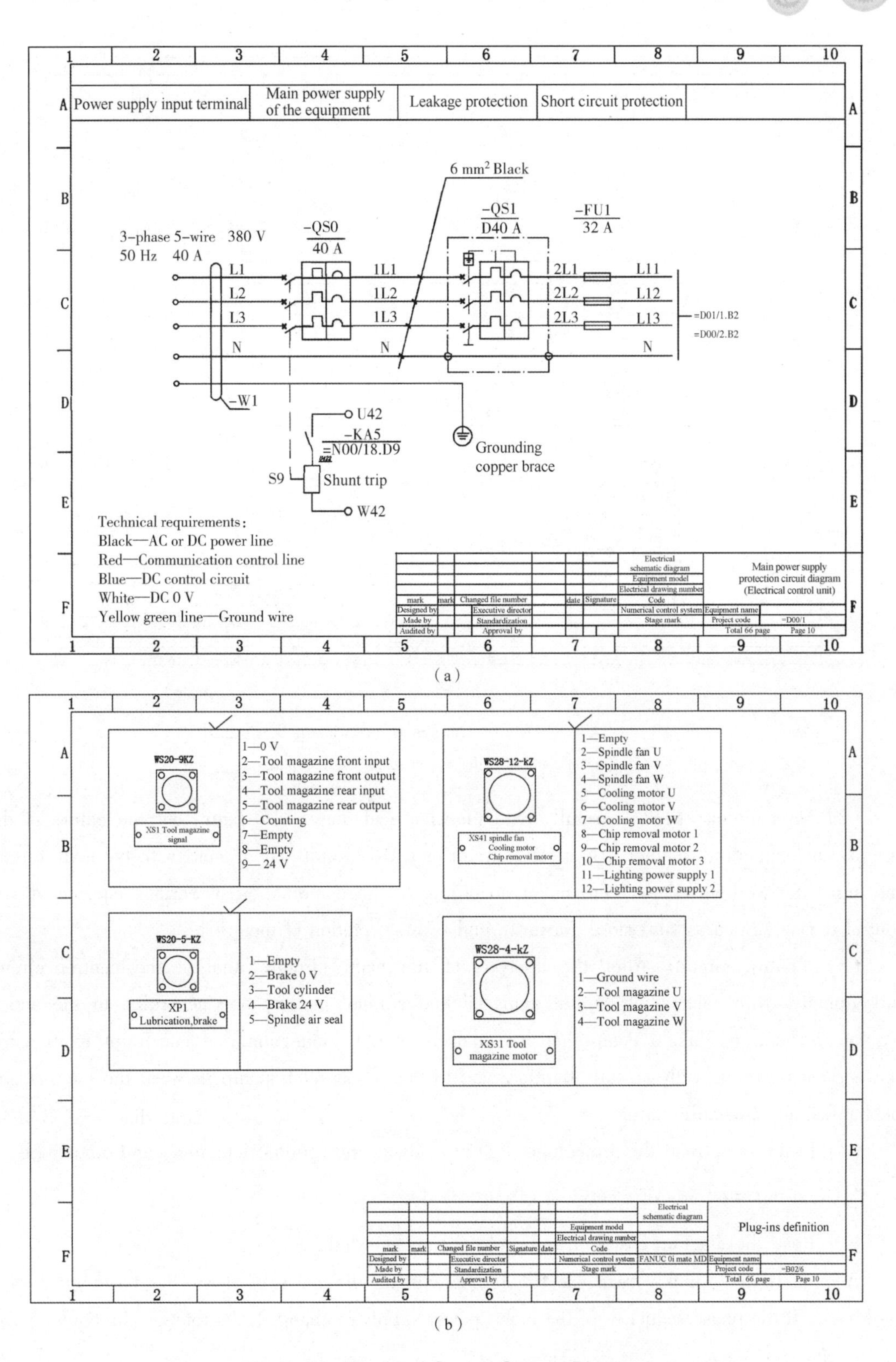

(a)

(b)

Fig. 3–24 Electrical Control System Diagrams

(a) Electrical schematic diagram of main power protection; (b) Plug-in definition diagram for some electrical plug-ins;

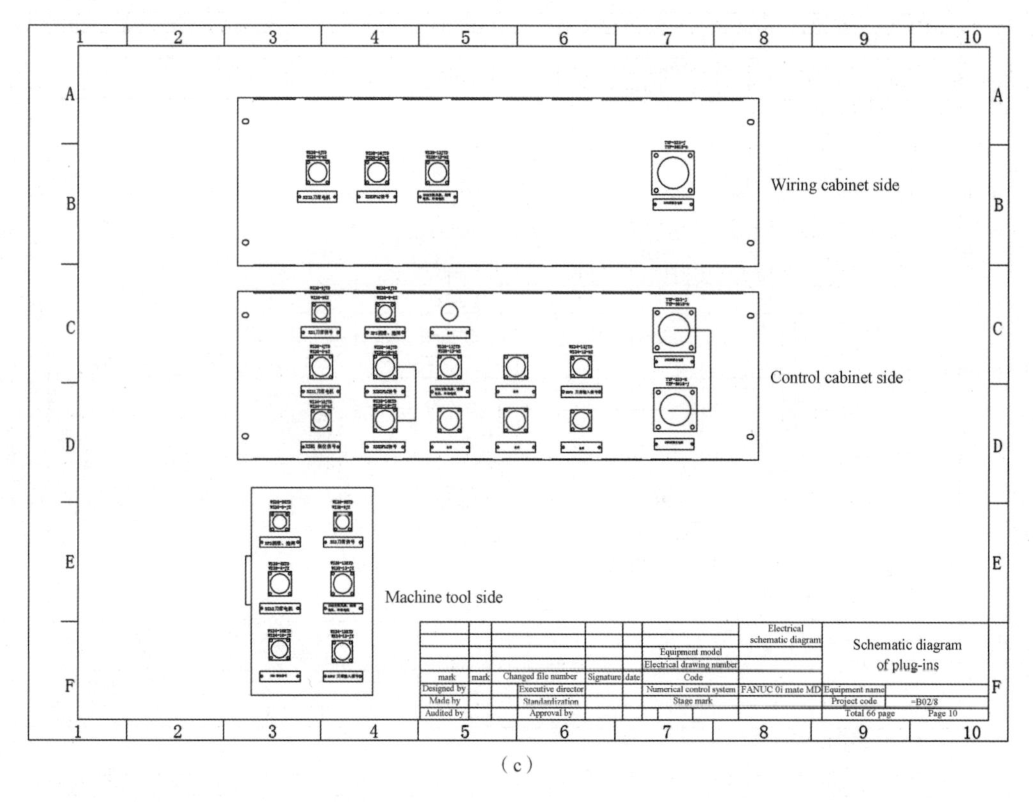

(c)

Fig. 3-24 Electrical Control System Diagram (Continued)

(c) Plug-in connection diagram of some electrical plug-ins

(2) Main circuit: First, carefully read the electrical diagram to figure out the nature of the circuit, understanding whether it is an AC circuit or a DC circuit. Then, start with the main circuit and judge the working state of the motor according to the combination of various equipment and control devices. Such as start-stop, forward, and reverse rotation of motors.

(3) Control circuit: After the analysis of the main circuit, analyze the control circuit subsequently, with each small circuit analyzed and studied one by one according to the action sequence. Then, conduct a comprehensive analysis of the connection and constraint relationship between each circuit, with special attention paid to the action relationship between the circuits and mechanical and hydraulic parts.

(4) In the end, read the protection, lighting, signal indication, detection, and other parts.

2. Reading methods of electrical schematic diagrams

(1) Read the diagram based on basic electrician knowledge.

Accurately and quickly read electrical diagrams on the basis of mastering basic electrician knowledge. If the phase sequence of the motor power supply is changed, its rotation direction control can be changed.

(2) Read the diagram based on typical circuits.

Typical circuits are common basic circuits, such as those related to motor starting, braking,

and sequential control. No matter how complex a circuit is, it almost always consists of several basic circuits. Therefore, familiarity with various typical circuits is the basis for understanding more complex electrical diagrams.

(3) Read the diagram based on drawing requirements.

When preparing electrical diagrams, in order to strengthen the standardization, universality and demonstrativeness of diagrams, some rules and requirements must be followed so that the diagrams can be accurately read by using these drawing requirements.

3. Electrical schematic diagrams

An electrical schematic diagram refers to a diagram showing the relationship and working principle among various electrical components of the circuit with graphic symbols, text symbols, item codes, etc.

It indicates the transmission of current from the power supply to the load and the action principle and interrelationship of each electrical component, regardless of the actual installation position and actual wiring of each electrical component.

(1) Text symbols.

A text symbol is a character code used to represent the name, function, state and characteristics of electrical equipment, devices and components. For example, FR stands for the thermal relay.

(2) Graphic symbols.

A graphic symbol is a figure, mark, or character used to represent a piece of equipment or a concept. For example, "~" indicates AC and *R* indicates resistor/resistance. The national standard *Graphical Symbols for Electrical Diagrams—Part 1: General information* (GB/T 4728.1—2018) specifies the preparation method of graphic symbols in electrical diagrams. The national electrical diagram preparation standard *Preparation of Documents Used in Electrotechnology—Part 1: Rules*(GB/T 6988.1—2024) provides specific rules for diagrams, graphs, and tables used in electrical diagrams, which will officially come into force from March 1, 2025, The electrical schematic diagrams are generally divided into power circuits, main circuits and auxiliary circuits. This book only gives examples of the power protection circuits, as shown in Fig. 3-24(a).

4. Plug-in definition diagrams

The installation positions of electrical equipment and components are detailed in the layout of electrical elements, as shown in Fig. 3-24 (b). Codes of electrical components in the layout shall be the same as those of all components on the list of relevant circuits and electrical components.

5. Installation and wiring diagrams

The installation and wiring diagram is used to show the wiring relationship between units of electrical equipment, as shown in Fig. 3-24(c). The diagram, indicating the relative positions of external components of electrical equipment and electrical connections between those external components, is the basis for actual installation and wiring.

II. Sheet subdivision of electrical control system diagram

To locate and identify the position of an component or device on a drawing for easy reading, it

is often necessary to subdivide the sheet. Sheet subdivision method: At the border of the drawing, starting from the opposite upper left corner of the title bar, the vertical direction is numbered with Latin letters in capital and the horizontal direction is numbered with Arabic numerals in turn, thus dividing the sheet into several areas. See Fig. 3−25 for the sheet subdivision example.

	1	2	3	4	5	6	
A							A
B		□X					B
C				□Y			C
D							D
E					Fig. 08		E
	1	2	3	4	5	6	

Fig. 3−25 Sheet Subdivision Example

A coordinate system has been established on the drawing after sheet subdivision. The position of an item and its connecting line on the electrical diagram is uniquely determined by this "coordinate system" and marked with "drawing number/row, column or area number".

In Fig. 3−25, element X, located in area B2, can be marked as 08/B2; element Y, located in area C4, can be marked as 08/C4.

In a simpler electrical schematic diagram of the machine tool, it is not necessary to subdivide the sheet in the vertical direction, but it is only necessary to number areas at the border in the horizontal direction on the lower part of the sheet. The border in the horizontal direction on the upper part of the sheet is set as a purpose column, and the functions or purposes of corresponding circuit components below this column are indicated with letters to help understand the functions of parts in the electrical schematic diagram and the working principle of the whole circuit, as shown in Fig. 3−25.

III. Drawing rules for electrical schematic diagrams

(1) The power circuit is drawn as a horizontal line. The phase sequence lines L1, L2 and L3 of the three-phase AC power supply are drawn in turn from top to bottom, while the center wire and protective ground wire are drawn below the phase lines in turn. Terminal "+" of the DC power supply is on the top and terminal "−" is on the bottom.

(2) The main circuit is on the left side of the circuit diagram and is vertical to the power circuit. It comprises the main fuse, the main contact of the contactor, the thermal elements of the thermal relay, and the motor. It passes through a larger working current of the motor.

(3) Auxiliary circuits generally include a control circuit for controlling the working state of the main circuit, an indicating circuit for displaying the working state of the main circuit, and a lighting circuit for locally lighting the machine tool. An auxiliary circuit comprises the contacts of the master switch, contactor coil and auxiliary contacts, relay coil and contacts, indicators and lighting lamps. The operation sequence is generally expressed from left to right and from top to bottom. The current passing through an auxiliary circuit is small, generally not exceeding 5 A.

(4) In the circuit diagram, contacts of electrical components are all drawn according to their normal positions when the circuit is not energized or electrical components are not subject to external force.

(5) In the circuit diagram, elements of the same electrical components are not drawn together according to their actual positions, but drawn in different circuits according to their functions in the circuits. However, the actions of these elements are interrelated and marked with the same letter symbols.

(6) In case of numbering in the main circuit, U11, V11 and W11 will be numbered sequentially according to the phase sequence at the outgoing terminal of the power switch. Then, the numbers shall be increased from top to bottom and from left to right after passing through each electrical component. In case of numbering in one auxiliary circuit, the numbers shall be marked in sequence from top to bottom and from left to right based on the "equipotential" principle and be increased successively after passing through each electrical components.

(7) The circuit diagram is divided into several areas by function. Generally, one area is formed by a circuit or circuit branch; areas are numbered with Arabic numerals from left to right in turn and are marked in the area column at the lower part of the circuit diagram.

(8) The purpose of each circuit in the circuit diagram in the electrical operation of the machine tool is marked with letters in the purpose column at the upper part of the circuit diagram.

(9) Draw two vertical lines below the letter symbol KM of each contactor coil in the circuit diagram, and three columns (left, middle and right) are thus formed, corresponding to the main contact, normally open contact and normally closed contact. Fill in the corresponding column with the area number where the corresponding contact controlled by the contactor coil acts according to regulations. For spare but unused contacts, mark "×" or do not mark any symbol in the corresponding column.

(10) Draw a vertical line below the letter symbol of each relay coil in the circuit diagram, and two columns (left and right) are thus formed, corresponding to the normally open contact and normally closed contact. Fill in the corresponding column with the area number where the corresponding contact controlled by the relay coil acts according to regulations. For spare but unused contacts, mark "×" or do not mark any symbol in the corresponding column.

(11) The number below the letter symbol of the contactor or relay contact in the circuit diagram indicates the area number of the contactor or relay coil.

IV. Typical Circuit Analysis

1. Inching control

(1) Inching control circuit is commonly used for the adjustment of the machine tool spindle or workbench, e.g., running test and maintenance of the machine tool. Fig. 3-26 shows the electrical schematic diagram of inching control.

(2) Operating process.

Turn on the power switch QS first.

Press SB→the KM is energized→the KM main contact is closed→the motor M starts rotating after being energized.

Release SB→the KM coil is de-energized→the KM main contact is reset→the motor M stops rotating after being de-energized.

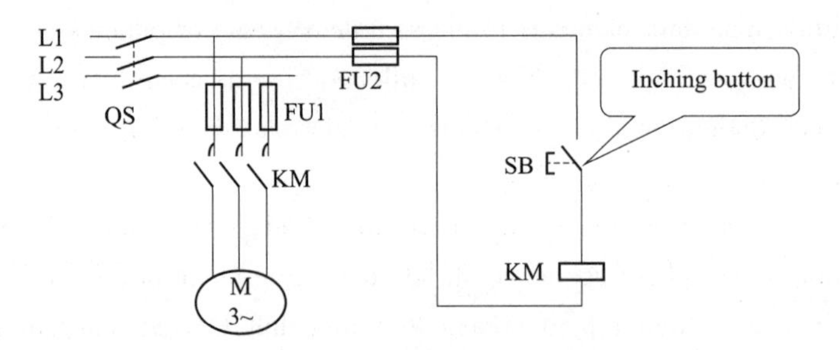

Fig. 3-26 Electrical Schematic Diagram of Inching Control

2. Continuous control

(1) The electrical schematic diagram of continuous control is shown in Fig. 3-27.

Protection link:

1) Short-circuit protection: FU1, FU2.

2) Overload protection: FR.

3) Undervoltage and no-voltage protection: KM self-locking link.

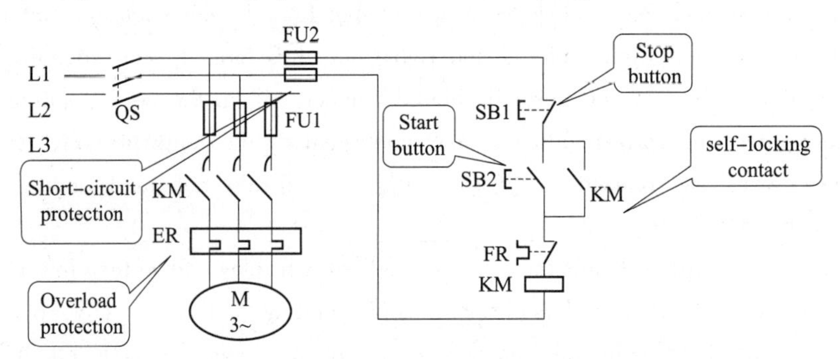

Fig. 3-27 Electrical Schematic Diagram of Continuous Control

(2) Operating process.

Turn on the power switch QS first.

Press SB2→ the KM coil is energized → { the KM main contact is closed →the motor M starts rotating.; the KM auxiliary contact is closed and self-locked. }

Press SB1→the KM coil is de-energized→the motor M stops rotating.

3. Forward and reverse rotation control of the motor

(1) Introduction to forward and reverse rotation control of the motor.

Such control is used for the forward and backward movement of the machine tool workbench or forward and reverse rotation of the spindle, etc. According to the motor working principle, as long as any two of the three-phase power incoming lines of the motor are exchanged, the rotation direction of the motor can be changed. This requirement can be achieved by using two contactors. When the forward contactor works, the motor rotates forward; when the reverse contactor works, the motor rotates reversely by exchanging any two of the lines connecting the motor to the power supply.

Electrical schematic diagram of forward and reverse rotation control is shown in Fig. 3-28.

(2) Operating process.

1) Forward rotation control.

The KM1 self-locking contact is closed.

Press SB2 ⟶ the KM1 coil is energized ⟶ { the KM1 main contact is closed ⟶ the motor M rotates forward. ; The KM1 interlocking contact is open. }

Press SB1 ⟶ the KM1 coil is de-energized ⟶ the motor M stops rotating.

2) Reverse rotation control.

Press SB3 ⟶ the KM2 coil is energized ⟶ the motor M rotates reversely.

4. Y-△ reduced voltage start of the motor

(1) The electrical schematic diagram of Y-△ reduced voltage start of the motor is shown in Fig. 3-29.

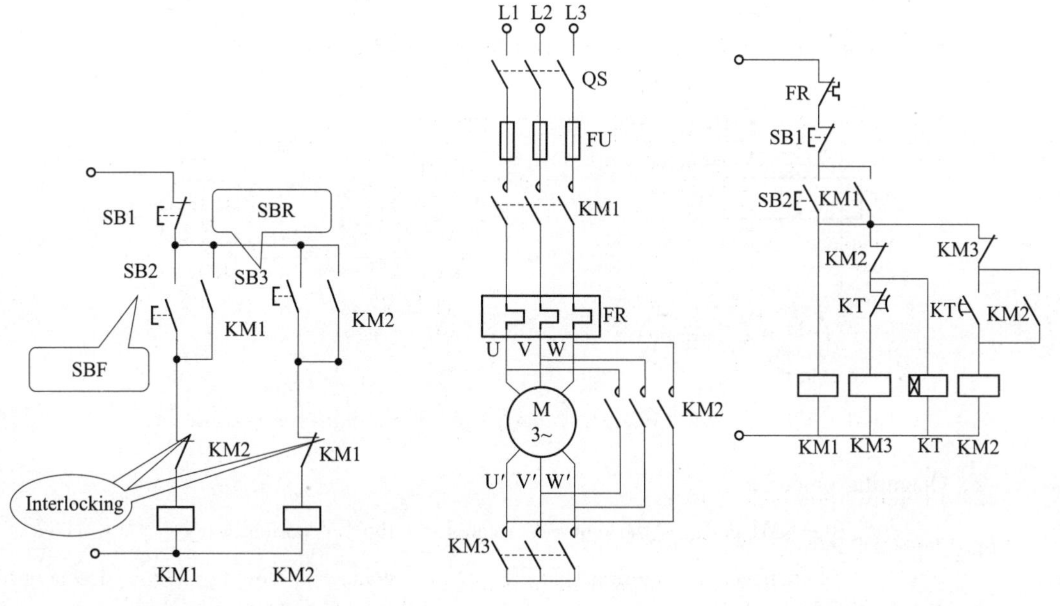

Fig. 3-28 Electrical Schematic Diagram of Forward and Reverse Rotation Control of the Motor

Fig. 3-29 Electrical Schematic Diagram of Y-△ Reduced Voltage Start of the Motor

(2) Operating process.

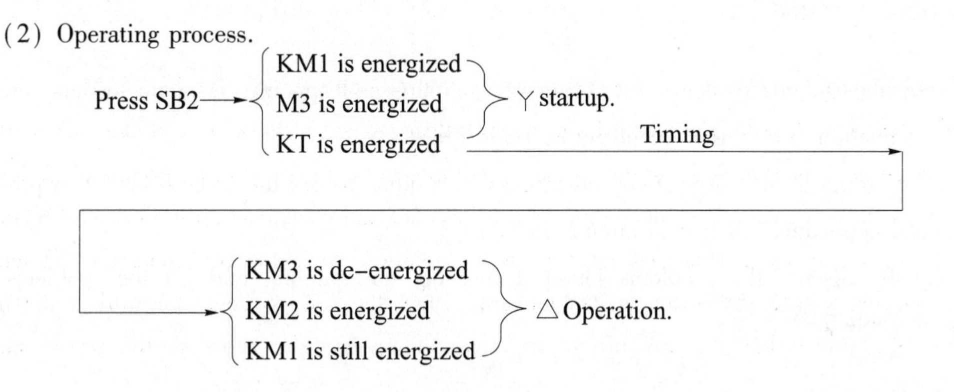

5. Motor reverse braking

(1) Electrical schematic diagram of motor reverse braking.

Reverse braking is a motor braking mode, which makes the motor generate retarding antitorque by reversing the phase sequence to brake the motor. Reverse braking features strong braking force, rapidness, simple control circuit and low equipment investment. However, reverse braking has poor accuracy, generates strong braking force during operation, and is easy to cause damage to transmission parts. Fig. 3-30 shows electrical schematic diagram of the motor reverse braking.

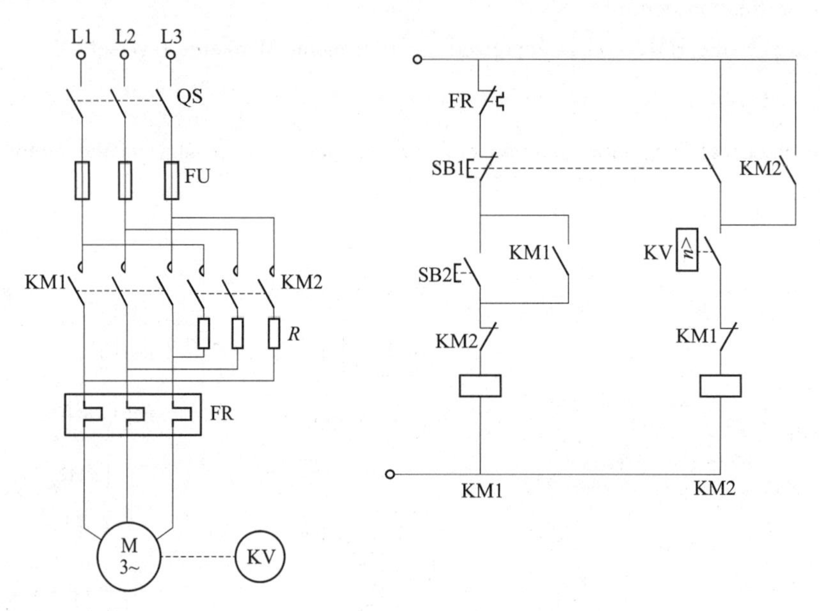

Fig. 3-30 Electrical Schematic Diagram of the Motor Reverse Braking

(2) Operating process.

Press SB2 ⟶ the KM1 coil is energized ⟶ the motor M rotates forward ⟶ the KV normally open contact is closed ⟶ KM1 is de-energized.

Press SB1 ⟶ { $n \approx 0$, KV is reset ⟶ KM2 is de-energized (end of braking). / KM2 is energized (start of braking). }

Task Implementation

(1) Divide students randomly into observation groups and appoint the group leaders. Fill the personal information of the group members in Table 3-2.

(2) According to the observation and discussion results, record the names, operating process, functions and applications of typical circuits in Table 3-2.

(3) Fully discuss the problems found during the observation, and fill the problems and solutions in Table 3-2.

Table 3-2 Task Work Order for Typical Circuit Analysis

Task Work Order					
Name:		Class:		Student ID:	
List of materials, tools, measuring instruments, and equipment required					
S/N	Name		Quantity	Model or specification	Remark
Operating process(Observation)					
S/N	Name	Operating process	Main function		Application
Problems			Solutions		

Task Evaluation

S/N	Evaluation contents	Total point	Score
1	Be proficient in the classification of electrical control system diagrams	25	
2	Be able to correctly draw electrical system diagrams	25	
3	Be able to analyze typical circuits with electrical schematic diagrams	25	
4	Have a professional spirit of effective communication, expression and solidarity	25	

Task III Design the Tool Change Circuit of Tool Magazines

Learning Objectives

(1) To understand the tool change process and working principle of the hat-type tool magazines.

(2) To master the circuit structure of the hat-type tool magazines.

(3) To master the circuit installation and commissioning methods of the hat-type tool magazines.

(4) To develop an ability to independently analyze the fault diagnosis of the hat-type tool magazine circuits.

(5) To master the engineering working methods, cultivate a rigorous work style, and abide by the 5S management system.

Task Description

Observe the tool change circuit unit of the hat-type tool magazines in groups, and fill the tool change process, working principle and circuit structure of the tool change circuit of the tool magazine in the task work order. Make sure to be careful and pay attention to the relevant operation specifications of the practical training site during observation.

Knowledge Link

I. Tool change process of hat-type tool magazines

1. Tool change actions of hat-type tool magazines

(1) Turn the tool magazine to the tool change coordinates, as shown in Fig. 3-31(a).

(2) The spindle stops exactly.

(3) The tool magazine moves forward (grasping the old tool), as shown in Fig. 3-31(b).

(4) Release the spindle tool.

(5) The Z-axis moves upward (giving way to the rotation space of the tool magazine), as shown in Fig. 3-31(c).

(6) The tool magazine rotates (for tool selection), as shown in Fig. 3-31(d).

(7) The Z-axis moves downward (to the tool change position), as shown in Fig. 3-31(e).

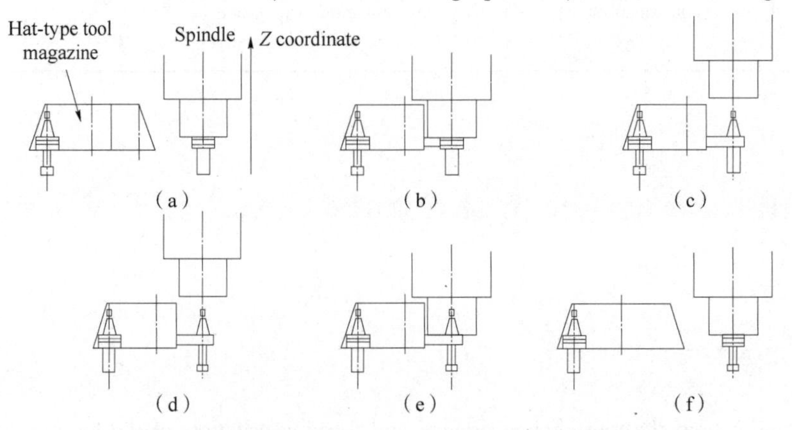

Fig. 3-31 Tool Change Process of Hat-type Tool Magazines

(8) Clamp the spindle tool (grasping the new tool).

(9) The tool magazine rotates backward (end of tool change), as shown in Fig. 3-31(f).

2. Positioning accuracy of the spindle

The required positioning accuracy of the spindle is very high during the tool change process of the hat-type tool magazine, which is determined by the structure of the standard tool BT600 and the spindle. The standard tool BT600 has two symmetrical positioning keyways, while there are two mating keys on the corresponding spindle. This requires that the spindle is always located in the same position during tool change. The requirement above cannot be achieved by the PMC alone but can be achieved under cooperation between the spindle and the CNC program. When the system detects that the program executes the M6TXX signal, the CNC program sends out tool change preparation and spindle positioning signals. The PMC controls the spindle frequency converter to position the spindle. The spindle encoder feeds the spindle position back to the CNC program, and then the CNC program detects whether the spindle is positioned to the required range. After the spindle has been positioned, the CNC program sends out a tool change start signal and transmits it to the PMC. The control flow chart is shown in Fig. 3-32.

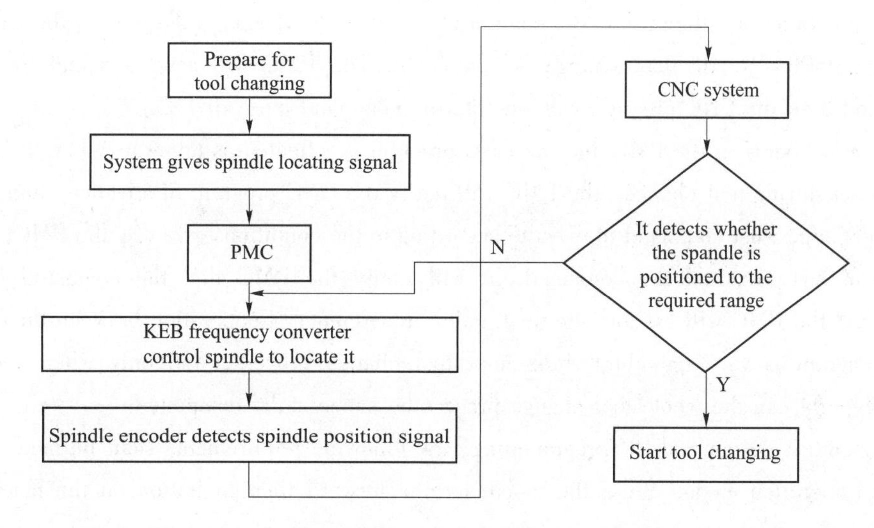

Fig. 3-32 Control Flow Chart

3. Cooperation Between the PMC and the CNC program

The PMC mainly controls the forward and reverse rotation of the tool magazine, the forward/backward movement of the tool magazine, the actions of the tool releasing/clamping valve and the sequence of tool change actions throughout tool change. As ascending, descending and positioning of the spindle are all controlled by the CNC program, cooperation between the PMC and the CNC program becomes not only a key point but also a difficult point throughout the tool change process.

Throughout tool change, the action sequence is controlled by the PMC. The D memory in the PMC is used, and different numbers memorized in the D memory represent different actions in the tool change process, as shown in Table 3-3.

Table 3-3 Tool Change Actions and Numbers

D800	Sequence of tool change actions	D800	Sequence of tool change actions
1	Start tool changing	5	The tool magazine swings in place
2	The cylinder moves forward in place	6	The *Z*-axis tool change point is in place
3	Tool feeding is completed	7	Tool clamping is completed
4	Tool position signal reading is completed	8	Tool change is completed

Table 3 - 3 and Fig. 3 - 31 indicate that the NC program and the PMC should act in a cooperative way when D800 = 3 and D800 = 5. When D800 = 3, the action is shown in Fig. 3-31 (b), and the next action is shown in Fig. 3-31(c). The *Z*-axis ascends, which requires cooperation between the CNC program and the PMC. Therefore, after the tool has been released, the PMC will send a signal to the CNC program, and then the CNC program will control the *Z*-axis to ascend. When D800=5, the action is shown in Fig. 3-31(d), and the next action is shown in Fig. 3-31 (e). The axis descends to the tool change point, which requires cooperation between the CNC program and the PMC. Therefore, after the cutter head swings in place, the PMC will send a signal to the CNC program, and then the CNC program will control the *Z*-axis to descend to the tool change point. When D800 = 8, the tool change is completed. The PMC will send a signal to the CNC program, and then the CNC program will execute the command after M6TXX.

Cooperation between the PMC and the CNC program is reflected as follows: When the *Z*-axis is required to act during tool change, the PMC will notify the CNC program in advance, and the CNC program will judge what command to execute according to the conditions given by the PMC; after the CNC program has executed the command, it will notify the PMC that the command has been executed, and the PMC will execute the next action in sequence. Cooperation between the PMC and the CNC program is very important throughout tool change process, and only when cooperation proceeds properly can the whole tool change process be successfully completed.

During circuit design and PC programming, the following requirements shall be met.

Manual operation mode: Press the tool magazine forward rotation button on the machine tool operation panel, and the tool magazine will rotate forward; Press the tool magazine reverse rotation button on the machine tool operation panel, and the tool magazine will rotate reversely.

Automatic operation mode: Enter the tool change command (e.g., "M6 T1") in the program, and press the cycle start button to complete the tool change process.

II. Design of tool change circuits of a tool magazine

The tool change circuits of the tool magazines consist of a main circuit, a control circuit and the PMC I/O circuit.

1. Tool change main circuit of a tool magazine

The electrical schematic diagram of the tool change main circuit of a tool magazine is shown in Fig. 3-33.

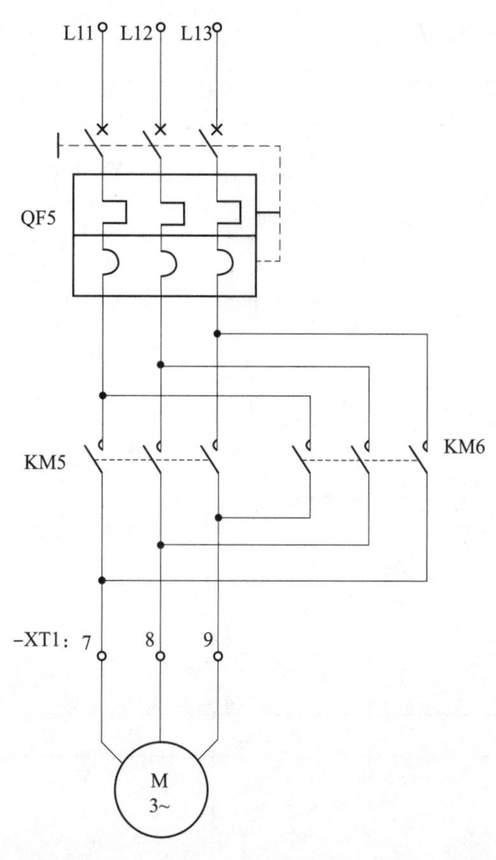

Fig. 3-33 Tool Change Main Circuit of a Tool Magazine

KM5-AC contactor for forward rotation of tool magazine motor;

KM6-AC contactor for reverse rotation of tool magazine motor

2. Design the tool change control circuit of a tool magazine

The electrical schematic diagram of the tool change control circuit of a tool magazine is shown in Fig. 3-34.

3. Design the tool change I/O circuit of a tool magazine

The electrical schematic diagram of the tool change I/O circuit of the tool magazine is shown in Fig. 3-35.

Task Implementation

(1) Divide students randomly into observation groups and appoint the group leaders. Fill the personal information of the group members in Table 3-4.

(2) According to the observation and discussion results, fill the tool change process, working principle, circuit structure, and a pplications during the design of the tool change circuit of the tool magazine in Table 3-4.

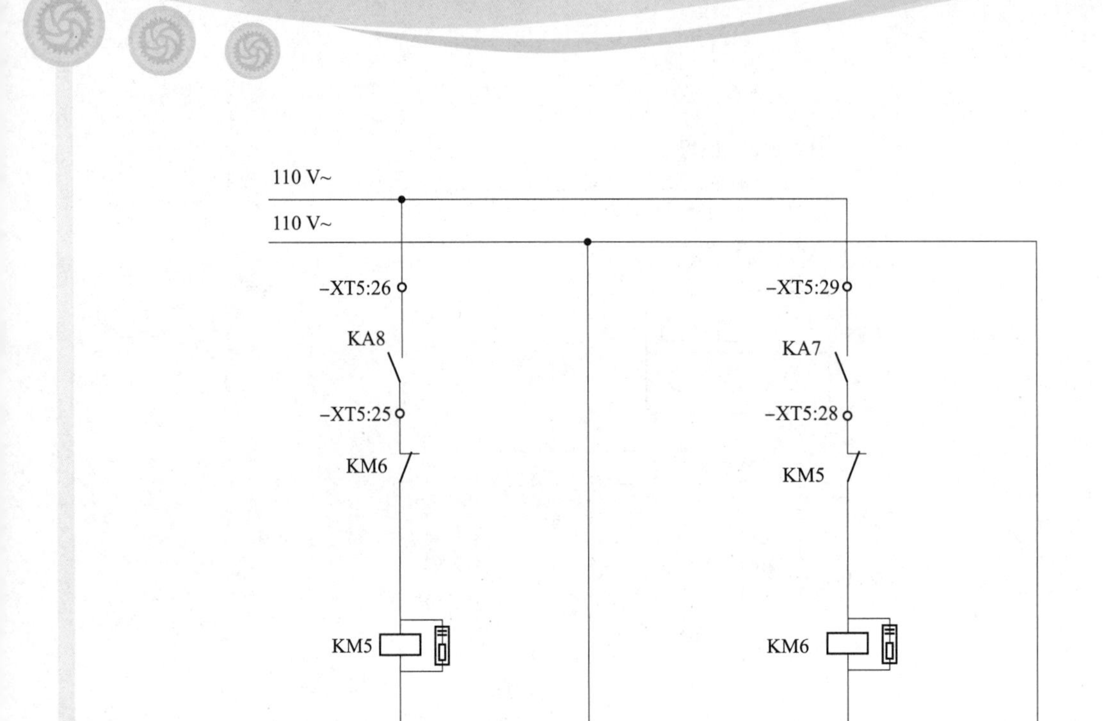

(a)

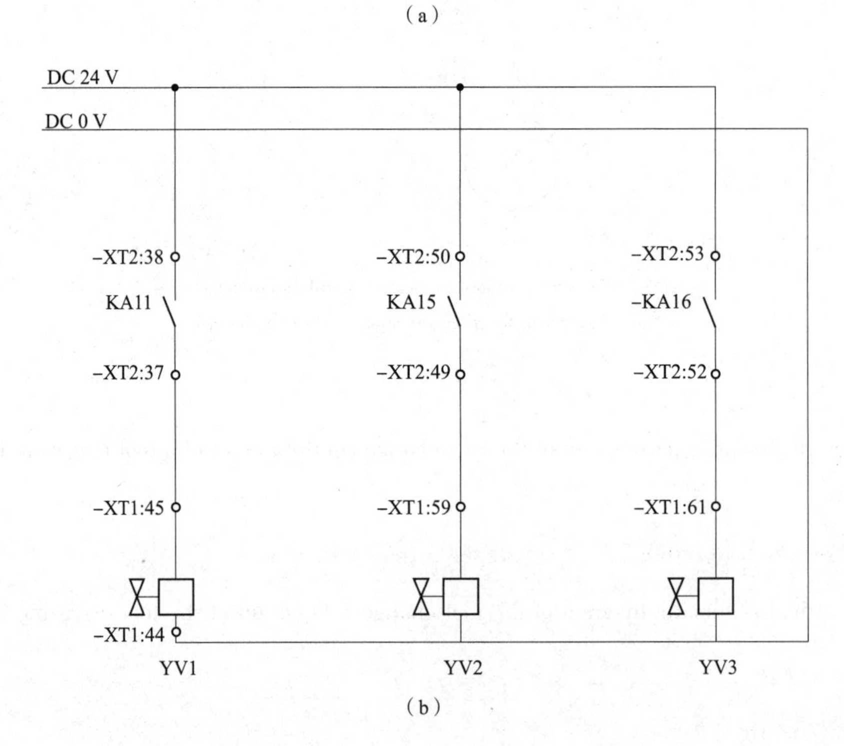

(b)

Fig. 3-34 Tool Change Control Circuit of a Tool Magazine

(a) Interlocking circuit for forward and reverse rotation of the tool magazine motor; (b) Circuit for spindle tool releasing, tool magazine approaching spindle and the tool magazine moving away from spindle

(3) Fully discuss the problems found during the observation, and fill the problems and solutions in Table 3-4.

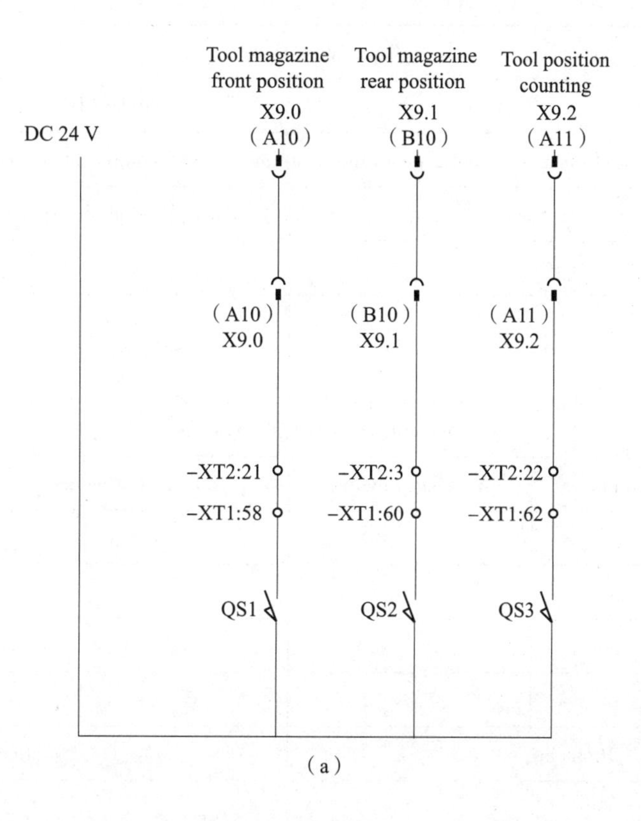

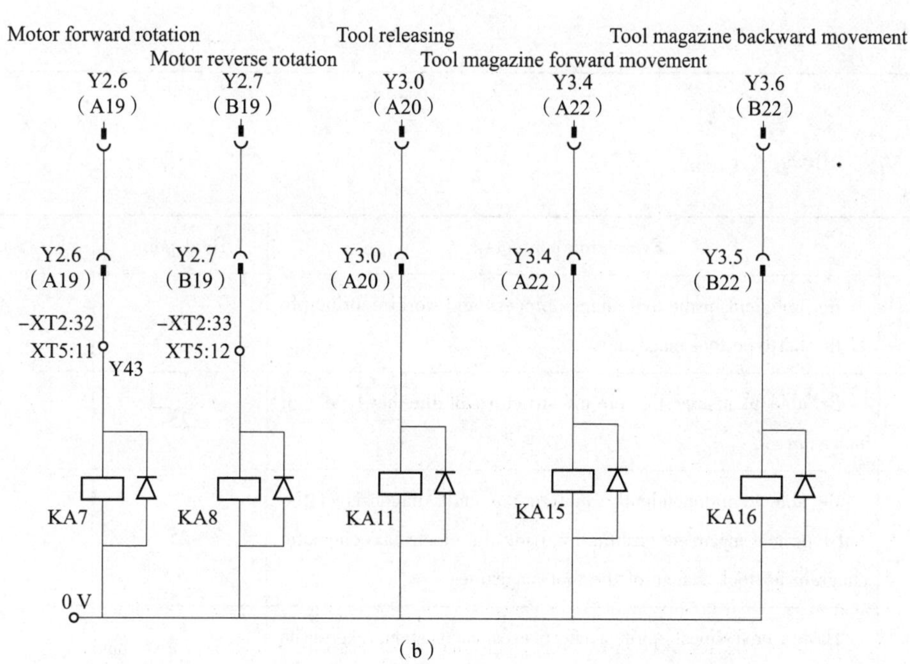

Fig. 3-35 Tool Change I/O Circuit of a Tool Magazine

(a) PMC input signal; (b) PMC output signal

Table 3-4 Task Work Order for Design of Tool Change Circuit of a Tool Magazine

<table>
<tr><td colspan="5">Task Work Order</td></tr>
<tr><td colspan="2">Name:</td><td colspan="2">Class:</td><td>Student ID:</td></tr>
<tr><td colspan="5">List of materials, tools, measuring instruments, and equipment required</td></tr>
<tr><td>S/N</td><td>Name</td><td>Quantity</td><td>Model or specification</td><td>Remark</td></tr>
<tr><td></td><td></td><td></td><td></td><td></td></tr>
<tr><td></td><td></td><td></td><td></td><td></td></tr>
<tr><td></td><td></td><td></td><td></td><td></td></tr>
<tr><td colspan="5">Operating process(Observation)</td></tr>
<tr><td>S/N</td><td>Tool change process</td><td>Working principle</td><td>Circuit structure</td><td>Application</td></tr>
<tr><td></td><td></td><td></td><td></td><td></td></tr>
<tr><td></td><td></td><td></td><td></td><td></td></tr>
<tr><td></td><td></td><td></td><td></td><td></td></tr>
<tr><td></td><td></td><td></td><td></td><td></td></tr>
<tr><td colspan="3">Problems</td><td colspan="2">Solutions</td></tr>
<tr><td colspan="3"></td><td colspan="2"></td></tr>
</table>

Task Evaluation

S/N	Evaluation contents	Total point	Score
1	Be proficient in the tool change process and working principle of the hat-type tool magazines	25	
2	Be able to master the circuit structure of the hat-type tool magazines	25	
3	Be able to independently analyze the fault diagnosis of the hat-type tool magazine circuits by using the electrical schematic diagram for tool change of the tool magazines	25	
4	Have a professional spirit of effective communication, expression and solidarity	25	

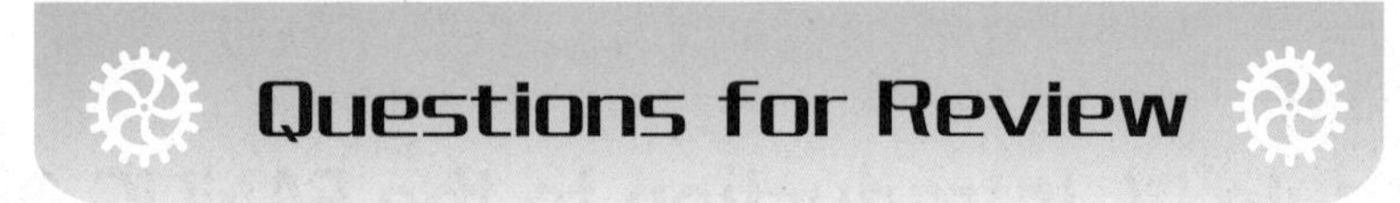

Questions for Review

(1) Briefly describe the classification of commonly used electrical components in CNC machine tools.

(2) Briefly describe the performance and selection methods of commonly used electrical components in CNC machine tools.

(3) Briefly describe the classification of electrical control system diagrams.

(4) Briefly describe the drawing principles of electrical system diagrams.

(5) Briefly describe the tool change process and working principle of hat-type tool magazines.

(6) Briefly describe the circuit structure of the hat-type tool magazine.

Module IV Introduction to the FANUC 0i-F CNC System

FANUC 0i-F(including FANUC Series 0i MF and FANUC Series 0i TF) CNC system is a new product rolled out by FANUC in 2017, which has been fully promoted and applied so far. The system, established based on the high-end 31iB system platform, perfectly presents the high efficiency and convenience required for machine tool operation in the support of its powerful intelligent functions. The αi and βi servo amplifiers connected to the system are also fully upgraded to B series high-speed and intelligent servo amplifiers to drive the spindle motor and servo motor of the upgraded B series, meeting high-speed and high-precision machining requirements. The 0i-F system has a highly integrated internal structure. It is connected with the spindle and servo unit of αi-B series and βi-B series through the latest FSSB high-speed interface; meanwhile, it is connected with the I/O unit by using the upgraded I/O link interface for intelligent and high-speed communication. The mainboard integrates the circlip function, and the system is equipped with FROM/SRAM boards for installing the system's cooling fan that can detect the speed and the battery used to save the system SRAM data in case of power failure. In addition, two slots can be expanded at most to meet the automated and networking functional requirements of the current machine tool industry to the greatest extent. This module comprehensively analyzes the hardware connection, basic parameters, PMC, data transmission and backup, and others of the FANUC 0i MF CNC system; at the same time, it explains in detail the specific contents of servo hardware connection, parameter setting and PMC programming of FANUC 0i MF CNC system.

Task I Hardware Connection of the FANUC 0i-F CNC System

Learning Objectives

(1) To master the hardware structure and interfaces of the FANUC 0i-F CNC system.

(2) To understand the integrated hardware connection of the FANUC CNC system.

(3) To understand the basic composition of the FANUC CNC system.

(4) To develop an ability to identify the definitions of hardware interfaces of the FANUC 0i–F CNC system.

(5) To master the structure and interfaces of the FANUC 0i–F CNC system mainboard.

(6) To master the engineering working methods, cultivate a rigorous work style, and abide by the 5S management system.

Task Description

Observe the hardware connection of the FANUC 0i – F CNC system in groups, and fill the names, definition description, main functions, and etc. of interfaces on the FANUC 0i – F CNC system mainboard in the task work order. Make sure to be careful and pay attention to the relevant operation specifications of the practical training site during observation.

Knowledge Link

I. Typical hardware structure and interfaces of FANUC CNC systems

1. Structure and interfaces of CNC system mainboard

The structure and interfaces of the FANUC 0i–F CNC system mainboard are shown in Fig. 4–1. Two fans are provided on the upper part of the mainboard for its heat dissipation. CP1 at the lower right of the mainboard is linked with a DC 3 V lithium battery, which serves as a backup battery for the memory. The part machining program, tool offset, system parameters, etc, made by users are stored in the CMOS memory of the control unit. When the main power supply of the system is cut off, these data are memorized by the lithium battery. Therefore, when the battery voltage drops to a certain extent and a "BAT" alarm pops up on the display, the battery shall be replaced in time to prevent data loss.

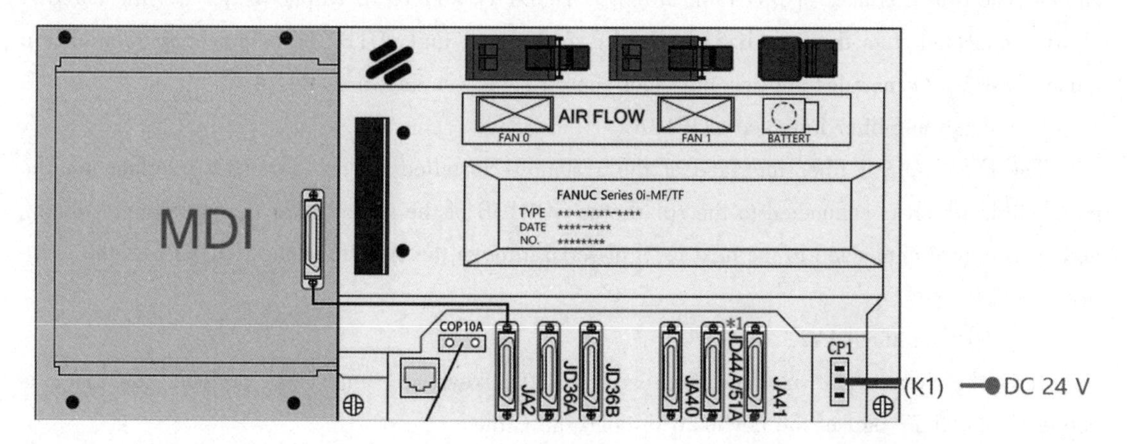

Fig. 4–1 Structure and Interfaces of the FANUC 0i–F CNC System Mainboard

The FANUC 0i-F mainboard has the following interfaces:

(1) Power interface CP1.

When the mainboard of the CNC system control unit works normally, an external DC 24 V power supply has to be provided. The external AC 200 V power supply is rectified by the switching mode power supply and changed into a DC 24 V power supply, which is input through the CP1 interface for the mainboard to work.

(2) Serial spindle/encoder JA41.

If measuring signals (like digital measuring instrument signals or tool measuring signals on the machine tool) are used on the machine tool, up to 8 high-speed input points can be connected when pursuing high-speed and accurate measurement.

(3) Analog spindle/jump signal JA40.

When the machine tool uses an analog spindle, the speed command of the spindle is converted into an analog quantity of ±10 V, and JA40 output is connected to the speed command port of the frequency converter. However, in order to reduce the detection alarm of circuits caused by external interference, shielded cables and shielding layer grounding treatment shall be selected for the command. At the same time, if the spindle encoder feedback is adopted for the analog spindle, the spindle encoder needs to feedback to JA41 of the mainboard interface, which is different from connecting to the spindle driver when the serial spindle is connected.

(4) I/O module communication JD44A/JD51A.

The motion control of each coordinate axis of the CNC machine tool, i.e., G and F commands in the user's machining program, is realized by the CNC system; while the sequential logic actions of the CNC machine tool, i.e., M, S and T commands in the user's machining program, are controlled by PMC, including spindle speed control, tool selection, workbench replacement, turntable indexing, and workpiece clamping and loosening. Communication is established between CNC and these input and output signals from the machine tool side through I/O Link. According to the number of PMC control points, it is necessary to connect multiple I/O modules through I/O Link connecting cables. The two interfaces of I/O Link are JD51A(JD1A) and JD1B respectively, and the cable is always connected from the unit JD51A(JD1A) to the next unit JD1B. The connection relationship among CNC, I/O modules and machine tool control signals is shown in Fig. 4-2.

(5) Servo amplifier interface COP10A.

The FSSB optical fiber interface of the system is installed on the COP10A interface on the mainboard, which is connected to the spindle or COP10B of the servo driver through optical fibers, and then in turn connected to the next level of servo through the spindle or the COP10A on the servo driver.

(6) MDI interface JA2.

It is the connection interface between the MDI keyboard and CNC system, and JA2 is connected to CK27 port of the keyboard through the cable.

(7) Embedded Ethernet CD38A.

CD38A is connected to the Ethernet port of the computer to complete the transmission and

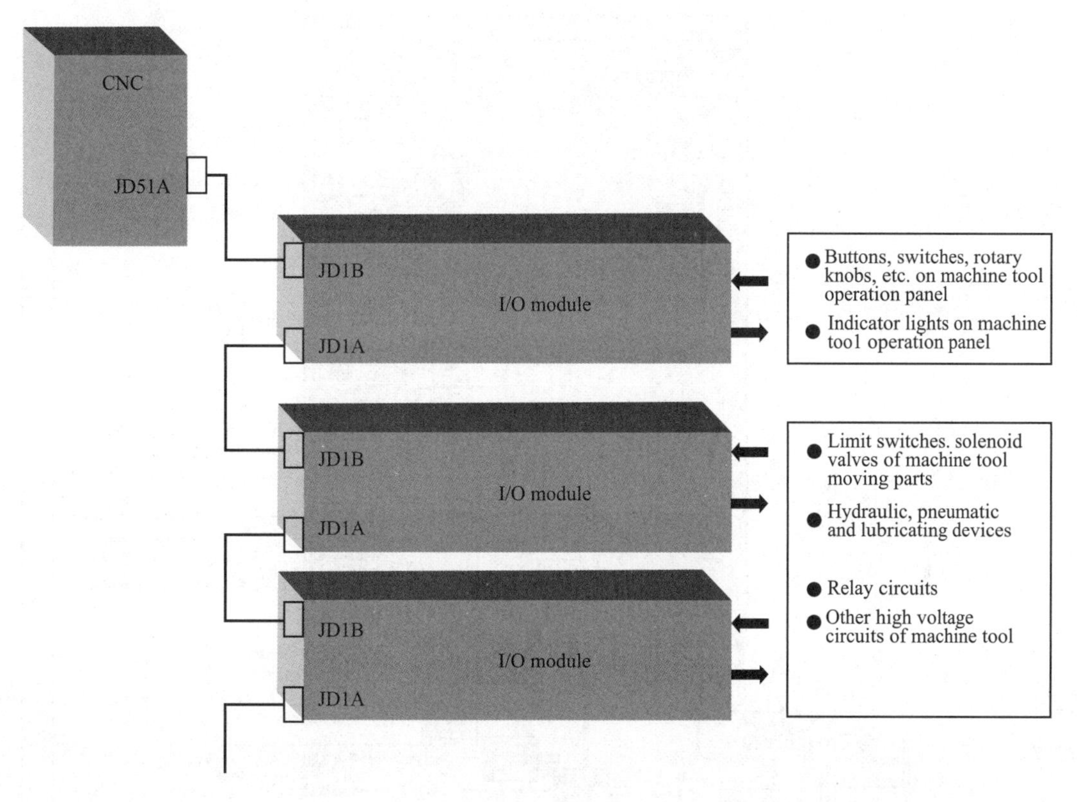

Fig. 4-2 Connection of CNC, I/O Modules and Machine Tool Control Signals

acquisition of machine tool data and information. The other communication interface is the RS232 serial communication interface. There are two interfaces, namely JD36A (port 1) and JD36B (port 2). Ports 1 and 2 can be set in the parameters, and the frequently used port is JD36A (port 1).

(8) Touch screen JD36B.

When the touch screen is used, this interface will be connected to the touch screen and not used as the RS232 communication port.

2. Servo amplifiers

(1) Function of a servo amplifier.

To machine workpieces of various shapes and meet the shape, position, and surface quality accuracy required by part drawings, a relative motion shall be done between the tool and the workpiece according to a given feed speed, a given feed direction, and a certain cutting depth. This relative motion is driven by one or more servo motors. The servo amplifier receives the feed motion command of the servo axis from the control unit CNC and drives the servo motor after the command is converted and amplified to realize the required feed motion.

(2) Definition and connection of βi-B servo interface.

The power module, spindle module and servo module constitute an integrated device, and the connection among them is shown in Fig. 4-3.

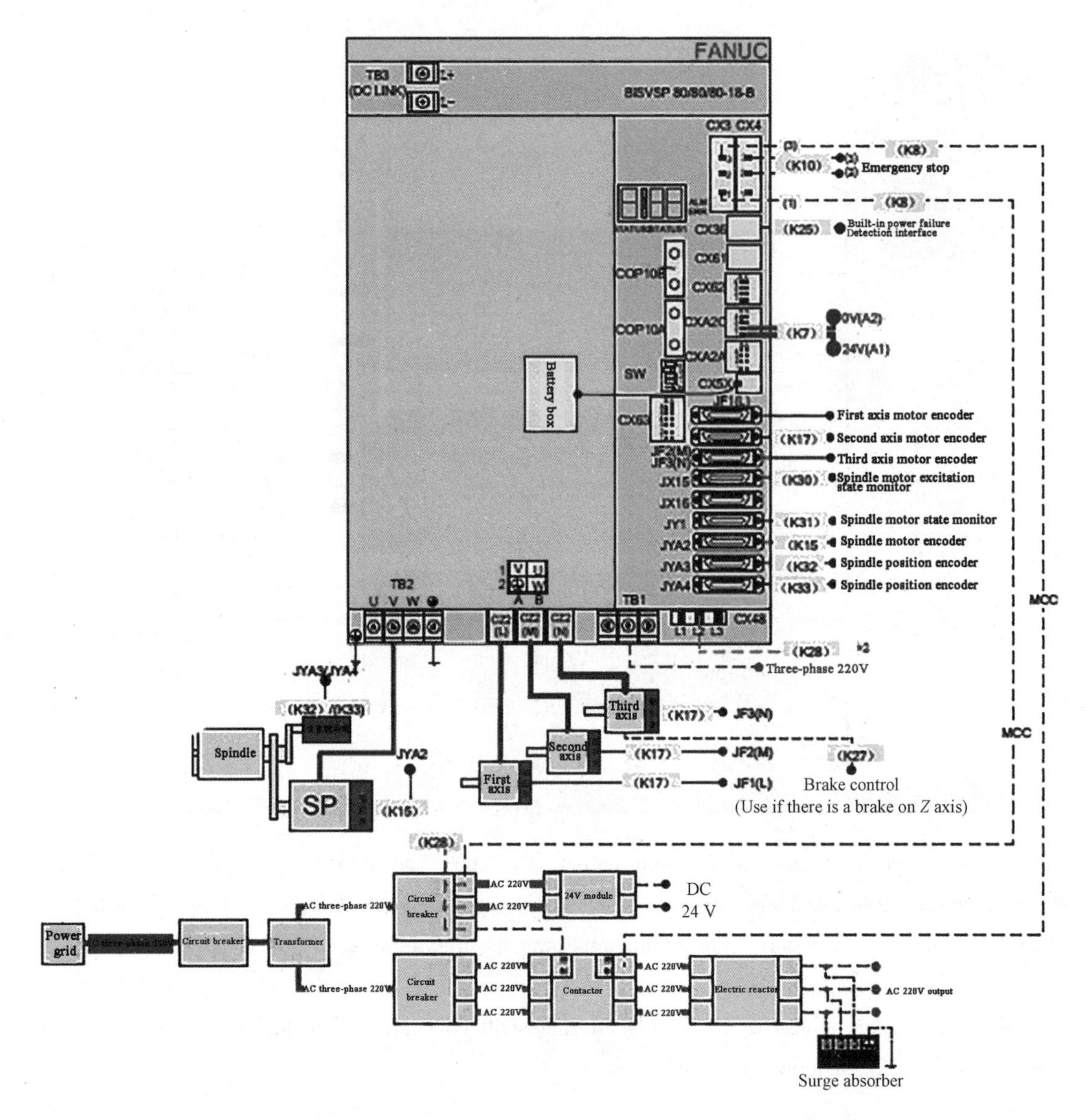

Fig. 4-3 Schematic Diagram of βi-B Servo Connection

3. I/O devices interfaces and connection of CNC system

(1) Common I/O devices of FANUC system.

The I/O devices commonly used in the FANUC system generally include the built-in I/O module, external I/O module, junction box I/O module, machine tool operation panel I/O card and system I/O module, as shown in Fig. 4-4.

(2) Definition of I/O devices interfaces.

The interfaces of I/O board is shown in Fig. 4-5.

POWER: power indicator;

CP1: DC24V power input;

CB104: DI/DO-1;

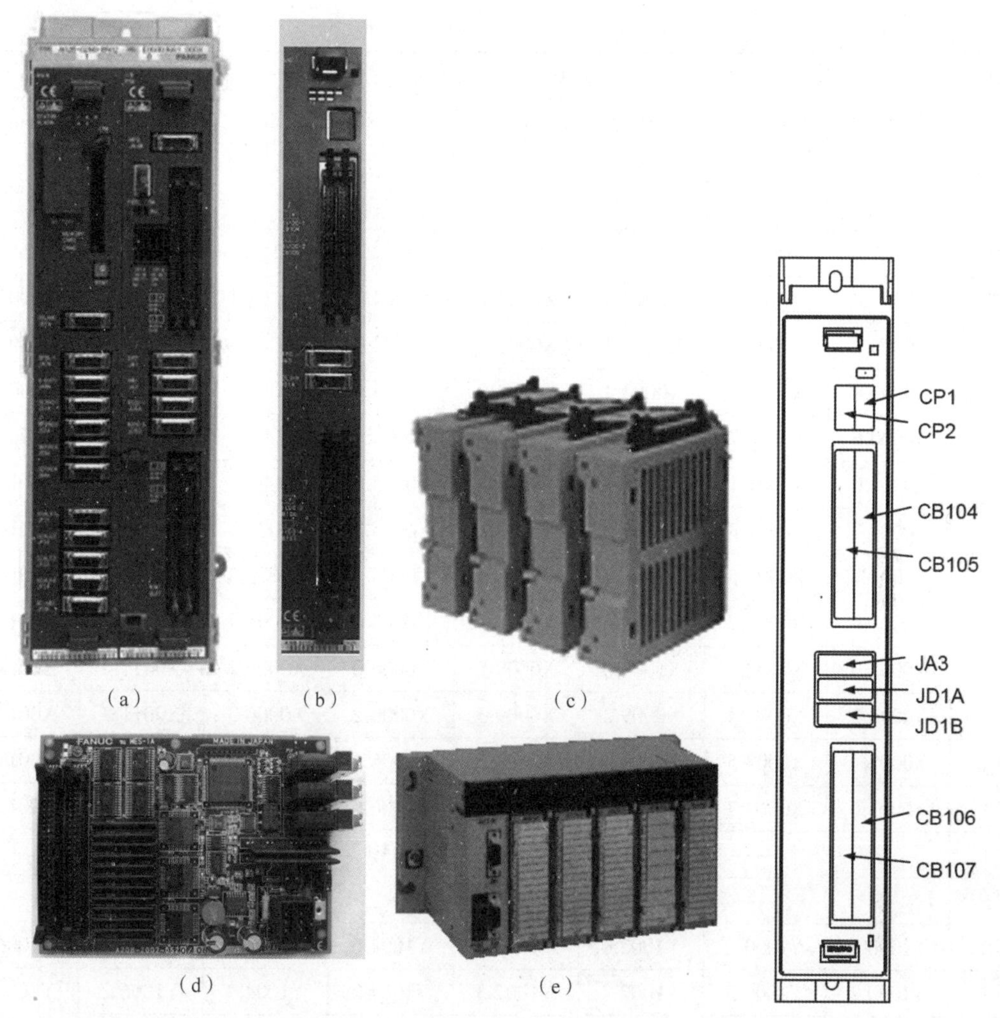

Fig. 4-4 Common I/O Devices of FANUC System

(a) Built-in I/O module; (b) External I/O module; (c) Junction box I/O module; (d) Machine tool operation panel I/O card; (e) System I/O module

Fig. 4-5 I/O Board Interfaces

CB105: DI/DO-2;

JA3: MPG electronic handwheel;

JD1A: I/O Link;

JD1B: I/O Link (connecting with system board JD51A/JD44A);

CB106: DI/DO-3;

CB107: DI/DO-4.

(3) Description of I/O board interface definition.

1) CB104-CB107: Each digital input/output interface has 24 input points and 16 output points.

2) CB104-CB107 has 96 input points and 64 output points.

If X and Y addresses are allocated from 0, the pin definition of CB104-CB107 interfaces is shown in Table 4-1.

Table 4-1 Pins Definition

Interface	CB104		CB105		CB106		CB107	
Pin	HIROSE 50 pin		HIROSE 50 pin		HIROSE 50 pin		HIROSE 50 pin	
Voltage terminal	A	B	A	B	A	B	A	B
01	0 V	+24 V	0 V	+24 V	0 V	+24 V	0 V	+24 V
02	X0000.0	X000.1	X0003.0	X0003.1	X0004.0	X0004.1	X0007.0	X0007.1
03	X0000.2	X0000.3	X0003.2	X0003.3	X0004.2	X0004.3	X0007.2	X0007.3
04	X0000.4	X0000.5	X0003.4	X0003.5	X0004.4	X0004.5	X0007.4	X0007.5
05	X0000.6	X0000.7	X0003.6	X0003.7	X0004.6	X0004.7	X0007.6	X0007.7
06	X0001.0	X0001.1	X0008.0	X0008.1	X0005.0	X0005.1	X0010.0	X0010.1
07	X0001.2	X0001.3	X0008.2	X0008.3	X0005.2	X0005.3	X0010.2	X0010.3
08	X0001.4	X0001.5	X0008.4	X0008.5	X0005.4	X0005.5	X0010.4	X0010.5
09	X0001.6	X0001.7	X0008.6	X0008.7	X0005.6	X0005.7	X0010.6	X0010.7
10	X0002.0	X0002.1	X0009.0	X0009.1	X0006.0	X0006.1	X0011.0	X0011.1
11	X0002.2	X0002.3	X0009.2	X0009.3	X0006.2	X0006.3	X0011.2	X0011.3
12	X0002.4	X0002.5	X0009.4	X0009.5	X0006.4	X0006.5	X0011.4	X0011.5
13	X0002.6	X0002.7	X0009.6	X0009.7	X0006.6	X0006.7	X0011.6	X0011.7
14	—	—	—	—	COM4	—	—	—
15	—	—	—	—	—	—	—	—
16	Y1000.0	Y1000.1	Y1002.0	Y1002.1	Y1004.0	Y1004.1	Y1006.0	Y1006.1
17	Y1000.2	Y1000.3	Y1002.2	Y1002.3	Y1004.2	Y1004.3	Y1006.2	Y1006.3
18	Y1000.4	Y1000.5	Y1002.4	Y1002.5	Y1004.4	Y1004.5	Y1006.4	Y1006.5
19	Y1000.6	Y1000.7	Y1002.6	Y1002.7	Y1004.6	Y1004.7	Y1006.6	Y1006.7
20	Y1001.0	Y1001.1	Y1003.0	Y1003.1	Y1005.0	Y1005.1	Y1007.0	Y1007.1
21	Y1001.2	Y1001.3	Y1003.2	Y1003.3	Y1005.2	Y1005.3	Y1007.2	Y1007.3
22	Y1001.4	Y1001.5	Y1003.4	Y1003.5	Y1005.4	Y1005.5	Y1007.4	Y1007.5
23	Y1001.6	Y1001.7	Y1003.6	Y1003.7	Y1005.6	Y1005.7	Y1007.6	Y1007.7
24	DOCOM	DOCOM	DOCOM	DOCOM	DOCOM	DOCOM	DOCOM	DOCOM
25	DOCOM	DOCOM	DOCOM	DOCOM	DOCOM	DOCOM	DOCOM	DOCOM

Task Implementation

(1) Divide students randomly into observation groups and appoint the group leaders. Fill the personal information of the group members in Table 4-2.

(2) Based on the results of observation and discussion, fill the names, definition description,

main functions and etc of interfaces on the FANUC 0i-F CNC system mainboard in Table 4-2.

(3) Fully discuss the problems found during the observation, and fill the problems and solutions in Table 4-2.

Table 4-2　Task Work Order for Mainboard interfaces of the FANUC 0i-F CNC System

<table>
<tr><td colspan="6">Task Work Order</td></tr>
<tr><td colspan="2">Name:</td><td colspan="2">Class:</td><td colspan="2">Student ID:</td></tr>
<tr><td colspan="6">List of materials, tools, measuring instruments, and equipment required</td></tr>
<tr><td>S/N</td><td colspan="2">Name</td><td>Quantity</td><td>Model or specification</td><td>Remarks</td></tr>
<tr><td></td><td colspan="2"></td><td></td><td></td><td></td></tr>
<tr><td></td><td colspan="2"></td><td></td><td></td><td></td></tr>
<tr><td></td><td colspan="2"></td><td></td><td></td><td></td></tr>
<tr><td colspan="6">Operating process(Observation)</td></tr>
<tr><td>S/N</td><td>Interface name</td><td>Definition description</td><td colspan="2">Main function</td><td>Application</td></tr>
<tr><td></td><td></td><td></td><td colspan="2"></td><td></td></tr>
<tr><td></td><td></td><td></td><td colspan="2"></td><td></td></tr>
<tr><td></td><td></td><td></td><td colspan="2"></td><td></td></tr>
<tr><td></td><td></td><td></td><td colspan="2"></td><td></td></tr>
<tr><td colspan="3">Problems</td><td colspan="3">Solutions</td></tr>
<tr><td colspan="3"></td><td colspan="3"></td></tr>
</table>

Task Evaluation

S/N	Evaluation contents	Total point	Score
1	Be able to clearly explain structure and interfaces of the FANUC 0i-F CNC system	25	
2	Be able to correctly identify the definitions of the hardware interfaces of the FANUC 0i-F CNC system	25	
3	Be able to master the structure and interfaces of the FANUC 0i-F CNC system mainboard	25	
4	Have a professional spirit of effective communication, expression and solidarity	25	

Task II Basic System Parameter Setting

Learning Objectives

(1) To understand the types and expressions of system parameters.

(2) To master the display and search of system parameters.

(3) To master the setting of basic parameters.

(4) To understand the meaning of alarms that occur during commissioning and be able to eliminate them.

(5) To possess a way of thinking and methods of troubleshooting through the commissioning of CNC system parameters.

(6) To master the engineering working methods, cultivate a rigorous work style, and abide by the 5S management system.

Task Description

Observe the basic parameters of FANUC 0i-F CNC system in groups, and fill the parameter No., brief description, setting description, machine tool setpoints and etc. of various system parameters on the parameter setting page into the task work order. Make sure to be careful and pay attention to the relevant operation specifications of the practical training site during observation.

Knowledge Link

I. Overview of system parameters

1. Definitions and functions of CNC system parameters

The parameters of the CNC system are a series of data used by the CNC system to match the machine tool and CNC functions. The parameters of the FANUC 0i-F CNC system can be divided into system parameters and PMC parameters. The function of system parameters is defined by FANUC. PMC parameters are the data used in the PMC program of CNC machine tools, such as the data of the timer, the counter and the latching relay. These parameters are defined by the machine tool manufacturer. These two types of parameters are the prerequisites for the normal operation of CNC machine tools.

2. Typical parameter expression

(1) Bit type and bit (mechanical group/path/axis/spindle) parameters.

A bit parameter uses an 8-bit binary number to represent a state in which the bit of the

parameter is 0 or 1. Bit 1 corresponds to bit 0 and bit 8 corresponds to bit 7.

0000					EIA	NCR	ISP	CTV	TVC

Data number — Data #0-#7 indicating bit

(2) Other parameters.

There are other expressions of parameters in addition to bit parameters.

1023	Servo axis number of each axis

Data number — Data

II. Display of parameter page and editing of parameters

1. Call and display of system parameters

(1) On the MDI keyboard of the CNC device, press the function key [SYSTEM] once, and then click the soft key [PARAMETER] to enter the parameter display page.

(2) Use the paging key or cursor movement key on the MDI keyboard to locate the desired parameters page by page.

(3) Alternatively, input the parameter number through the MDI keyboard first, and then click the soft key [SEARCH NUMBER] so as to display the page where the specified parameters are located. The cursor is at the position of the specified parameters at the same time.

2. System parameter editing

(1) Parameter rewriting state.

After the read-only parameter setting of the CNC system is completed, the program is in the read-only state. In this state, it is not allowed to change the parameters. To modify or adjust the parameters, you should put them in a writable state, i.e. remove the read-only state. The operation steps are as follows.

1) Set the CNC system to MDI mode or emergency stop mode.

2) Press the function key [OFFSET SETTING] on the MDI keyboard for several times, or click the soft key [SETTING] after pressing the function key [OFFSET SETTING] once to display the homepage of "SETTING", as shown in Fig. 4-6.

3) Click the cursor to the "WRITE PARAMETER" line.

4) Click the soft key [OPERATION].

5) Enter "1" and then click the soft key [INPUT] to make "WRITE PARAMETER = 1", so that the parameters are in a writeable state.

(2) Conventional parameter setting method.

The conventional parameter setting steps are as follows.

1) Enter the parameter setting page, place the cursor at the position of the parameter to be set, enter data, and then click the soft key [INPUT]. The entered data will be set to the parameter position specified by the cursor.

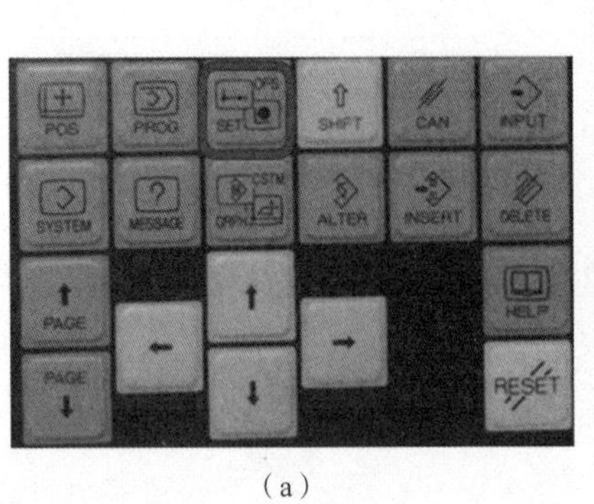

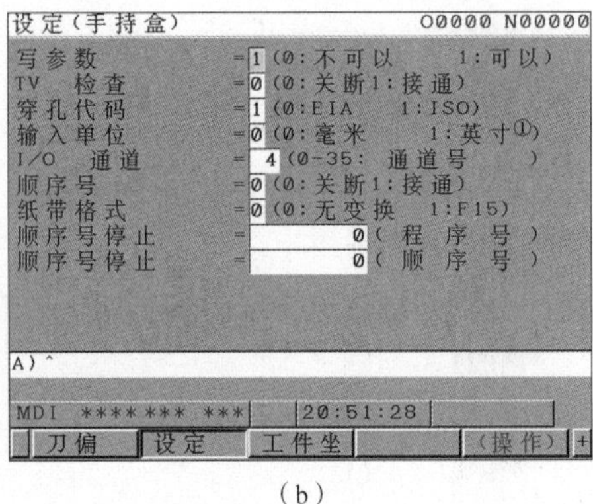

(a) (b)

Fig. 4-6 Interfaces of Parameter Writing Allowed

(a) Function keys; (b) "SETTING" page

2) For bit parameters: Click the soft key [ON: 1] to set the cursor position to 1; click the soft key [OFF: 0] to set the cursor position to 0.

3) After inputting the parameter value, click the soft key [+INPUT] to add the input value to the original value.

4) After inputting the parameter value, click the soft key [INPUT] to enter a new parameter value, as shown in Fig. 4-7.

5) After inputting the parameter, you can also press the function key INPUT on the MDI keyboard to complete the parameter writing operation.

III. Setting of system parameters

Call of parameter setting page to set the system parameters.

Press the function key [SYSTEM] in the emergency stop state, and then click the soft key [+] several times until the soft key [PARAMETER ADJUSTMENT] appears. Select the soft key [PARAMETER ADJUSTMENT] to display the parameter setting help page, as shown in Fig. 4-8. The items in the Fig. 4-8 are the parameter setting and adjustment steps. Set the parameters in turn according to the items.

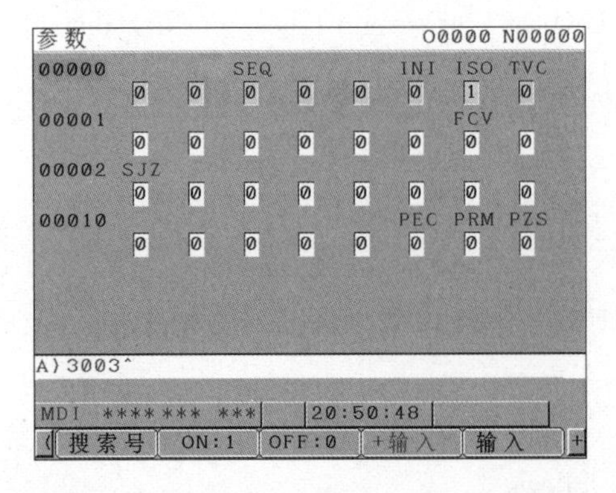

Fig. 4-7 Parameter Value Input page

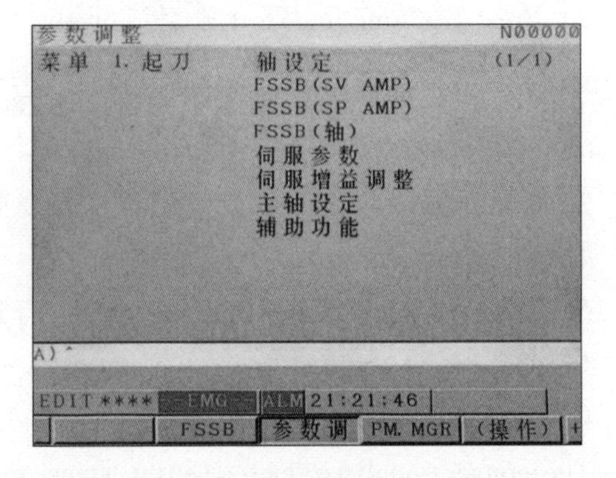

Fig. 4-8 Parameter Setting Page

① 英寸为英制长度单位,符号为 in, 1 in=2.54cm.

On the parameter setting page, complete the input and commissioning of the following parameters in turn.

(1) Axis setting.

There are the following groups in the axis setting, and each group of parameters needs to be set

1) Basic group.

The settings of basic group parameters are shown in Table 4-3.

Table 4-3 Basic Parameter Group

Group	Parameter No.	Brief description	Setting description	Machine tool setpoint		
				X	Y	Z
Basic	1001#0	The minimum unit of movement for a linear axis. 0: Metric system (machine tool of metric system); 1: British system (machine tool of British system)	Machine tool of metric system in general	0	0	0
	1013#1	Set the minimum input increment and the minimum command increment. 0: IS - B (0.001 mm, 0.001°, 0.0001 in); 1: IS - C (0.0001 mm, 0.0001°, 0.00001 in)	Generally set to 0	0	0	0
	1005#0	When the reference point is not established, specify movement commands other than G28 in automatic operation, P/S 224 alarms or not. 0: Alarms (No.224); 1: No alarm	It is generally set to 0 for machine tool safety	1	1	1
	1005#1	Whether the setting function of returning to the reference point without a stopper is effective. 0: invalid (each axis); 1: valid (each axis)	0 for setting with stopper; 1 for setting without stopper	1	1	1
	1006#0	Set a linear axis or a rotation axis. 0: Linear axis; 1: Rotation shaft	—	0	0	0

Continued

Group	Parameter No.	Brief description	Setting description	Machine tool setpoint		
				X	*Y*	*Z*
Basic	1006#3	Set whether the movement type of each axis is specified by radius or diameter. 0: By radius; 1: By diameter	The *X* axis of the lathe is set to 1	0	0	0
	1006#5	Set the return direction of each axis to the reference point. 0: Positive direction; 1: Negative direction	Movement direction of rear axle after disengaging the stopper	1	0	0
	1008#0	Set whether the rotation axis circulation function is valid. 0: invalid; 1: valid	Set whether the coordinates are cyclically recirculated	0	0	0
	1008#2	Relative coordinate values. 0: It is not displayed circularly according to the movement amount of each revolution; 1: It is displayed circularly according to the movement amount of each revolution	—	0	0	0
	1020	Programming name of each axis	*X* axis: 88; *Y* axis: 89; *Z* axis: 90	88	89	90
	1022	Properties of axes	*X* axis: 1; *Y* axis: 2; *Z* axis: 3	1	2	3
	1023	Servo axis number of each axis	Determine the relationship between CNC axis and servo motor	1	2	3
	1815#1	Separate position encoder. 0: Not used; 1: Used	Set to 1 when connected with the grating ruler or separate rotary encoder	0	0	0

Continued

Group	Parameter No.	Brief description	Setting description	Machine tool setpoint		
				X	Y	Z
Basic	1815#4	Position of mechanical position and absolute position detector when absolute position detector is used. 0: Inconsistent; 1: Consistent	When the absolute position detector is used, it is set to 0 during initial debugging and 1 when the mechanical position coincides with the position of the absolute position detector by moving the machine tool	1	1	1
	1815#5	Absolute position detector. 0: Not used; 1: Used	Set to 1 when using the absolute position detection function, requiring hardware support (absolute encoder)	1	1	1
	1825	Servo loop gain for each axis	3,000-8,000, the mutually interpolated axes must be set consistently	5,000		
	1826	To-position width of each axis	20-50	50		
	1828	Maximum permissible positional deviation when each axis is moving	500-10,000 Setpoint = Fast moving speed / (60 × loop gain)	10,000		
	1829	Maximum permissible positional deviation when each axis stops	50-2,000	2,000		

2) Spindle group.

Parameters of the spindle group are shown in Table 4-4.

Table 4-4 Spindle Parameter Group

Group	Parameter No.	Brief description	Setting description	Setpoint
Spindle	3716#0	Specifies the spindle motor type. 0: Simulation; 1: Serial	—	1
	3717	Set the number for each spindle motor	—	1

3) Coordinate group.

Parameters of the coordinate group are shown in Table 4-5.

Table 4-5 Coordinate Parameter Group

Group	Parameter No.	Brief description	Setting description	Setpoint
Coordinate system	1240	Coordinate the value of the first reference point of each axis in the machine coordinate system	Coordinates of the reference point in the mechanical coordinate system	0
	1241	Coordinate the value of the second reference point of each axis in the mechanical coordinate system	Establish the coordinates of the reference point in the machine coordinate system. In this machine tool, the value Z is the point where the tool magazine changes tool	0 0 -101
	1260	Movement per revolution of rotation	It is generally set to 360, 000, indicating that the coordinates rotate 360° for one revolution of the rotating shaft	0
	1320	Coordinate value of positive direction boundary of storage-type stroke detection 1 for each axis	Set after returning to the reference point. The datum is the machine tool coordinate system	999, 999.000
	1321	Coordinate value of negative direction boundary of storage-type stroke detection 1 for each axis	Set after returning to the reference point. The datum is the machine tool coordinate system	-999, 999.000

4) Feed rate group.

The parameters of the feed rate group are shown in Table 4-6.

Table 4-6 Feed Rate Group

Group	Parameter No.	Brief description	Setting description	Setpoint
Coordinate system	1401#6	Whether the fast no load running is effective. 0: invalid; 1: valid	—	0

Continued

Group	Parameter No.	Brief description	Setting description	Setpoint
Coordinate system	1410	No load run speed and feed speed of manual linear and circular interpolation	Generally, the speed is set in mm/min	3,000
	1420	Fast running speed of each axis		3,000
	1421	F0 speed at fast running override of each axis		500
	1423	Feed speed during manual continuous feed (JOG feed) of each axis		3,000
	1424	The manual fast running speed of each axis		5,000
	1425	F1 speed of each axis returning to the reference point		500
	1428	Reference point approach speed		3,000
	1430	Maximum cutting speed		6,000

(2) FSSB setting items.

The CNC system is connected to multiple servo amplifiers through the high-speed serial servo bus FSSB with an optical fiber, which realizes photoelectric isolation, improves reliability and anti-interference performance, and also requires fewer connecting cables. After the connection of the CNC system and servo hardware is completed, it needs to be activated by setting parameters through FSSB during the first commissioning.

1) FSSB (SV AMP) setting.

Enter the parameter adjustment page, click the soft key [OPERATE] to move the cursor to "FSSB (SV AMP)", and click soft key [SELECT]. The parameter setting page will appear. After setting relevant items, click the soft key [OPERATE] and then press the soft key [SELECT]. If the CNC system fails to detect the servo module through FSSB, no servo-related information will appear on the parameter page, it is necessary to check for hardware problems. FB(SV AMP) setting page is shown in Fig. 4-9.

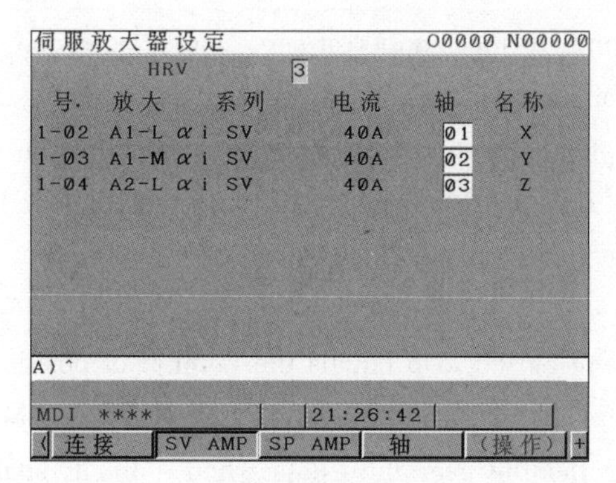

Fig. 4-9 FSSB (SV AMP) Setting Page

2) FSSB (Axis) setting.

Enter the parameter adjustment page, click the soft key [OPERATE] to move the cursor to "FSSB (axis)", and click the soft key [SELECT] to display the parameter setting page (no data modification is required when the CNC machine tool is connected in a semi-closed loop). After setting relevant items, click the soft key [SETTING]. FBBS(axis) setting page is shown in Fig. 4-10.

(3) Initialization setting of servo.

In the emergency stop state, enter the parameter setting page. Click the soft key [OPERATE] to move the cursor to "SERVO SETTING", click the soft key [SELECT] to display the servo parameter setting page, and click the soft key [SWITCH] to enter the initialization setting page. Set on this page.

1) Axis initialization setpoint.

The system automatically sets the default value of initialization setting bit parameters for each axis to 00000010. When it is necessary to initialize servo parameters, the initialization setting bit corresponds to 00000000 for each axis, as shown in Fig. 4-11.

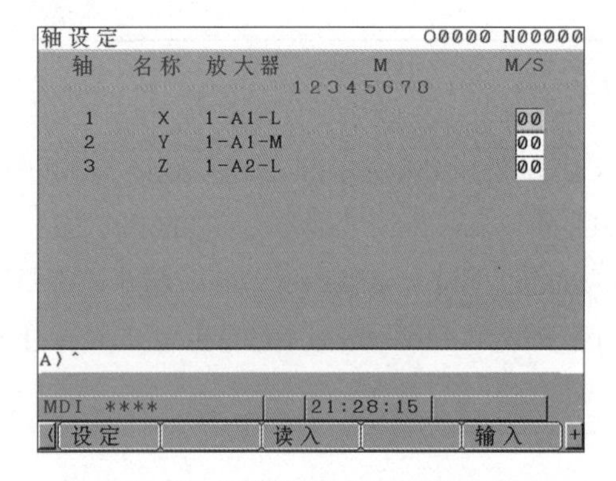

Fig. 4-10 FSSB (Axis) Setting Page

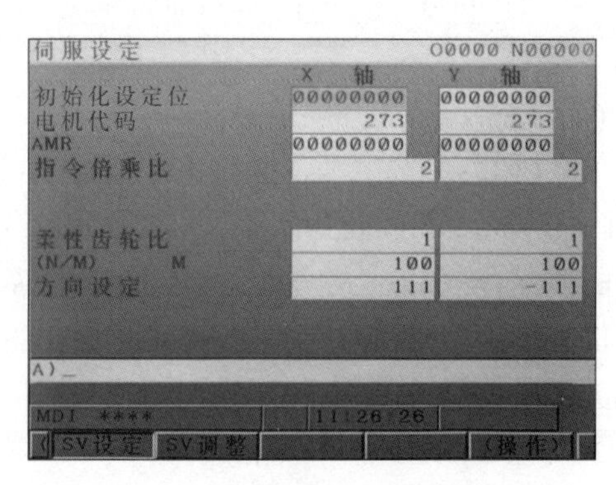

Fig. 4-11 Parameter Setting Page

2) Motor code.

Refer to the *Parameter Specifications for Servo Motor* according to the motor model on the servo motor nameplate and the servo drive amplifier nameplate, find the corresponding motor code, and enter the code value.

3) AMR.

This parameter corresponds to the number of servo motor poles and is set to 00000000.

4) Command multiplication ratio.

Set the command multiplication ratio of the movement amount from the CNC system to the servo system. Setpoint = (command unit/detection unit) × 2, and the command unit is usually equal to the detection unit, so this parameter is set to 2.

5) Flexible gear ratio.

Flexible gear ratio in semi-closed loop equals the number of position pulses required per motor revolution/1, 000, 000, where the denominator refers to the amount of shaft movement per motor revolution. For example, the flexible gear ratio data when a linear shaft motor is directly coupled with a ball screw by 1: 1 is shown in Table 4-7.

Table 4-7 Example of Flexible Gear Ratio in Semi-closed Loop

Detection unit/ μm	Lead of ball screw/mm					
	6	8	10	12	16	20
1	6/1 000	8/1 000	10/1 000	12/1 000	16/1 000	20/1 000
0.5	12/1 000	16/1 000	20/1 000	24/1 000	32/1 000	40/1 000
0.1	60/1 000	80/1 000	100/1 000	120/1 000	160/1 000	200/1 000

6) Direction setting.

111: The motor shaft rotates clockwise when viewed from the pulse encoder.

-111: The motor shaft rotates counterclockwise when viewed from the pulse encoder.

(4) Spindle setting.

In emergency stop state, first press the emergency stop button of the machine tool, press the function key [SYSTEM], then click the soft key [+] several times until the soft key [SPINDLE SETTING] appears on the spindle setting page, and click the soft key [SWITCH], as shown in Fig. 4-12.

1) Click the soft key [CODE] to display the motor code list page(Fig. 4-13), and then click the soft key [CODE] will be displayed when the cursor is located at the motor item. Go back to the previous page from the motor list page, click the soft key [BACK]. When switching to the motor list page, the motor name and amplifier name corresponding to the motor code are displayed. Move the cursor to the code number you want to set, and click the soft key [SELECT] to complete the input.

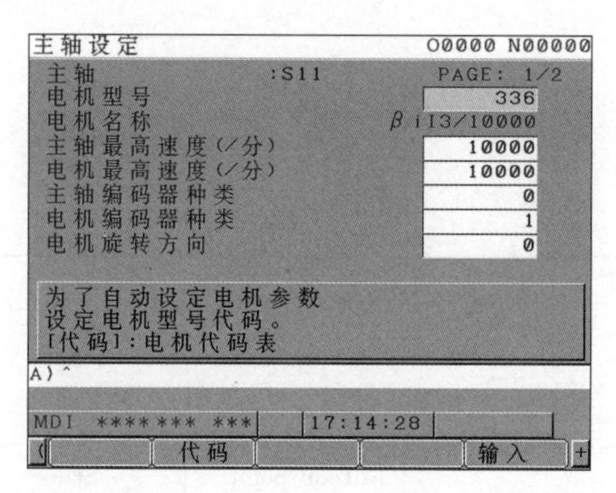

Fig. 4-12 Spindle Setting Page

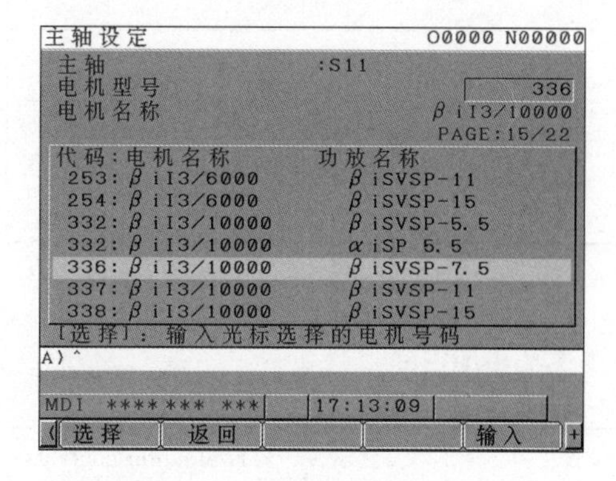

Fig. 4-13 Motor Code List Page

Task Implementation

(1) Divide students randomly into observation groups and appoint the group leaders. Fill the personal information of the group members in Table 4-8.

(2) According to the results of observation and discussion, record the parameter No., brief

description, setting description, machine tool setpoints and etc. of various system parameters in Table 4-8.

(3) Fully discuss the problems found during the observation, and fill the problems and solutions in Table 4-8.

Table 4-8 Task Work Order for System Basic Parameter Setting

<table>
<tr><td colspan="6">Task Work Order</td></tr>
<tr><td colspan="2">Name:</td><td colspan="2">Class:</td><td colspan="2">Student ID:</td></tr>
<tr><td colspan="6">List of materials, tools, measuring instruments, and equipment required</td></tr>
<tr><td>S/N</td><td colspan="2">Name</td><td>Quantity</td><td>Model or specification</td><td>Remark</td></tr>
<tr><td></td><td colspan="2"></td><td></td><td></td><td></td></tr>
<tr><td></td><td colspan="2"></td><td></td><td></td><td></td></tr>
<tr><td></td><td colspan="2"></td><td></td><td></td><td></td></tr>
<tr><td colspan="6">Operating process(Observation)</td></tr>
<tr><td>S/N</td><td>Parameter No.</td><td>Brief description</td><td colspan="2">Setting description</td><td>Machine tool setpoint</td></tr>
<tr><td></td><td></td><td></td><td colspan="2"></td><td></td></tr>
<tr><td></td><td></td><td></td><td colspan="2"></td><td></td></tr>
<tr><td></td><td></td><td></td><td colspan="2"></td><td></td></tr>
<tr><td></td><td></td><td></td><td colspan="2"></td><td></td></tr>
<tr><td colspan="3">Problems</td><td colspan="3">Solutions</td></tr>
<tr><td colspan="3"></td><td colspan="3"></td></tr>
</table>

Task Evaluation

S/N	Evaluation contents	Total point	Score
1	Be able to clearly explain the types and expressions of system parameters	25	
2	Be able to complete the setting of basic parameters	25	
3	Be able to understand the meaning of alarms that occur during commissioning and eliminate them.	25	
4	Have a professional spirit of effective communication, expression and solidarity	25	

Task III Introduction to CNC System PMC

Learning Objectives

(1) To understand the principles of PMC and the I/O address assignment.

(2) To master the PMC program structure.

(3) To master the tool change procedures of hat-type tool magazine.

(4) To master the PMC programming for the tool change function of hat-type tool magazine.

(5) To possess the PMC program design and debugging capabilities.

(6) To master the engineering working methods, cultivate a rigorous work style, and abide by the 5S management system.

Task Description

Observe the PMC programs of FANUC 0i-F CNC system by groups, fill the signals, addresses, descriptions, resource type and etc. of the PMC programs of each part in the task work order. Make sure to be careful and pay attention to the relevant specifications of the practical training site during observation.

Knowledge Link

I. Introduction to PMC principle and I/O address assignment

1. Definition of PMC

Programmable logic controller PLC is called programmable machine controller PMC in the FANUC CNC system. Regardless of different names, PLC and PMC are the same in nature. Unless otherwise required, they will be called PMC for consistency. At present, PMC is built in CNC products of FANUC, and independent PMC equipment is not necessary. PMC has become an important constituent part of CNC systems.

2. Introduction to PMC program structure

In general, the PMC program of FANUC 0i-F CNC system consists of level 1 program, level 2 program, level 3 program and subprograms, as shown in Fig. 4-14.

(1) Level 1 program.

The PMC level 1 program is executed once in each ladder diagram execution cycle from the beginning of the program to the END1 command, featuring real-time signal sampling and fast

response of output signals. It mainly processes short pulse signals, such as the command signal of torque limit HIGH, machine tool ready signal, emergency stop signal, spindle stop signal, interlocking signal of each axle, and brake control signal. In the level 1 program, the program should be kept as short as possible so that PMC procedure execution time can be shortened. When there is no input signal, it is only necessary to write END1 function command.

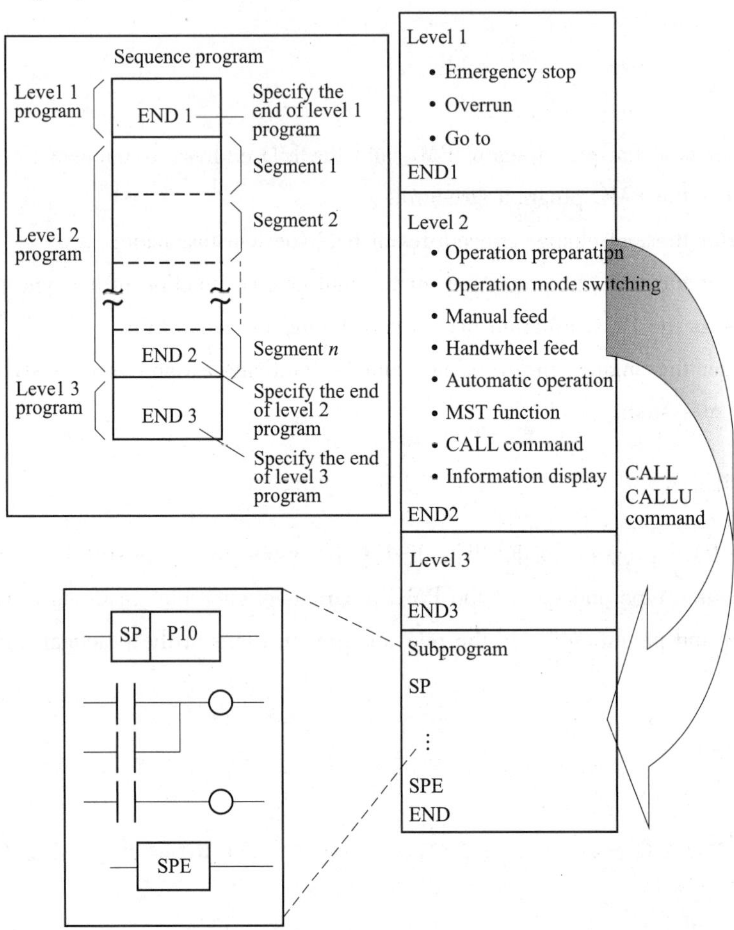

Fig. 4-14 PMC Program Structure of FANUC 0i-F CNC System

(2) Level 2 program.

The PMC level 2 program is a program after the END1 command and before the END2 command. Level 2 program usually comprises operation mode selection, spindle function, feedrate override and other programs. For conciseness and clear distinction of functions, function programs can also be written in subprograms and called in level 2 program.

(3) Level 3 program.

Level 3 program of PMC is the program after END2 command and before END3 command. Level 3 program mainly processes low-speed response signals, and is usually used for processing PMC program alarm signals. When compiling a sequence program, you can choose whether to use level 3 program. It is not used herein.

(4) Subprogram.

Subprogram is the program after the END3 command and before END command. Program segments having specific functions that are used many times are usually used as subprograms. Commands used in the main program determine the execution state of specific subprograms. When the main program calls and executes the subprogram, the subprogram executes all commands till the end, and then the system will return to the main program which calls the subprogram. Subprograms are used for dividing programs into segments and blocks for easier management. During program debugging and maintenance, these areas and the whole program are simply debugged and troubleshooted by using smaller program blocks. Subprogram block is called only when necessary, which makes the use of PMC more efficient, because the PMC program processing time can be shortened as all subprogram blocks may not need to execute scanning every time.

3. Sequence program of FANUC CNC systems

The sequence program of an FANUC CNC system consists of level 1 program and level 2 program, as shown in Fig. 4-15.

Level 1 program only processes short pulse signals, for examples, emergency stop, overrun of each feed coordinate axis, machine tool interlock signal, deceleration for returning to reference point, skip, feed pause signal, counting of CNC center large tool magazine, etc.

Level 2 program comprises main contents of CNC machine tool functions, such as processing of operation mode, auxiliary function, tool change, etc.

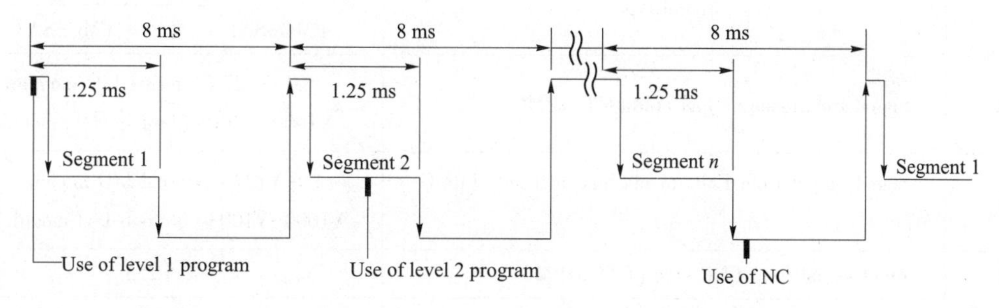

Fig. 4-15 Sequence Program of an FANUC CNC System

4. I/O address assignment

Addresses are used to distinguish signals. Different addresses correspond to the input and output signals on machine tool side, the input and output signals on CNC side, the internal relay, the counter, the latching relay and the data table, as shown in Fig. 4-16.

In Fig. 4-16, the PMC-related input and output signals indicated by continuous lines are transmitted via the receiving circuit and the drive circuit of I/O board; the PMC-related input and output signal indicated by dotted lines are transmitted only in memories, for example, transmitted in RAM. The state of all these signals can be displayed on CRT.

(1) Address format and signal type.

Address is expressed by address number and bit number in the format shown below.

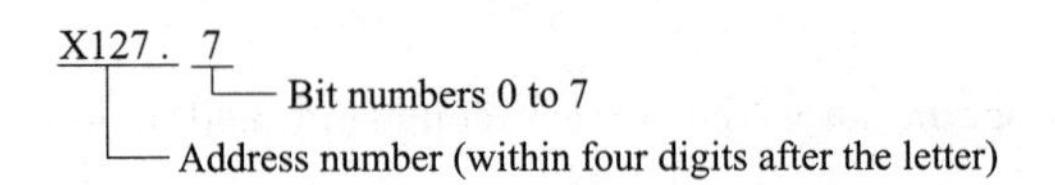

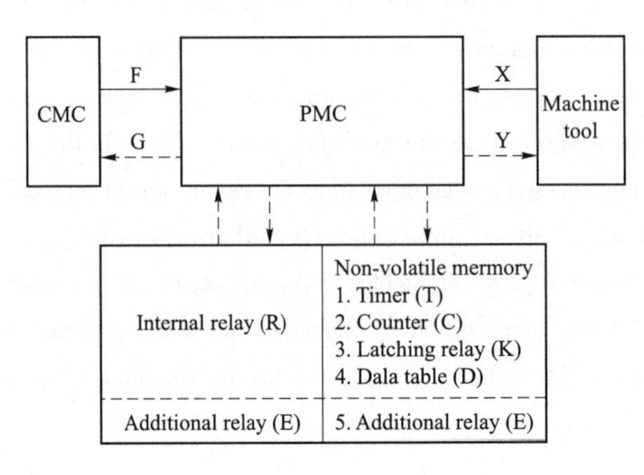

Fig. 4-16 I/O Address Assignment

A letter must be specified in the beginning of address number to indicate the signal type shown in Table 4-9. When specifying the address of byte unit in a function command, the bit number can be omitted, for example, X127.

Table 4-9 Signal Type of PMC Address

Letter	Signal type	Remarks	
		PMC-SA1	PMC-SA3
X	Signal from machine tool side (MT→PMC)	X0-X127 (external I/O module) X1000-X1003 (built-in I/O module)	
Y	Signal output from PMC to machine tool side (PMC→MT)	Y0-Y127 (external I/O module) Y1000-Y1003 (built-in I/O module)	
F	Input signal from CNC side (CNC→PMC)	F0-F255	
G	Signal output from PMC to CNC (PMC→NC)	G0-G255	
R	Internal relay	R0-R999 R9000-R9099	R0-R1499 R9000-R9117
A	Request signal for information display	A0-A24	
C	Counter	C0-C79	
K	Latching relay	K0-K19	
T	Variable timer	T0-T79	
D	Data table	—	D0-D1859
L	Mark number	—	L1-L9999
P	Subprogram	—	P1-P512

(2) Assignment method of I/O address.

The operation steps of CNC milling machine maintenance training equipment are as follows. Press the function key [SYSTEM] twice on the operation panel, then press [PMCMCNF], and then press [MODULE], put the cursor to X0 and input "0.0.1.OC02I", then put the cursor to Y0 and input [0.0.1.OC02O], so as to assign the I/O address X0-X13, Y0-Y13, as shown in Fig. 4-17.

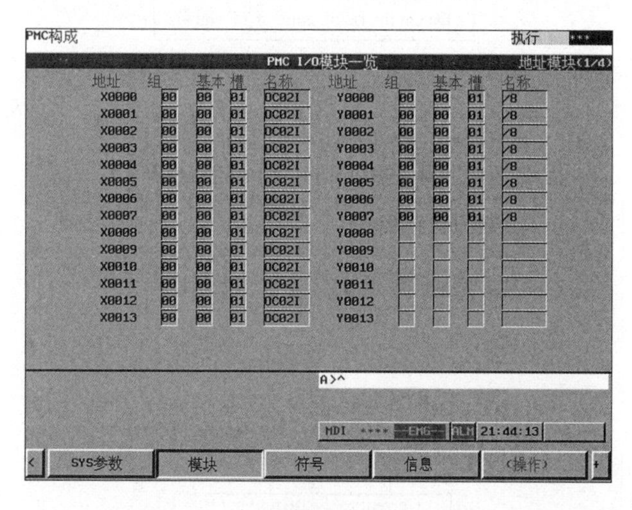

Fig. 4-17 I/O Address Assignment

II. PMC programming for tool change of hat-type tool magazine

1. Working principle of hat-type tool magazine

1) Control requirements of tool magazine.

The tool magazine of FANUC 0i Mate MF CNC milling machine maintenance training equipment is a hat-type tool magazine using 12 tools. Therefore, this part introduces the control of the hat-type tool magazine of vertical CNC center; implementation of other types of CNC centers may refer to this method.

Tool magazine control comprise manual control and program automatic control. Manual control is mainly used for installation, commissioning or maintenance of tool magazine, mainly including manual tool selection, spindle tool clamping, release and other operations. Program automatic control is mainly used for the automatic tool change control during production. The general automatic tool change process of tool magazines is shown in Fig. 4-18.

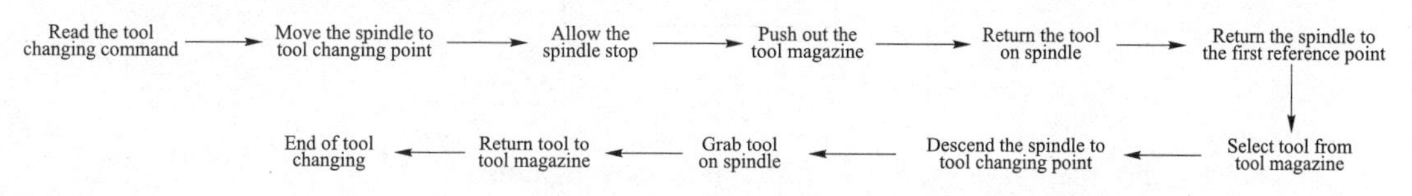

Fig. 4-18 Automatic Tool Change Control Process of Tool Magazines

2) Tool change procedures and thoughts.

For automatic tool change, it is necessary to consider whether the T command is consistent with the spindle tool number, whether there is a tool on spindle, whether the tool case number of tool magazine is consistent with the spindle tool number. The tool change procedures are shown in Fig. 4-19.

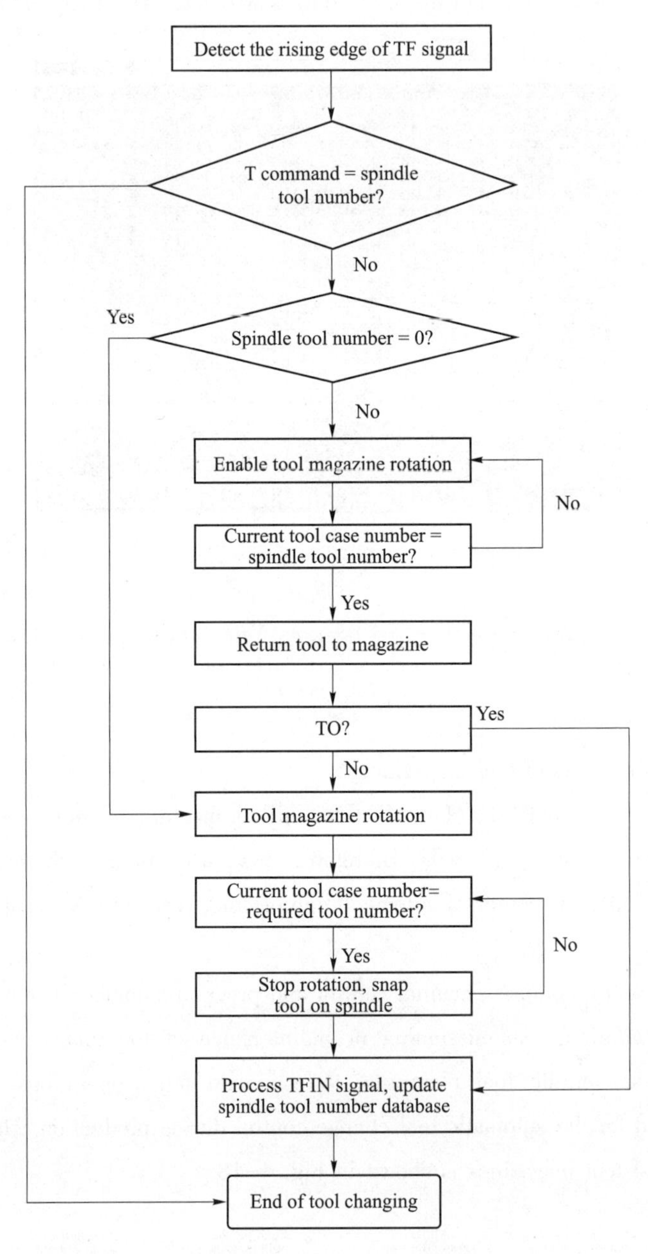

Fig. 4-19 Tool Change procedures

2. Tool change macro program of hat-type tool magazine

(1) Compilation of tool change macro program.

Compile tool change macro program "O9001" as below according to Fig. 4-47. In EDIT

mode, press program button [PROG], input O9001, then press BG to edit program, as shown below.

O9001; (tool change macro program number)

IF[#1001EQ1] GOTO 40; (if spindle tool number is consistent with command tool number, go to N40, macro variable #1001 corresponds to G54.1 in PMC program; EQ means "=")

#199=#4003; (G90, G91 mode)

#198=#4006; (G20, G21 mode)

IF[#1003EQ1] GOTO 20; (determine whether there is a tool on spindle, if not, go to N20 directly, macro variable #1003 corresponds to G54.3 in PMC program)

G21 G91 G30 P2 *Z*0 M19; (*Z*-axis moves to spindle orientation at the second reference point)

M81; (determine whether the spindle tool number is consistent with the current tool number of tool magazine, if not, the tool magazine rotates until it is consistent with the spindle tool number)

M80; (tool magazine moves forward, close to the spindle)

M82; (release tool and blow air)

G91 G28 *Z*0; (*Z*-axis moves to the first reference point)

IF[#1002EQ1] GOTO 10; (whether the command is T0 or not? Macro variable #1002 corresponds to G54.2 in PMC program)

M83; (at the spindle end, the tool magazine rotates to the tool position specified by machining program)

G91 G30 P2 *Z*0; (*Z*-axis moves to the second reference point)

N10 M84; (tool clamping)

M86; (tool magazine moves backward, away from spindle)

GOTO 30;

N20 G21 G91 G28 *Z*0 M19; (*Z*-axis moves to the spindle orientation at the first reference point)

M83; (at the end away from spindle, the tool magazine rotates to the tool position specified by machining program)

M80; (tool magazine moves forward, close to the spindle)

M82; (release tool and blow air)

G91 G30 P2 *Z*0; (*Z*-axis moves to the second reference point)

M84; (tool clamping)

M86; (tool magazine moves backward, away from spindle)

N30 G#199G#198; (mode recovery)

N40 M99; (subprogram ends)

Press BG to end editing.

(2) Macro program calling and relevant system parameters.

Use command "M06" to call the tool change macro program to realize the automatic tool change control of tool magazine. The relevant system parameter settings are shown in Fig. 4-10.

Table 4-10 Relevant System Parameters Setting for Automatic Tool Change of Tool Magazine Realized by Calling Tool Change Macro Program

Parameter No.	Meaning	Set value	Description
6071	Use M command to call tool change macro program O9001	6	Specify M06 to call macro program O9001
3202#4	Set the macro program to allow display, edit and delete	0	When it is set to "0", it means that programs " O9000 - O9999 " can be displayed, edited or deleted

(3) Variables in macro program corresponding to G address of PMC program.

The direct relation between macro program and PMC program is the contents of macro program that save and recover system modes and determine jumping conditions, etc. Therefore, system variables will be used.

1) Input signal variables (G54.1-G54.3).

The description of variables of macro program corresponding to PMC program input signals is shown in Table 4-11.

Table 4-11 Variables of Macro Program Corresponding to PMC Program Input Signals

Signal/Function	Address	Corresponding variable
Spindle tool number is consistent with command tool number	G54.1	#1001
Determine whether there is any tool on spindle	G54.3	#1003
Whether the command is T0	G54.2	#1002

In the Table 4-11, the operation result of PMC program determines whether G54.1, G54.2 and G54.3 are "0" or "1".

2) System mode information variables (#4003, #4006).

Due to the increment programming used in macro program, the system mode of main program must be protected before executing the macro program, and the system mode of main program must be recovered after execution of the macro program, so the system mode information variables will be used. System modes in main program mainly comprise Metric/British system programming and absolute/increment value programming modes, to which the corresponding system variables are "#4003" (corresponding to the programming coordinate mode value G90(absolute)/G91(increment) currently used by the system) and "#4006" (corresponding to the programming unit system G20 (British system)/G21 (Metric system) currently used by the system).

3. PMC program of hat-type tool magazine

(1) Spindle orientation.

In the tool change process, the spindle orientation means that the spindle needs to stop at a

certain angle, otherwise the tool will impact the locating block at the front end of spindle. Spindle orientation function is mainly realized by PMC program and system parameter settings. The relevant system parameter settings are shown in Table 4-12 and Table 4-13. The reference PMC program is shown in Fig. 4-20.

Table 4-12 Parameters and Settings Relevant to Spindle Orientation

Parameter No.	Meaning	Set value	Description
4077	Set the spindle exact stop position data	Spindle exact stop position data	Save the spindle exact stop position data
4038	Set the spindle orientation speed	Spindle orientation speed	Save the spindle orientation speed

Table 4-13 Resources and Allocation for Realizing Spindle Orientation Control Function

Resource type	Signal/Function	Address	Description
Input	Spindle orientation	X5.4	Spindle orientation button on operation panel
	Air pressure	X8.3	Air pressure detection signal
Output	Spindle orientation indicator	Y5.4	Spindle orientation button indicator on operation panel
Internal signal	Reset signal	F1.1	System in reset state (RST signal)
	Handle selection signal	F3.1	Handle feed selection confirmation signal
	JOG selection signal	F3.2	Manual continuous feed selection confirmation signal
	MDI selection signal	F3.3	Manual data input selection confirmation signal
	DNC selection signal	F3.4	DNC operation selection confirmation signal
	Automatic selection signal	F3.5	Memory operation selection confirmation signal
	Spindle orientation end signal	F45.7	End of spindle orientation
	Spindle orientation signal	G70.6	Control request signal from PMC to CNC

(2) Determine whether T command equals to spindle tool number.

In the tool change process, it is necessary to determine whether the T command equals to spindle tool number. Resources and allocation for realizing the function are shown in Table 4-14. The reference PMC program is shown in Fig. 4-21.

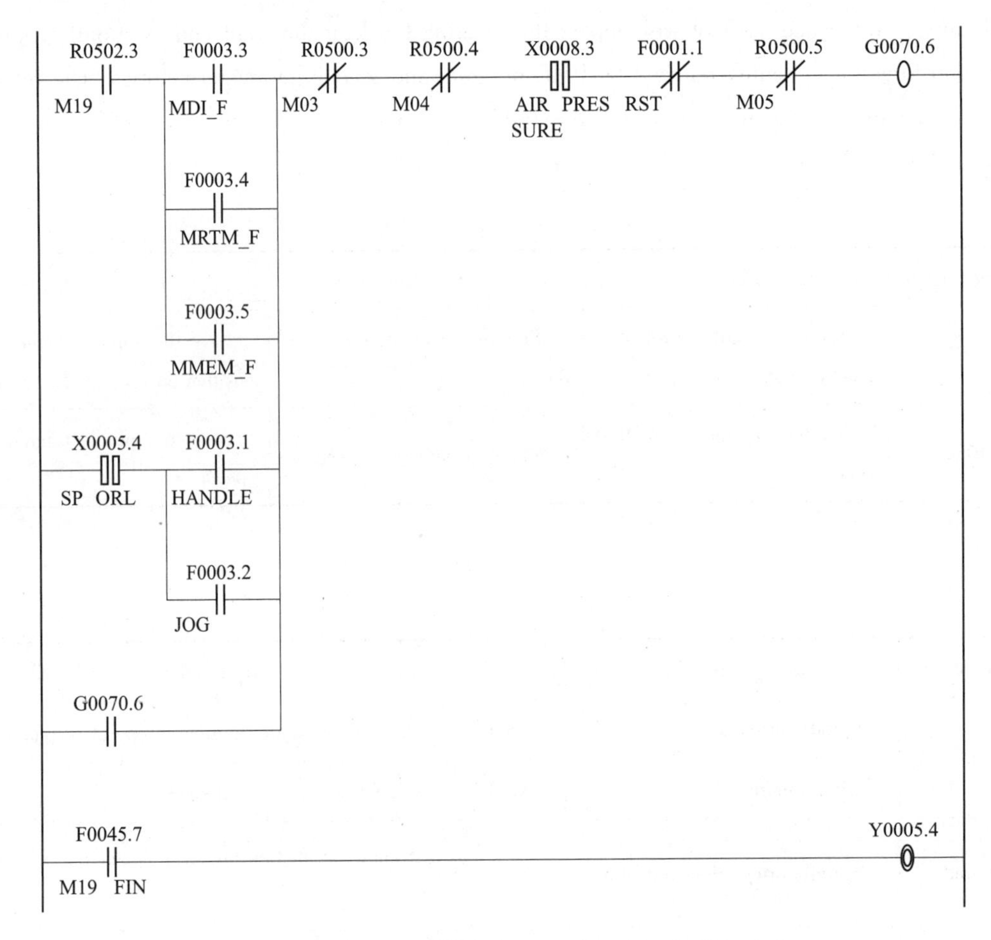

Fig. 4-20 Reference PMC Program for Spindle Orientation

Table 4-14 Resources and Allocation for Determining Whether T command Equals to Spindle Tool Number

Resource type	Signal/Function	Address	Description
Internal signal	Normally "1" signal	R9091.1	PMC normally "1" signal
	Reset signal	F1.1	System in reset state (RST signal)
	T command selection signal	F7.3	T command selection signal from CNC to PMC
	T command decipher code register	F26	T command decipher code from CNC to PMC
	T command decipher result signal	R26	Save the T command decipher data
	Decipher result signal of current tool number in tool magazine	R10	Save the decipher data of current tool number in tool magazine
	Register of current tool number in the tool magazine	C10	It can be queried and modified through MDI
	Signal of T command equals to spindle tool number	G54.1	Corresponding variable #1001

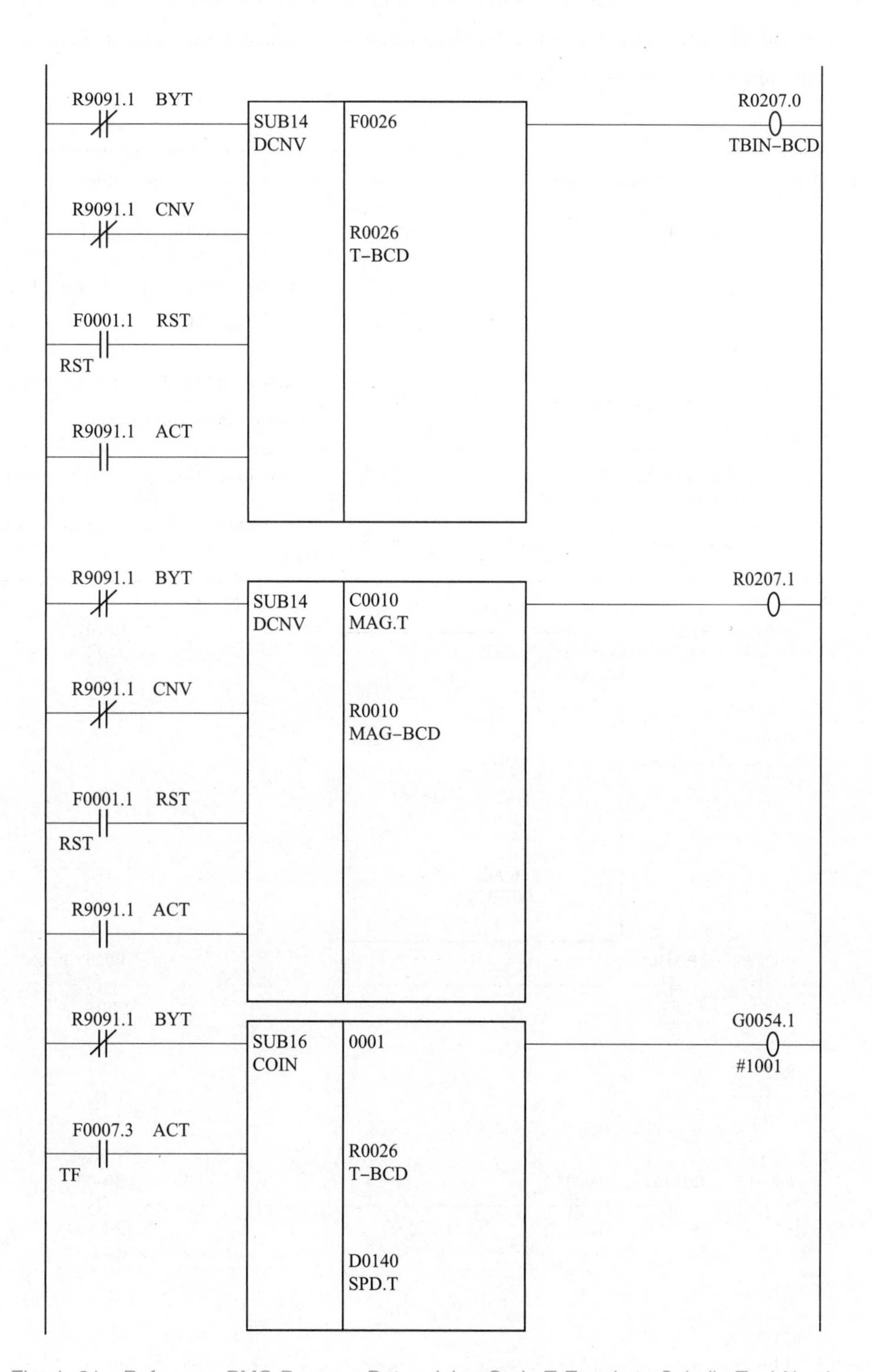

Fig. 4-21 Reference PMC Program Determining Code T Equals to Spindle Tool Number

(3) Determine the value range of T command.

The hat-type tool magazine used in this project can hold 12 tools. Therefore, PMC program

needs to determine the value range of T command, and the input tool number cannot exceed T12. The resources and allocation for determining the value range of T command are shown in Table 4–15. The reference PMC program is shown in Fig. 4–22.

Table 4–15 Resources and Allocation for Value Range of T Command

Resource type	Signal/Function	Address	Description
Internal signal	Normally "1" signal	R9091.1	PMC normally "1" signal
	T command error signal	R207.3	When R207.3 is "1", it means T command exceeds the value range, that is $T>12$
	T command error cancel signal	R430.3	When R430. 3 is "1", it means T command error cancel signal
	Reset signal	F1.1	System in reset state (RST signal)
	T command selection signal	F7.3	T command selection signal from CNC to PMC

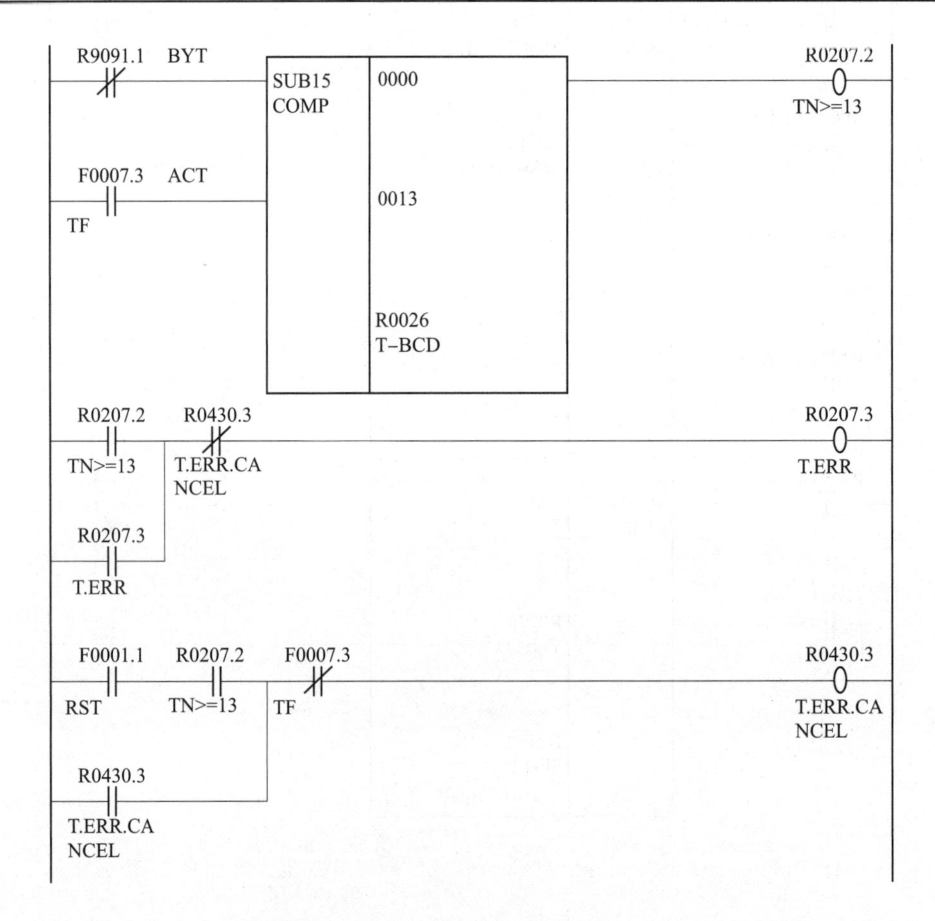

Fig. 4–22 Reference PMC Program for Determining Value Range of T Command

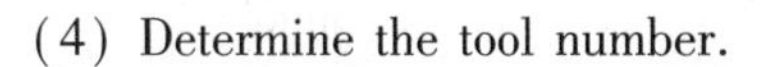

(4) Determine the tool number.

The hat-type tool magazine of this project requires determining whether there is any tool on spindle, determining whether the T command is zero and determining the current tool change point of tool magazine. Resource and allocation for determining the tool number are shown in Table 4 - 16. The reference PMC program is shown in Fig. 4-23.

Table 4-16 Resources and Allocation for Determining the Tool Number

Resource type	Signal/Function	Address	Description
Output	Tool magazine backward drive signal	Y0002.6	Control the backward rotation of tool magazine
Internal signal	Normally "1" signal	R9091.1	PMC normally "1" signal
	T command decipher result signal	R0026	Save the T command decipher data
	Count falling edge signal of tool magazine count sensor	R1000.0	When R1000.0 is "1", it means that the tool magazine count sensor has counted 1
	Decipher signal of tool magazine moving backward away from spindle	R0200.6	M86 signal
	T command selection holding signal	R0205.1	When R0205.1 is "1", it means T command selection holding
	T command selection signal	F0007.3	T command selection signal from CNC to PMC
	Spindle tool number register (data table)	D0140	Save the current tool number on spindle
	Signal of T command equals to zero	G0054.2	Corresponding variable #1002
	Signal of no tool on spindle	G0054.3	Corresponding variable #1003

(5) Select tool in the tool magazine by spindle tool number.

When executing the tool change command, if there is a tool on spindle, it is necessary to return the tool on spindle to the tool magazine first, and the tool will be selected in tool magazine by spindle tool number. Resource and allocation for selecting tool in tool magazine by spindle tool number are shown in Table 4-17. The reference PMC program is shown in Fig. 4-24.

(6) Manual forward and backward rotation of the tool magazine.

Press the tool magazine forward and backward button on machine tool operation panel to

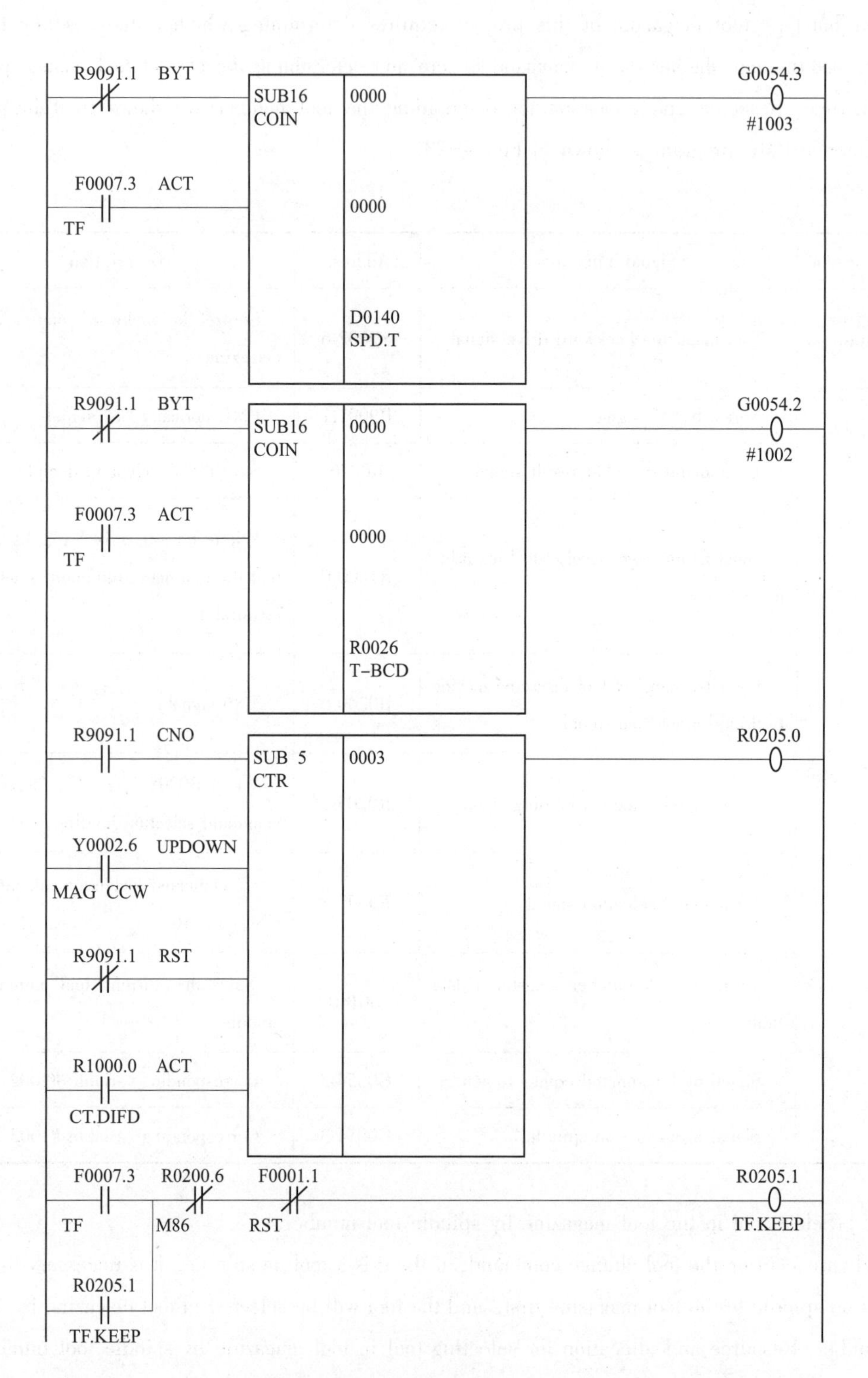

Fig. 4–23 Reference PMC program for Determining the Tool Number

realize the forward and backward rotation of tool magazine, to rotate it to the next tool position. Resources and allocation for manual forward and backward rotation of the tool magazine are shown in Table 4-18. The reference PMC program is shown in Fig. 4-25.

Table 4-17 Resources and Allocation for Selecting Tool in the Tool Magazine by Spindle Tool Number

Resource type	Signal/Function	Address	Description
Internal signal	Normally "1" signal	R9091.1	PMC normally "1" signal
	Decipher result signal of current tool number in the tool magazine	R0010	Save the decipher data of current tool number in the tool magazine
	Decipher signal for determining consistency between spindle tool number and current tool number in the tool magazine	R0200.1	M81 signal
	Rotation control signal for determining current tool number in the tool magazine and spindle tool number	R0205.4	When R0205.4 is "0", it means the tool magazine needs to rotate forward, when it is "1", it means the tool magazine needs to rotate backward
	Spindle tool number register (data table)	D0140	Save the current tool number on spindle

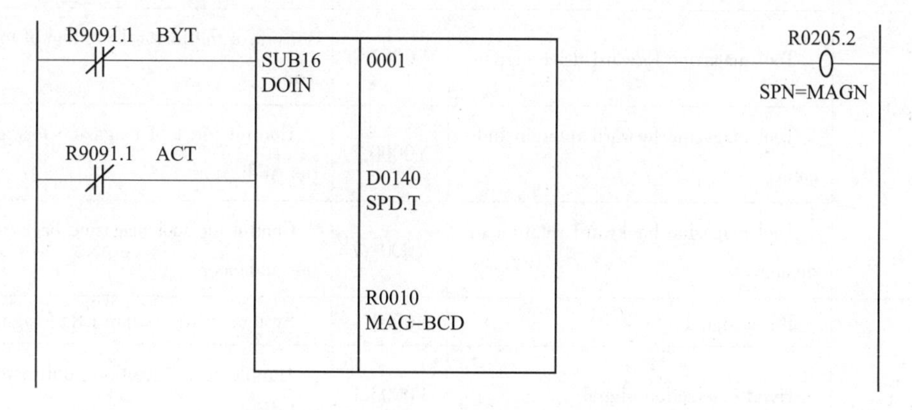

Fig. 4-24 Reference PMC Program for Selecting Tool in the Tool Magazine by Spindle Tool Number

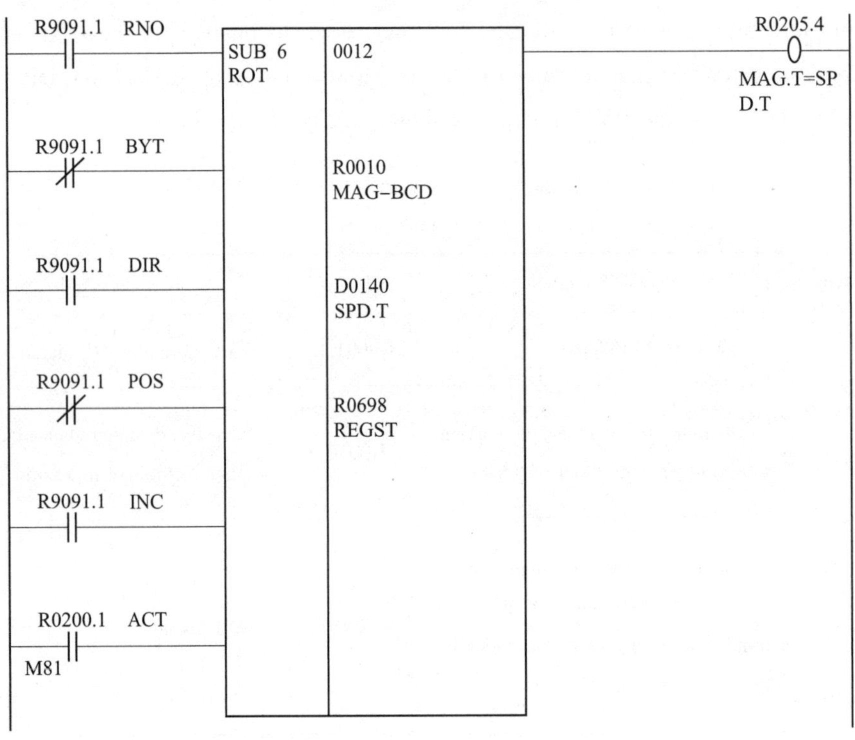

Fig. 4-24 Reference PMC Program for Selecting Tool in Tool Magazines by Spindle Tool Number(Continued)

Table 4-18 Resources and Allocation for Manual Forward and Backward Rotation of Tool Magazines

Resource type	Signal/Function	Address	Description
Input	Emergency stop signal	X0008.4	Machine tool emergency stop
Output	Tool magazine backward drive signal	Y0002.6	Control the backward rotation of tool magazine
	Tool magazine forward drive signal	Y0002.7	Control the forward rotation of tool magazine
	Tool magazine forward rotation indicator	Y0000.7	Control the tool magazine forward rotation indicator
	Tool magazine backward rotation indicator	Y0001.7	Control the tool magazine backward rotation indicator
Internal signal	Reset signal	F0001.1	System in reset state (RST signal)
	Handle selection signal	F0003.1	Handle feed selection confirmation signal
	JOG selection signal	F0003.2	Manual continuous feed selection confirmation signal

Continued

Resource type	Signal/Function	Address	Description
Internal signal	MDI selection signal	F0003.3	Manual data input selection confirmation signal
	DNC selection signal	F0003.4	DNC operation selection confirmation signal
	Automatic selection signal	F0003.5	Memory operation selection confirmation signal
	The third axis arrives at the first reference point	F0094.2	Confirmation signal of Z-axis arrives at the first reference point
	The third axis arrives at the second reference point	F0096.2	Confirmation signal of Z-axis arrives at the second reference point
	Decipher signal for determining consistency between spindle tool number and current tool number in tool magazine	R0200.1	M81 signal
	The tool magazine rotates to the tool position specified by machining program	R0200.3	M83 signal
	Signal of the tool magazine forward rotation at the second reference point	R0201.0	When R0201.0 is "1", it means the tool magazine forward rotation signal at the second reference point
	Signal of the tool magazine forward rotation at the first reference point	R0201.1	When R0201.1 is "1", it means the tool magazine forward rotation signal at the first reference point
	Signal of the tool magazine backward rotation at the second reference point	R0201.2	When R0201.2 is "1", it means the tool magazine backward rotation signal at the second reference point
	Signal of the tool magazine backward rotation at the first reference point	R0201.3	When R0201.3 is "1", it means the tool magazine backward rotation signal at the first reference point
	T command selection holding signal	R0205.1	When R0205.1 is "1", it means T command selection holding
	Signal for determining consistency between spindle tool number and current tool number in the tool magazine	R0205.2	When R0205.2 is "1", it means the spindle tool number equals to the current tool number in tool magazine
	Rotation control signal for determining current tool number in the tool magazine and spindle tool number	R0205.4	When R0205.4 is "0", it means the tool magazine needs to rotate forward; when it is "1", it means the tool magazine needs to rotate backward

Continued

Resource type	Signal/Function	Address	Description
Internal signal	Rotation control signal for determining current tool number in tool magazine and T command	R0206.2	When R0206.2 is "0", it means the tool magazine needs to rotate forward, when it is "1", it means the tool magazine needs to rotate backward
	Manual tool magazine forward rotation signal	R0206.3	When R0206.3 is "1", it means the manual tool magazine forward rotation signal
	Manual tool magazine backward rotation signal	R0206.4	When R0206.4 is "1", it means the manual tool magazine backward rotation signal
	Signal for determining consistency between current tool number in the tool magazine and T command	R0208.2	When R0208.2 is "1", it means the current tool number in tool magazine equals to T command

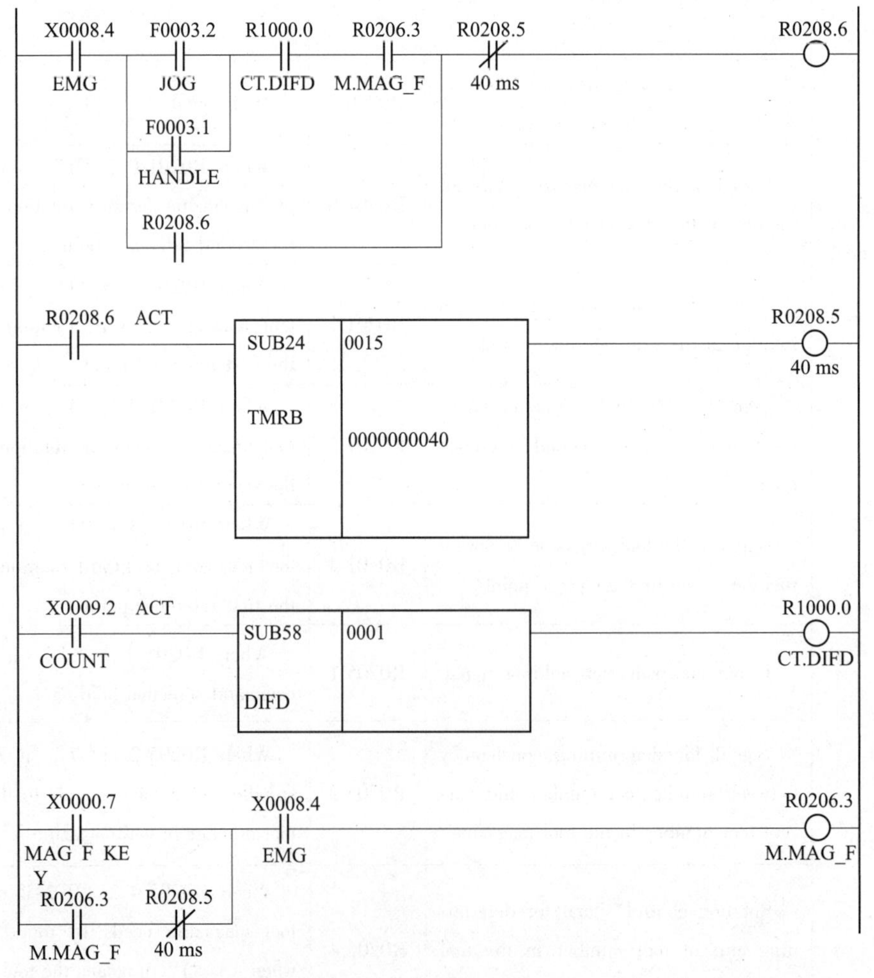

Fig. 4-25 Reference PMC Program for Manual Tool Magazine Forward and Backward Rotation

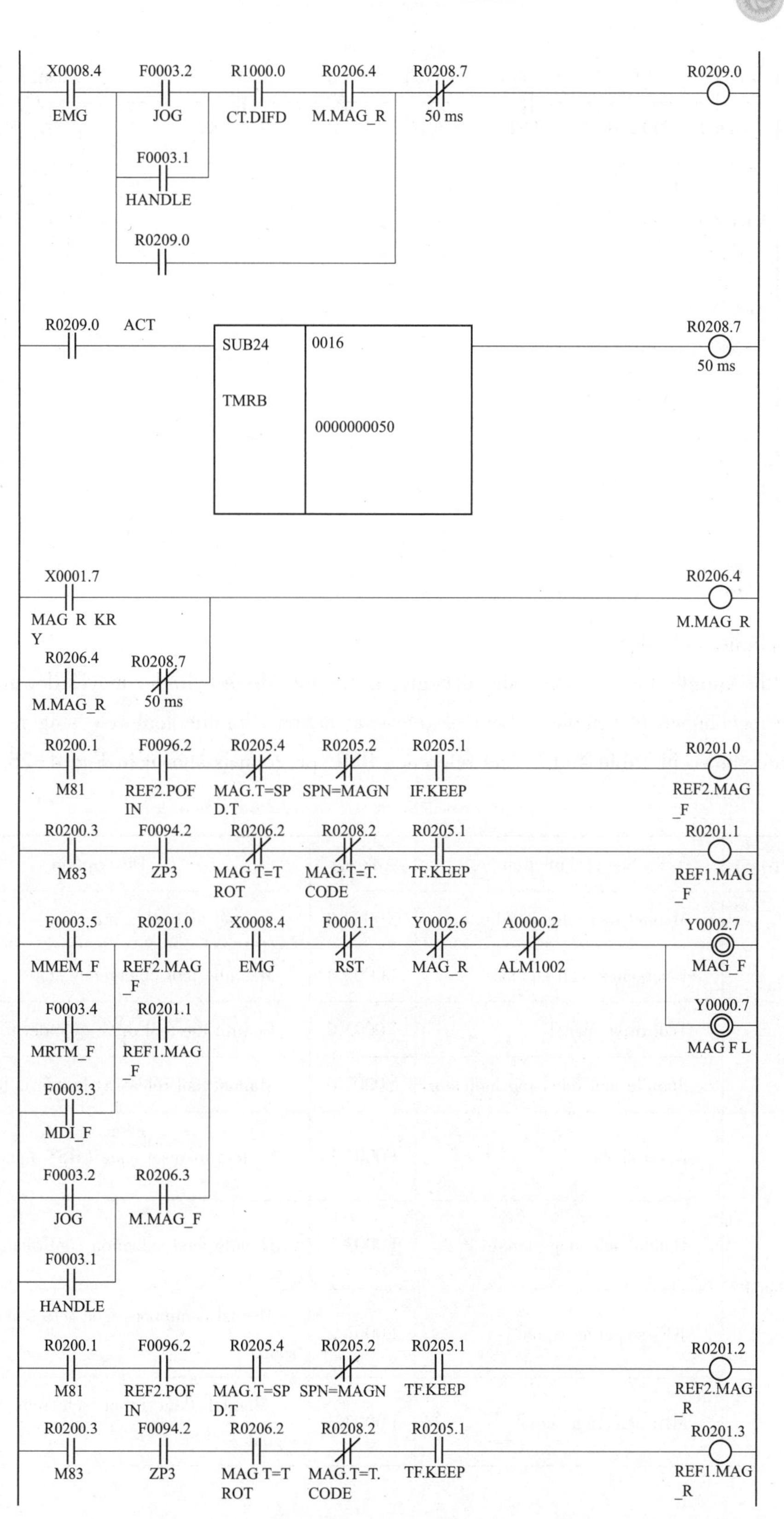

Fig. 4–25 Reference PMC Program for Manual Tool Magazine Forward and Backward Rotation(Continued)

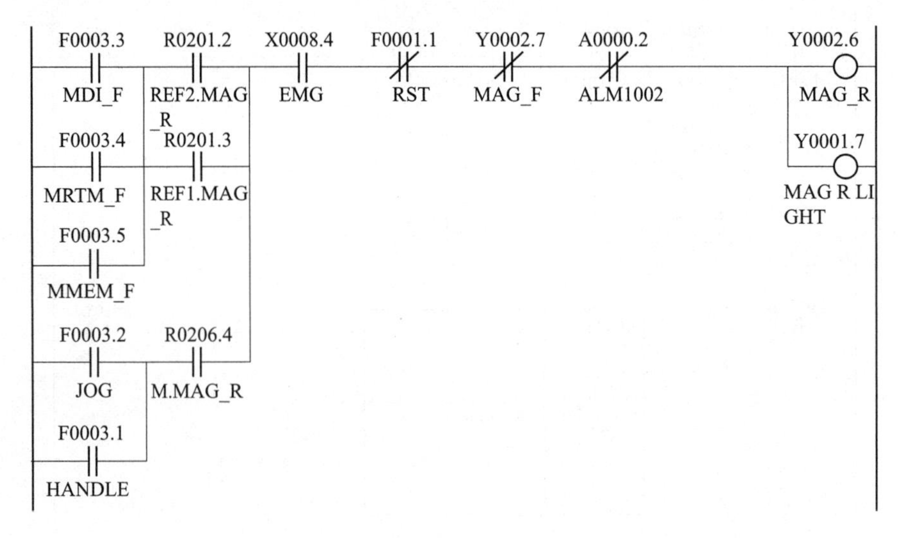

Fig. 4-25 Reference PMC Program for Manual Tool Magazine Forward and Backward Rotation(Continued)

(7) Spindle tool releasing.

When the spindle tool releasing signal is given, the tool drive cylinder moves down to push the tool pulling mechanism to complete the tool releasing action. Spindle tool releasing resources and allocation are shown in Table 4-19. The reference PMC program is shown in Fig. 4-26.

Table 4-19 Spindle Tool Releasing Resources and Allocation

Resource type	Signal/Function	Address	Description
Input	Manual tool releasing button	X0008.0	Manual tool releasing signal
	Emergency stop signal	X0008.4	Machine tool emergency stop
Output	Tool drive signal	Y0003.0	Control the tool drive cylinder to move down
	Spindle tool releasing indicator	Y0007.0	Manual tool releasing button indicator signal
Internal signal	Reset signal	F0001.1	System in reset state (RST signal)
	Handle selection signal	F0003.1	Handle feed selection confirmation signal
	JOG selection signal	F0003.2	Manual continuous feed selection confirmation signal
	MDI selection signal	F0003.3	Manual data input selection confirmation signal

Continued

Resource type	Signal/Function	Address	Description
Internal signal	DNC selection signal	F0003.4	DNC operation selection confirmation signal
	Automatic selection signal	F0003.5	Memory operation selection confirmation signal
	The third axis arrives at the second reference point	F0096.2	Z - axis arriving at the second reference point confirmation signal
	Tool releasing decipher signal	R0200.2	M82 signal
	Tool clamping decipher signal	R0200.4	M84 signal

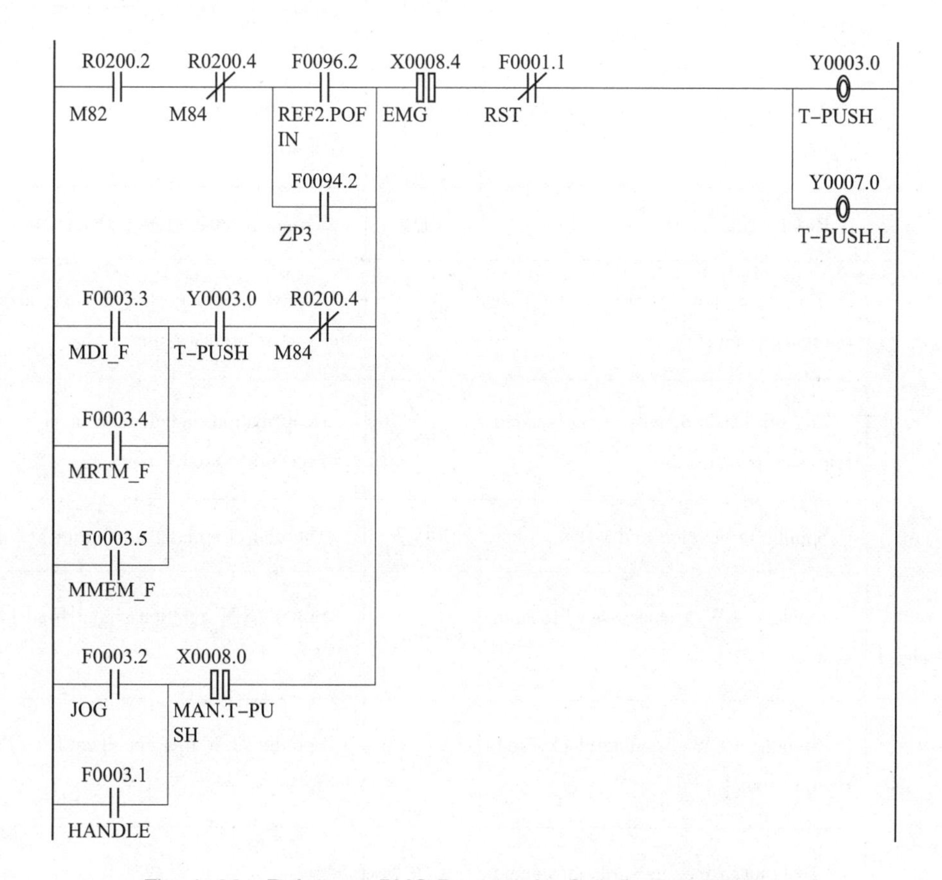

Fig. 4-26 Reference PMC Program for Spindle Tool Releasing

(8) Tool magazine forward and backward movements.

When the machine tool gives the tool change command, the tool magazine moves forward to approach the spindle, and moves back away from the spindle after tool change. Resources and allocation for the tool magazine front and back movement signals are shown in Table 4-20. The reference PMC program is shown in Fig. 4-27.

Table 4-20 Resources and Allocation for Tool Magazine Forward and Backward Movement Signals

Resource type	Signal/Function	Address	Description
Input	Emergency stop signal	X8.4	Machine tool emergency stop
	Tool magazine front position signal	X9.0	Tool magazine front position detection signal
	Tool magazine rear position signal	X9.1	Tool magazine rear position detection signal
Output	Tool magazine forward movement signal	Y3.4	Control the tool magazine to move forward
	Tool magazine backward movement signal	Y3.5	Control the tool magazine to move backward
Internal signal	Reset signal	F1.1	System in reset state (RST signal)
	The third axis arrives at the first reference point	F94.2	Confirmation signal of Z-axis arrives at the first reference point
	The third axis arrives at the second reference point	F96.2	Confirmation signal of Z-axis arrives at the second reference point
	Spindle orientation end signal	F45.7	The end of spindle orientation
	Spindle CW (clockwise) rotation signal	G70.4	Spindle CW rotation signal from PMC to CNC
	Spindle CCW (counterclockwise) rotation signal	G70.5	Spindle CCW rotation signal from PMC to CNC
	Tool magazine forward movement decipher signal	R200.0	M80 signal
	Tool magazine backward movement decipher signal	R200.6	M86 signal

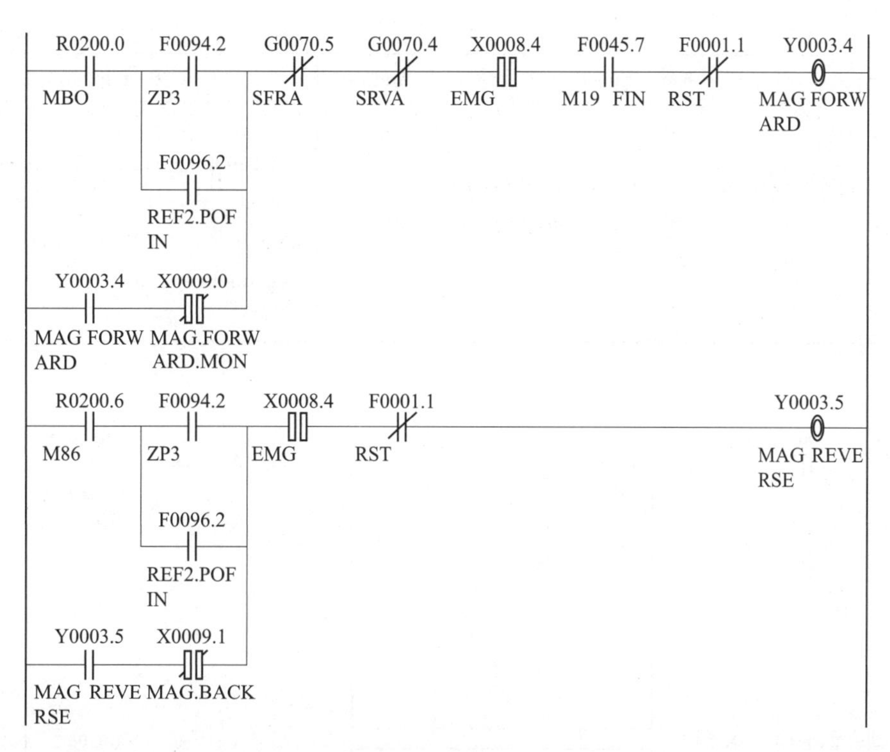

Fig. 4-27 Reference PMC Program for Tool Magazine Forward and Backward Movements

(9) Tool magazine rotation by T command.

When executing the tool change commands, if there is no tool on spindle, the tool magazine rotates to the specified position by T command and prepares to change the tool. Resource and allocation for tool magazine rotation by T command are shown in Table 4-21. The reference PMC program is shown in Fig. 4-28.

Table 4-21 Resources and Allocation for Tool Magazine Rotation by T command

Resource type	Signal/Function	Address	Description
Internal signal	Register of current tool number in the tool magazine	C0010	It can be queried and modified through MDI
	Decipher result signal of current tool number in the tool magazine	R0010	Save the decipher data of current tool number in the tool magazine
	T command decipher result signal	R0026	Save the T command decipher data
	Normally "1" signal	R9091.1	PMC normally "1" signal
	The tool magazine rotates to the tool position specified by machining program	R0200.3	M83 signal
	Signal for determining consistency between current tool number in tool magazine and T command	R0208.2	When R0208.2 is "1", it means the current tool number in the tool magazine equals to T command

Continued

Resource type	Signal/Function	Address	Description
Internal signal	Rotation control signal for determining current tool number in the tool magazine and T command	R0206.2	When R0206.2 is "0", it means the tool magazine needs to rotate forward; when it is "1", it means the tool magazine needs to rotate backward

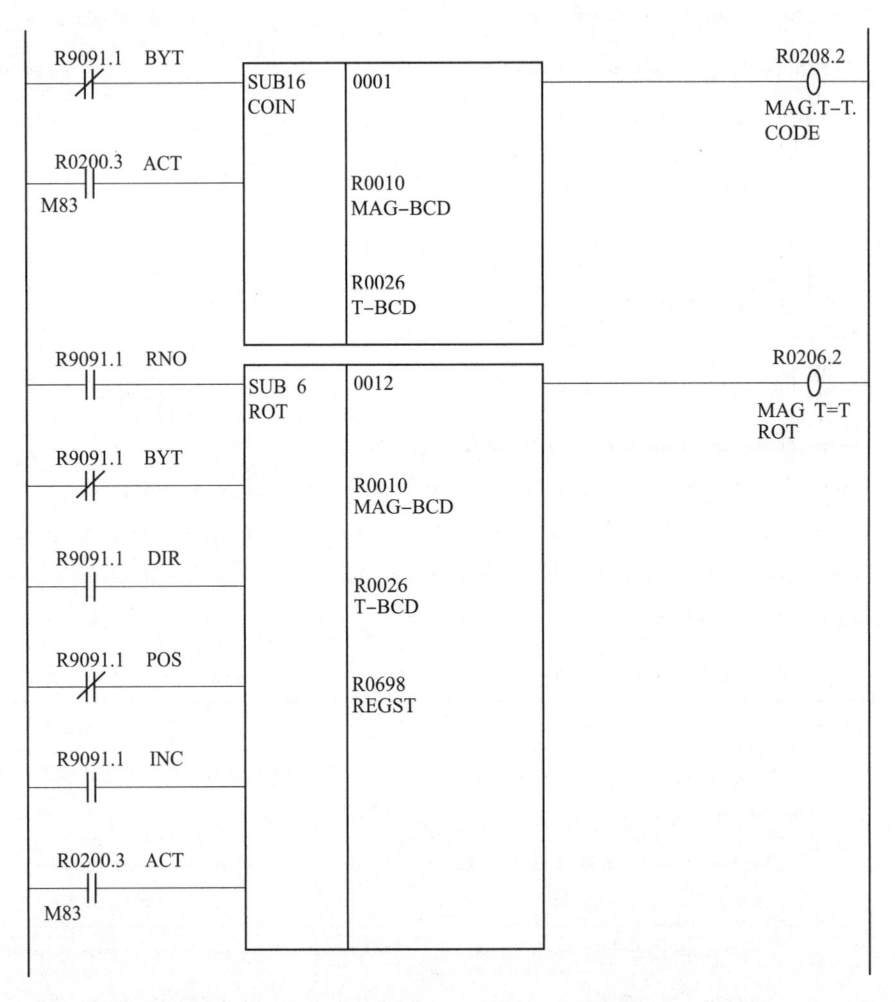

Fig. 4-28 Reference PMC Program for Tool Magazine Rotation by T command

(10) M and T commands completion signals.

When tool change is completed, M and T commands give completion signals. Resources and allocation for M and T commands completion signals are shown in Table 4-22. The reference PMC program is shown in Fig. 4-29.

Table 4-22 Resources and Allocation for M and T Commands Completion Signals

Resource type	Signal/Function	Address	Description
Input	Tool locking signal	X0008.2	Tool locking detection signal
	Tool magazine front position signal	X0009.0	Tool magazine front position detection signal
	Tool magazine rear position signal	X0009.1	Tool magazine rear position detection signal
Internal signal	Auxiliary function code signal	F0010	M command decipher signal
	Spindle orientation signal	G0070.6	Control request signal from PMC to CNC
	Normally "1" signal	R09091.1	PMC normally "1" signal
	Tool magazine forward movement decipher signal	R0200.0	M80 signal
	Decipher signal for determining consistency between spindle tool number and current tool number in the tool magazine	R0200.1	M81 signal
	Tool releasing decipher signal	R0200.2	M82 signal
	Tool magazine rotates to the tool position specified by machining program	R0200.3	M83 signal
	Tool clamping decipher signal	R0200.4	M84 signal
	Decipher signal of tool magazine moving backward away from spindle	R0200.6	M86 signal
	Signal for determining consistency between the spindle tool number and current tool number in tool magazine	R0205.2	When R0205.2 is "1", it means the spindle tool number equals to the current tool number in the tool magazine
	Rotation control signal for determining current tool number in the tool magazine and spindle tool number	R0205.4	When R0205.4 is "0", it means the tool magazine needs to rotate forward; when it is "1", it means the tool magazine needs to rotate backward
	Spindle orientation decipher signal	R0502.3	M19 signal

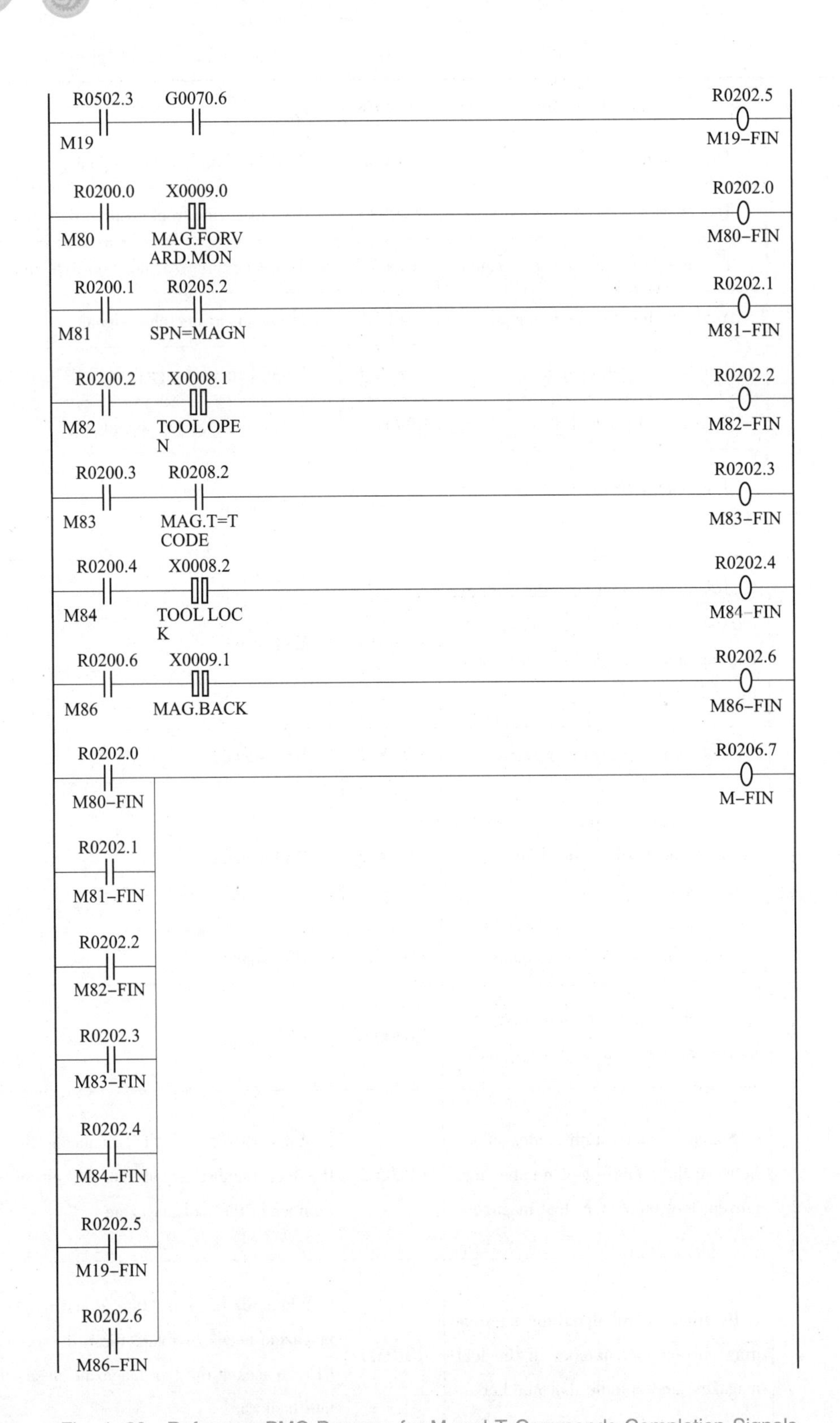

Fig. 4-29 Reference PMC Program for M and T Commands Completion Signals

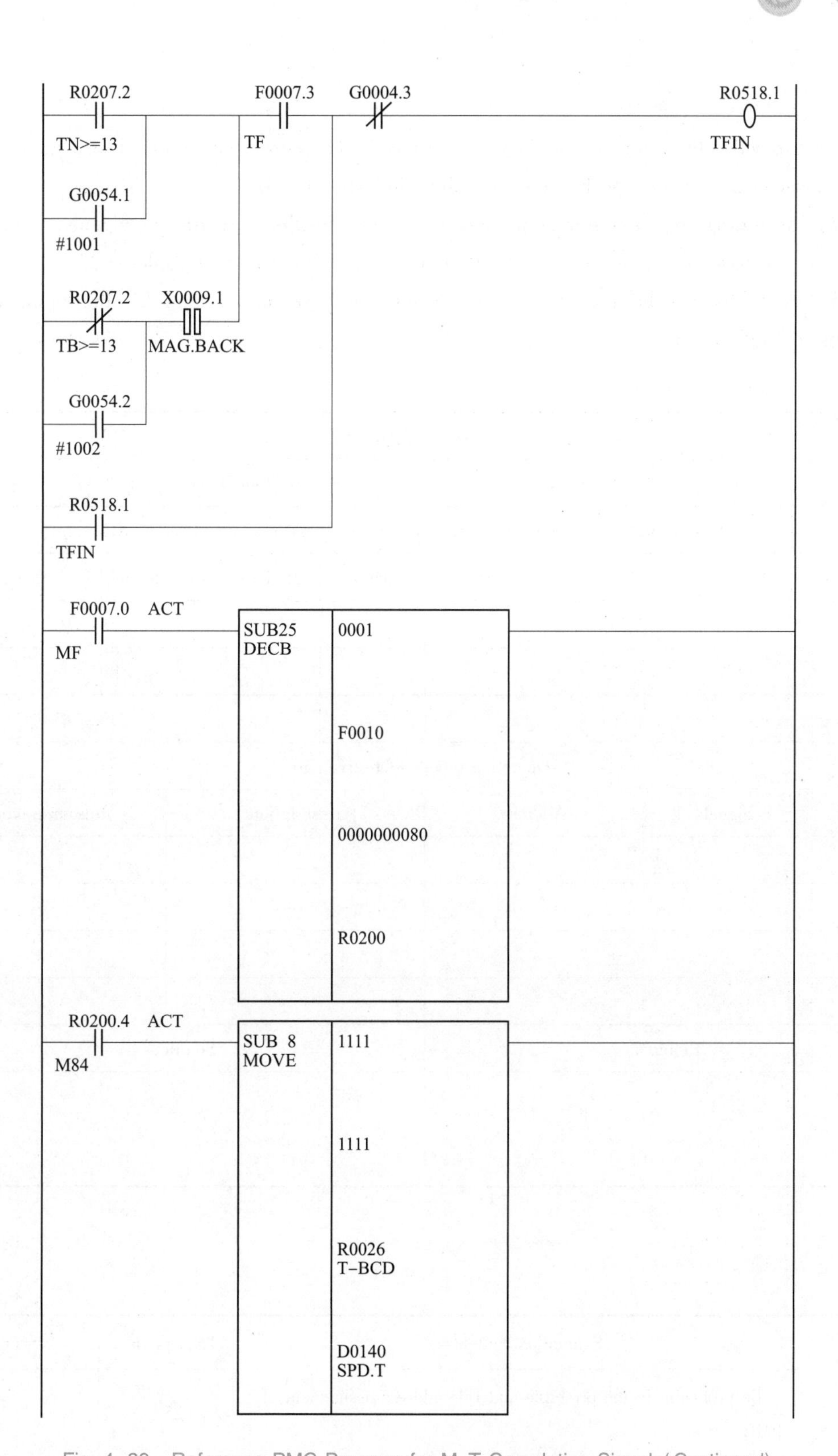

Fig. 4-29 Reference PMC Program for M, T Completion Signals(Continued)

Task Implementation

(1) Subject to the random grouping, divide into observation groups and select a group leader. Fill the personal information of the group members in Table 4-23.

(2) According to the observation and discussion results, record the signals, addresses, description, resource type, and etc. of the PMC program of each part in Table 4-23.

(3) Fully discuss the problems found during the observation, and fill the problems and solutions in Table 4-23.

Table 4-23 Task Work Order for PMC Program of CNC System

<table>
<tr><td colspan="7">Task Work Order</td></tr>
<tr><td colspan="3">Name:</td><td colspan="2">Class:</td><td colspan="2">Student ID:</td></tr>
<tr><td colspan="7">List of materials, tools, measuring instruments, and equipment required</td></tr>
<tr><td>S/N</td><td colspan="2">Name</td><td>Quantity</td><td colspan="2">Model or specification</td><td>Remark</td></tr>
<tr><td></td><td colspan="2"></td><td></td><td colspan="2"></td><td></td></tr>
<tr><td></td><td colspan="2"></td><td></td><td colspan="2"></td><td></td></tr>
<tr><td></td><td colspan="2"></td><td></td><td colspan="2"></td><td></td></tr>
<tr><td colspan="7">Operating process(Observation)</td></tr>
<tr><td>S/N</td><td>Signal</td><td>Address</td><td colspan="2">Description</td><td colspan="2">Resource type</td></tr>
<tr><td></td><td></td><td></td><td colspan="2"></td><td colspan="2"></td></tr>
<tr><td></td><td></td><td></td><td colspan="2"></td><td colspan="2"></td></tr>
<tr><td></td><td></td><td></td><td colspan="2"></td><td colspan="2"></td></tr>
<tr><td></td><td></td><td></td><td colspan="2"></td><td colspan="2"></td></tr>
<tr><td colspan="3">Problems</td><td colspan="4">Solutions</td></tr>
<tr><td colspan="3"></td><td colspan="4"></td></tr>
</table>

Task Evaluation

S/N	Evaluation contents	Total point	Score
1	Be proficient in the principle and I/O address assignment of PMC	25	
2	Master the PMC program structure	25	

Continued

S/N	Evaluation contents	Total point	Score
3	Master the PMC programming for the tool change function of hat-type tool magazines	25	
4	Have a professional spirit of effective communication, expression and solidarity	25	

Task IV Data Transmission and Backup of CNC Systems

Learning Objectives

(1) To master the data backup and recovery of systems.

(2) To master the transmission of FANUC LADDER-III PMC program.

(3) To possess the data transmission and backup capabilities of FANUC CNC systems.

(4) To master the engineering working methods, cultivate a rigorous work style, and abide by the 5S management system.

Task Description

Observe the data transmission and backup functions of FANUC 0i-F CNC systems in groups, and fill the names, main functions, applications and etc. of various data transmission methods in the task work order. Make sure to be careful and pay attention to the relevant operation specifications of the practical training site during observation.

Knowledge Link

I. Data backup and recovery of guidance interface

1. Backup and recovery of static radom access memory data

(1) Functions of static random access memory (SRAM).

SRAM is used to store the machining program, tool compensation, user macro variable, CNC parameters, PMC parameters and pitch error compensation of lead screw, and keep the data related to power failure of machine tool. It is powered by the system battery.

(2) Backup or recovery steps of SRAM data.

1) Insert the memory card when the CNC machine tool is shut down.

2) Click and hold two soft keys on the lower right of the display when the CNC system is

powered on to enter the guidance page of the system. The main menu of startup page is shown in Fig. 4-30.

3) Click [UP] or [DOWN], move the cursor onto "7.SRAM DATA UNILITY", and click [SELECT]. The "SRAM DATA BACKUP" page pops up. The main menu of startup page is shown in Fig. 4-31.

4) Click [UP] or [DOWN] to select function(s).

Backup data through the memory card: SRAM BACKUP.

Restore data in SRAM: RESTORE SRAM.

Restore the data that is automatically backed up: AUTO BKUP RESTORE.

```
SYSTEM MONITOR MAIN MENU

  1. END
  2. USER DATA LOADING
  3. SYSTEM DATA LOADING
  4. SYSTEM DATA CHECK
  5. SYSTEM DATA DELETE
  6. SYSTEM DATA SAVE
  7. SRAM DATA UTILITY
  8. MEMORY CARD FORMAT

* * * MESSAGE * * *
SELECT MENU AND HIT SELECT KEY.

[SELECT] [YES  ] [ NO  ] [ UP  ] [DOWN]
```

Fig. 4-30 Main Menu of Startup Page

```
SRAM DATA BACKUP

1. SRAM BACKUP       ( CNC→MEMORY CARD )
2. RESTORE SRAM      (MEMORY CARD→CNC )
3. AUTO BKUP RESTORE    ( F-ROM→CNC )
4. END

* * * MESSAGE * * *
SELECT MENU AND HIT SELECT KEY.

[SELECT] [ YES ] [ NO ] [ UP ] [DOWN]
```

Fig. 4-31 SRAM Data Backup Page

(3) After backing up or restoring the SRAM data, power off the CNC machine tool and unplug the memory card. The backup data in the SRAM_BAK.001 file is the machine command and is packaged, which cannot be opened on the computer.

2. Backup and recovery of the PMC program

(1) The PMC program of machine tool shall be modified during maintenance. At complex

maintenance sites, however, modifications can be remembered, and sometimes the PMC program may be modified incorrectly due to unfamiliar operations and mistakes. In order to restore the machine tool to the state before maintenance, it is better for us to backup parameters of the machine tool before maintenance. Backup and recovery steps of the PMC program are mainly introduced in this part.

(2) To back up the PMC program, click [UP] or [DOWN], move the cursor onto "6. SYSTEM DATA SAVE", and click [SELECT]. The "SYSTEM DATA SAVE" page pops up, as shown in Fig. 4-32.

(3) Restore the PMC program page is shown in Fig. 4-33.

Fig. 4-32 System Data Save Page

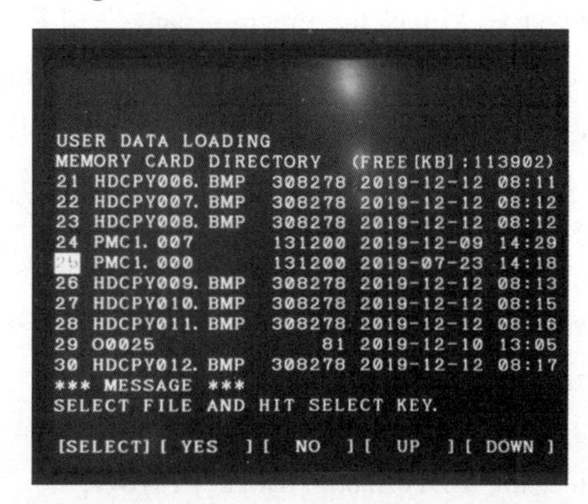

Fig. 4-33 Restore the PMC Program Page

II. Data backup in the work pages

1. Backup of system parameters

Select the MDI mode:

(1) Enable the write protection of system parameters;

(2) Click [SYSTEM]—[PARAMETER] in turn to set the parameter from 20 to 4;

(3) Select the EDIT mode. Click [SYSTEM]—[PARAMETER]—[OPERATE] —[>]—[FOUTPT]—[EXECUTE]—[COVER] in turn, wait the process to complete, and the "CNC-PARA.TXT" file is automatically generated in the PCMCIA card.

2. Backup of PMC parameters

Select the EDIT mode.Click [SYSTEM]—[>] for 3 times—[PMCMNT]—[I/O] in turn—move the cursor onto [MEMORY CARD, WRITE, PARAMETER, FILE NAME]. Click [OPERATE]—[FILE NAME] in turn, the "PMC1_PRM.000" file is automatically generates—[EXECUTE].

3. Backup of machining program (from machine tool to memory card)

Select the EDIT mode. Click [PROG]—[DIRECTORY]—[OPERATE]—[>]—[EQUIPMENT]—[CNCMEM] in turn.

Method 1: input all file names, e.g., O1000 and then click [FOUTPT]—[EXECUTE] in turn.

Method 2: input the file name to be backed up, e.g. O1000 and click [P SET]. Save the file name, e.g., O123 in the memory card, and then click [F SET]—[EXECUTE]. Save the changed file name to the memory card, e.g. change the file name of machine tool from O1000 into O123.

4. Input the machining program

Input the machining program:

Rotate the key switch on the operating panel of machine tool from "0" to "1".

Select the EDIT mode. Click [PROG]—[DIRECTORY]—[OPERATE]—[>]— [EQUIPMENT]—[M-CARD]— [>] in turn.

Method 1: "input the program name"—[F INPT]—[EXECUTE].

Method 2: Click [F INPT], "input the program name in the card"—[F SET]—"input the program name on the machine tool side"—[P SET]—[EXECUTE].

5. Input and output of all machining programs

(1) Output steps of all machining programs.

Select the EDIT mode. Click [PROG]—[DIRECTORY]—[OPERATE]—"input PROGRAM. ALL"—[FOUTPT]—[EXECUTE] in turn.

(2) Input steps of all machining programs.

Select the EDIT mode. Click [PROG]—[DIRECTORY]—[OPERATE]—"input O9999"—[FINPT]—[EXECUTE] in turn.

6. Data input/output in the "ALL I/O" page

Select the EDIT mode. Click the [SYSTEM]—[>] 3 times—[ALL I/O]. Select to enter the page shown in Fig. 4-34, and input/output the machining program, parameter, tool deviation, macro program, pitch error compensation, workpiece coordinate system and other data.

《 螺补 | 工件坐 | | | (操作) | +

Fig. 4-34 "ALL I/O" Page

(1) Output the pitch error compensation.

Select [PITCH ERROR COMPENSATION]—[OPERATE] to enter the parameter input and output page.

Select [FOUTPT]. If you select [EXECUTE] directly, the file name output is the default name PITCH.TXT in the system, as shown in Fig. 4-35.

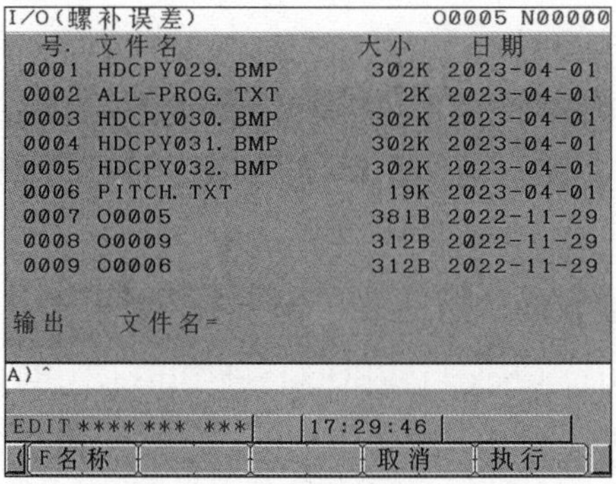

I/O(螺补误差) O0005 N00000

号.	文件名	大小	日期
0001	HDCPY029. BMP	302K	2023-04-01
0002	ALL-PROG. TXT	2K	2023-04-01
0003	HDCPY030. BMP	302K	2023-04-01
0004	HDCPY031. BMP	302K	2023-04-01
0005	HDCPY032. BMP	302K	2023-04-01
0006	PITCH. TXT	19K	2023-04-01
0007	O0005	381B	2022-11-29
0008	O0009	312B	2022-11-29
0009	O0006	312B	2022-11-29

输出 文件名=

A)^

EDIT **** *** *** | 17:29:46

《 F名称 | | | 取消 | 执行 |

Fig. 4-35 Select [FOUTPT]

(2) Input the pitch error compensation by F input mode.

Select the EDIT mode. Click the [SYSTEM]—[>]—[ALL I/O], enter the ALL I/O page. Select [PITCH ERROR COMPENSATION] —[OPERATE]—[F INPT]—“input the file No. 6”—[F SET]—[EXECUTE], as shown in Fig. 4-36.

(3) Input of pitch error compensation by N input mode.

Select the EDIT mode. Click[SYSTEM]—[>] —[ALL I/O], enter the ALL I/O page. Select [PARAMETER]—[OPERATE]—[N INPT]—“input the file name, PITCH.TXT”—[F NAME]—[EXECUTE], as shown in Fig. 4-37.

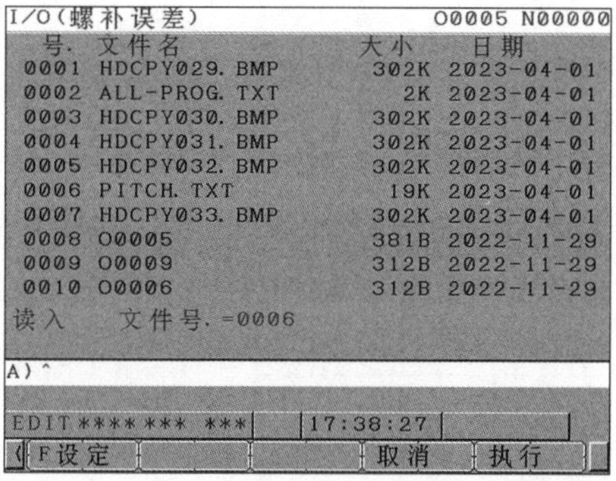

Fig. 4-36 F Input Mode

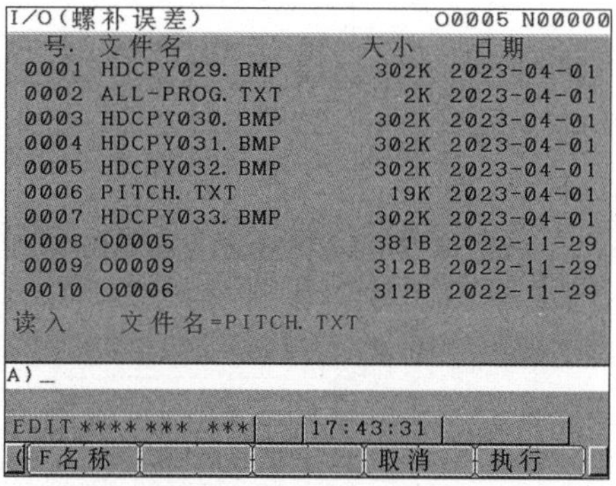

Fig. 4-37 N Input Mode

III Transmission of FANUC LADDER-III PMC program

1. Import PMC program through LADDER-III

The PMC LAD backed up through the memory card is called the PMC program in the format of memory card. As it adopts the machine language format, FAPT LADDER-III is unable to recognize and read it directly and then modify and edit it. Therefore, it must be imported for format conversion.

Program importing steps are listed below.

(1) Run FAPT LADDER-III, and create an empty file with the type the same with the backed-up PMC program in the M-CARD format.

(2) Select [Import] in the [File] menu, and the “Select import file type” dialog box pops up in the system, as shown in Fig. 4-38. Select the [Memory-card Format File] format based on the as prompted, and then click [Next].

(3) The [Specify import file name] dialog box pops up, as shown in Fig. 4-39. Select the file name to be imported in the corresponding path, e.g., D: \ desktop \ textbook \ 1. LAD, Click 【(Complete)】and import the file.

2. Export PMC program through LADDER-III

Similarly, when the PMC program edited on the computer cannot be saved into the memory card directly, it cannot be loaded into CNC system before its format is converted. When the program is exported, its format is converted. The operation steps are as follows.

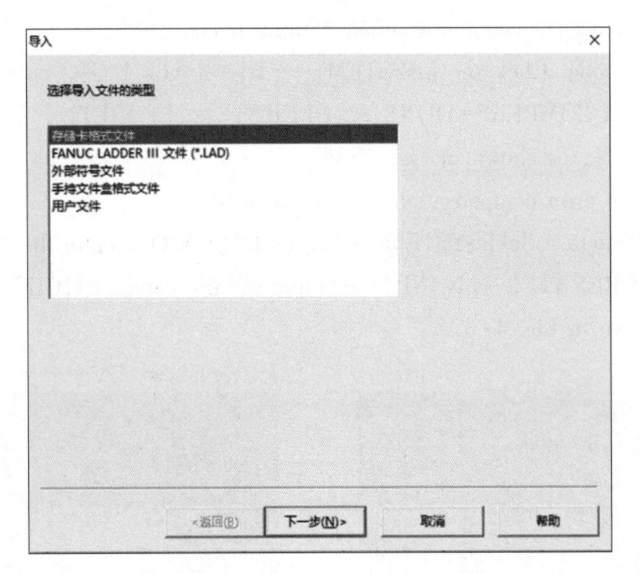

Fig. 4-38 Dialog Box of Selecting Type of File to Be Imported

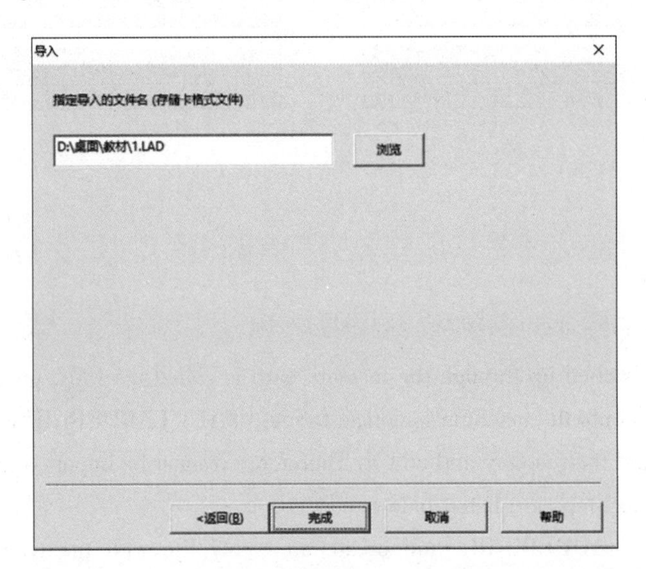

Fig. 4-39 Dialog Box of Specifying Name of File to Be Imported

(1) Open the PMC program with its format to be converted in FAPT LADDER-III.

(2) Select [Compile] in the [Tool] menu to compile the program. If no error is prompted, the program is compiled successfully; If any error is prompted, exit the program, compile it again, and save it.

(3) Select the [Export] option in the [File] menu.

(4) After selection of [Export], the [Select export file type] dialog box pops up, and the software prompts to select the type of file to be output. Select the [Memory-card Format File] format, and then click [Next].

(5) The [Specify export file name] dialog box pops up. Select the file path, input the name of file to be exported, and click [Save]. The [Specify export file name] dialog box pops up again, as

shown in Fig. 4-40. Click [Finish], and click [OK] after the [Export completed] information box pops up to complete the whole export process.

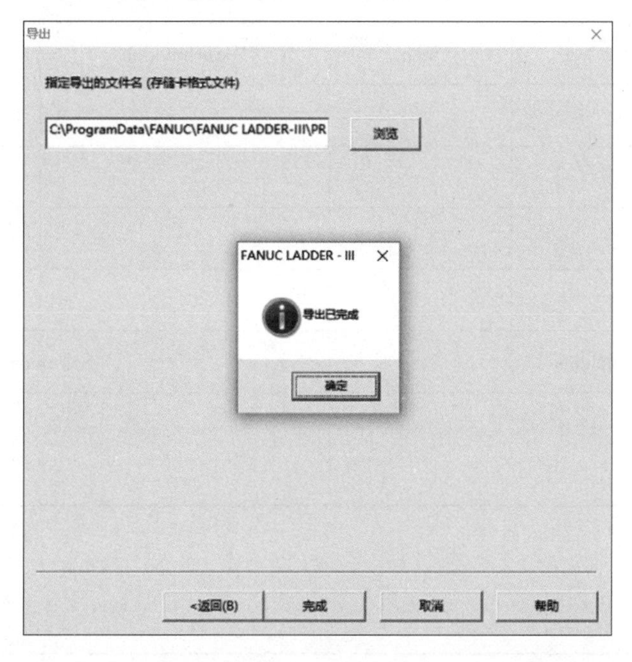

Fig. 4-40 Dialog Box of Specifying Name of File to Be Exported

Task Implementation

(1) Divide students randomly into groups and appoint the group leaders. Fill the personal information of the group members in Table 4-24.

(2) According to the observation and discussion results, record the names, steps main functions, applications and etc in Table 4-24.

(3) Fully discuss the problems found during the observation, and fill the problems and solutions in Table 4-24.

Table 4-24 Task Work Order for Data Transmission and Backup of CNC System

Task Work Order				
Name:	Class:		Student ID:	
List of materials, tools, measuring instruments, and equipment required				
S/N	Name	Quantity	Model or specification	Remark

Continued

Operating process(Observation)				
S/N	Name	Step	Main function	Application
Problems			Solutions	

Task Evaluation

S/N	Evaluation contents	Total point	Score
1	Be proficient in various system data backup and recovery methods	25	
2	Transmit the data through the FANUC LADDER-III PMC program	25	
3	Master the data transmission and backup capabilities of the FANUC CNC system	25	
4	Have a professional spirit of effective communication, expression and solidarity	25	

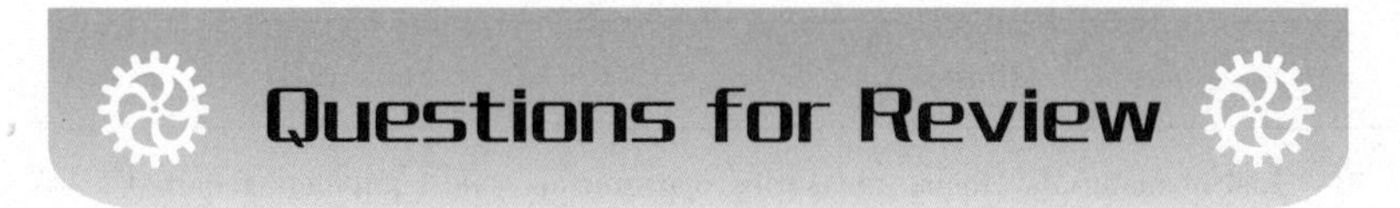

Questions for Review

(1) Please briefly describe definitions of servo interfaces of FANUC βi-B.

(2) Please briefly describe connecting definitions of interfaces of the FANUC 0i-F CNC system mainboard.

(3) Please briefly describe the expression of FANUC 0i MF CNC system parameters.

(4) Please briefly describe the initialization steps of system parameters.

(5) Please briefly describe the information and editing requirements of the level-1 PMC program.

(6) Please briefly describe the tool change process of the hat-type tool magazine in the vertical machining center.

(7) Please briefly describe the PMC program in the guidance page as well as the SRAM data backup and recovery steps.

(8) Please briefly describe the that data can be input and output in the ALL I/O page.

(9) Please briefly describe FANUC LADDER-Ⅲ PMC program transmission steps.

Module V Fault Diagnosis and Maintenance of the Servo System

If the CNC device serves as the "brain" of CNC machine tool and the command organization that sends "commands", the servo system functions as "four limbs" of the CNC device and the executing organization, which executes the motion commands sent by the CNC device faithfully and accurately.

Servo system is also called the "follow-up system", which specifically refers to the feedback control system with the mechanical displacement (or rotation angle), speed or acceleration as the controlled variable (output variable of system). It is used to accurately track the input displacement (or rotation angle) based on the output mechanical displacement (or rotation angle). The servo system is an important component of the CNC machine tool, which is generally the servo feed system of each moving coordinate axis. The servo system consists of a servo motor, a servo drive device, a position detection device, etc. It links the CNC system and the mechanical transmission parts of the machine tool. In the CNC machine tool, the servo system receives the feed pulse signal from the CNC device. The servo drive device amplifies the voltage and power. The servo motor drives the moving parts of the machine tool to move, and ensures the moving speed and accuracy. It converts the position command generated by the interpolation operation of the CNC system into the displacement of moving parts of machine tool accurately, which directly reflects the performance of machine tool coordinate axis tracking motion command and actual positioning. The excellent performance of the servo system largely determines the high efficiency and precision of the CNC machine tool. The servo system is thus an important part of the CNC machine tool. It includes mechanical transmission, electrical drive, detection, automatic control and other information, and involves strong current and weak current control.

The servo system of CNC machine tool includes the feed servo system and spindle servo (drive) system. The feed servo system is an automatic control system that takes mechanical displacement (position control) as the direct control target to ensure machining contour. The spindle servo system is mainly based on speed control, which generally only meets the functions of spindle speed regulation and forward and reverse rotation, and provides the torque and power required in the cutting process. When the machine tool shall be provided with the thread machining, exact stop, machining at the constant linear speed and other functions, the corresponding position control requirements shall be imposed on the spindle.

This module mainly introduces the servo systems of CNC machine tools, the common faults diagnosis and the maintenance servo spindle drive system and feed servo system of the CNC machine tools.

Task I Servo Systems of CNC Machine Tools

Learning Objectives

(1) To understand the classification of the servos systems of CNC machine tools.

(2) To understand the composition of the servo systems of CNC machine tools.

(3) To understand the requirements of servo systems of CNC machine tools.

(4) To master the basic principle of position control in the feed servo systems of CNC machine tools.

(5) To understand the drive device that controls the speed of feed servo system of CNC machine tools.

(6) To possess the ability to interpret the working principle of servo system control.

(7) To master the engineering working methods, cultivate a rigorous work style, and abide by the 5S management system.

Task Description

Observe the servo system of the YL569 CNC milling machine in groups, and fill the names, classification and composition, applications and etc of the servo systems of the CNC machine tools in the task work order. Make sure to be careful and pay attention to the relevant operation specifications of the practical training site during observation.

Knowledge Link

I. Classification and composition of servo systems

There are many ways to classify servo systems. servo systems are generally classified according to the regulation control mode.

1. Classification by regulation control mode

Servo systems can be divided into an open-loop system, closed-loop system and semi-closed-loop system according to the regulation control mode.

(1) Open-loop servo systems.

A stepping motor is often used in open-loop systems, which is directly controlled after the feed

pulse is power-amplified. The speed or position detection device and feedback signals are not available on the stepping motor shaft or workbench. The speed of the stepping motor is controlled by the frequency of pulses output by the CNC device, and the position of workbench is controlled by the number of output pulses, An open-loop servo system is shown in Fig. 5-1.

The open-loop servo system is simply structured, which runs stably, controls easily, costs low, and is also easy to use and maintain. However, its accuracy is not high, operation at low speed is unstable, and the torque when operating at high speed is small. It is generally used in economical CNC machine tools and general machine tool transformation.

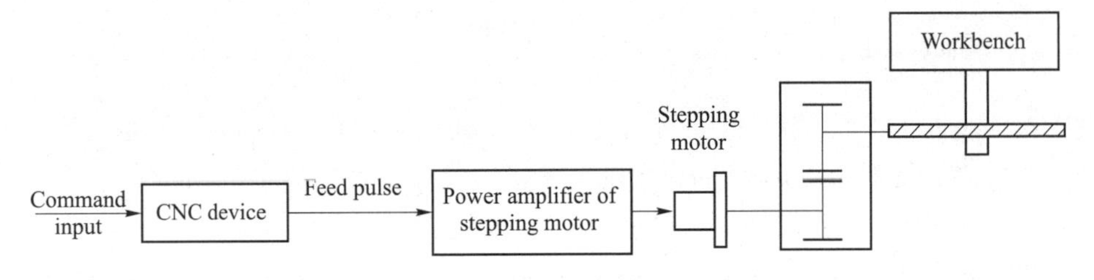

Fig. 5-1 An Open-Loop Servo System

(2) Closed-loop servo systems.

The DC or AC servo motors are often used in the closed-loop and semi-closed-loop servo systems. The speed and position detection components obtain feedback signals of the controlled objects, e.g., speed and position of workbench, and adjust the speed and position of servo motor based on them.

The position detection device of a closed-loop servo system is directly installed on the moving part of machine tools. For example, the linear position detection device is installed on the workbench, which feeds back the actual position detected into the CNC system, compares and calculates the positions, and generates control outputs, A closed-loop servo system is shown in Fig. 5-2.

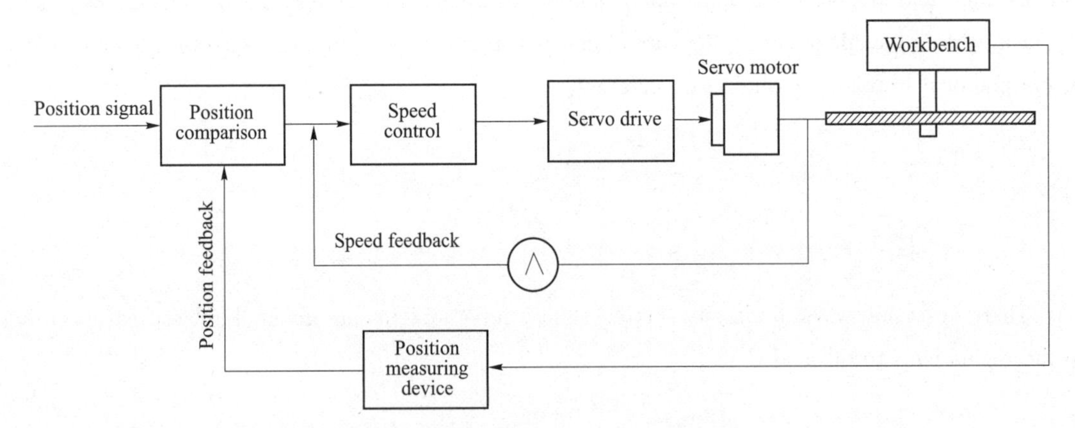

Fig. 5-2 A Closed-Loop Servo System

As all errors of the transmission chain are included during the position detection of a closed-loop servo system, high control accuracy can be achieved. However, it cannot be considered that the closed-loop servo system may impose the less strict requirements on the transmission mechanism.

That is because the transmission mechanism affects the dynamic characteristics of the system, which makes it difficult in commissioning and stability, causing the decrease of the position gain during the adjustment of the loop, and adversely influencing the following error and contouring error. Therefore, when the closed-loop mode is adopted, the rigidity of the machine tool must be increased, the friction characteristics of the sliding surface is improved and the transmission clearance is reduced to increase the position gain. A closed-loop servo system is mainly applied to large CNC machine tools with high precision requirements.

(3) Semi-closed-loop servo systems.

The position detection device of a semi-closed-loop servo system is generally installed on the motor shaft or at the ball screw shaft end. The linear displacement of moving part is measured indirectly based on the measured angular displacement, fed back into the CNC system, and compared with the position command in the system. The movement of moving part is controlled based on the compared difference, and the moving part stops moving until the difference is eliminated. The composition of a semi-closed-loop servo system is shown in Fig. 5-3.

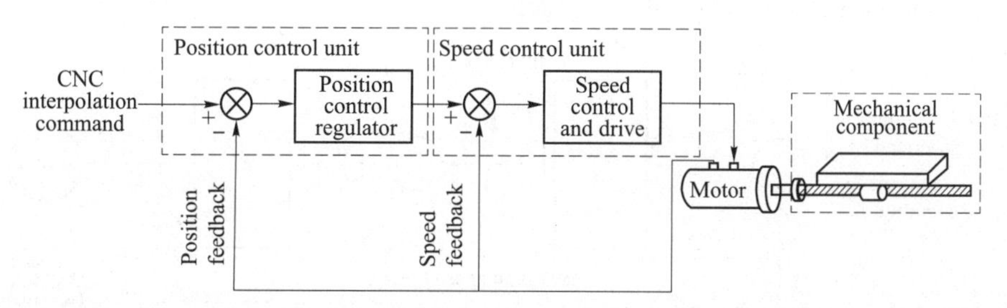

Fig. 5-3 A Semi-Closed-Loop Servo System

As not all errors of the ball screw nut pair and guide rail pair of the feed transmission chain in the semi-closed-loop servo system are contained in the position feedback, errors of the transmission mechanism still affect the position accuracy of the moving part. However, as the number of unstabilizing factors reduces during the feedback process, the system is easy to achieve a high position gain without oscillation, and thus has a good rapidity and high dynamic accuracy, being widely applied at present. In terms of errors of the transmission chain, e.g., cumulative errors of the reverse clearance and lead screw pitch can be compensated through parameter settings of the CNC system to improve the positioning accuracy of machine tools.

II. Position control

Position control and speed control are the important links of the feed servo systems of CNC machine tools. The position deviation is generated when the position command obtained by the position control device through the interpolation operation of computer is compared with the actual position of machine tool coordinate axis fed back by the position detection device, which is then converted into the voltage at a given speed. The speed control device controls the servo motor based on the speed voltage signals output by the position control device and the actual rotating speed fed back by the speed detection device, so as to drive the transmission part of machine tools. As the

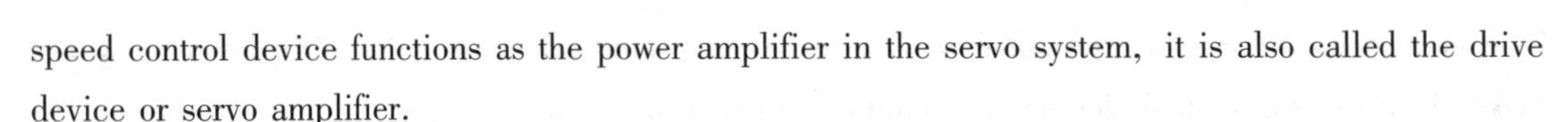

speed control device functions as the power amplifier in the servo system, it is also called the drive device or servo amplifier.

Basic principle of position control is shown below.

Position control is a closed or semi-closed-loop system. The position of the moving part can be detected by the detection component and sent to a computer for comparison. Therefore, the positioning accuracy of the moving part of the machine tool is high.

The schematic diagram of a position control system is shown in Fig.5-4. The system is mainly composed of the position control, speed control and position detection. The position control device is used to compare the instantaneous position command value P_s obtained through interpolation operation with the actual position detected, generate the position deviation ΔP, and convert it into the instantaneous speed command voltage U_{sn}. The speed control device is used to compare the instantaneous speed command voltage U_{sn} with the detected speed voltage U_{fn}, and amplify them into the armature voltage that drives the servo motor. The position detection device is used to convert the signal detected by the position detection component into the digital quantity P_f with the same command position magnitude for use in the position control.

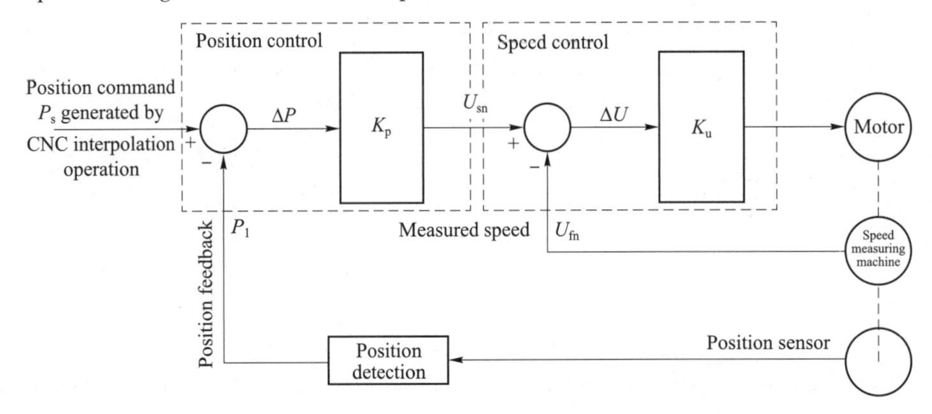

Fig. 5-4 Schematic Diagram of a Position Control System

III. Speed control

The speed control device is also called the drive device. The drive device in the CNC machine tool varies with the drive motor. The drive device of AC servo motor features other-controlled frequency conversion and self-controlled frequency conversion. The drive device of DC spindle motor is controlled by the AC thyristor, while the AC spindle motor is controlled by the universal frequency conversion and vector. Power devices commonly used in the drive device include the high-power transistor, power field effect transistor, insulated gate transistor, ordinary thyristor and disconnectable thyristor, which are used to amplify the power of control signals in the drive device.

The three-phase AC permanent magnet synchronous motor is mostly used as AC servo motor for feeding. The variable frequency speed regulation is often adopted, and the variable frequency speed regulation system can be divided into two categories: other-controlled variable frequency and self-controlled variable frequency. The other-controlled variable frequency speed regulation system uses an independent variable frequency device to provide variable voltage and frequency power supply to

the motor, while the self-controlled variable frequency speed regulation system uses a rotor position detector on the motor shaft to control frequency conversion.

(1) Frequency modulation principle.

The frequency of industrial electricity is fixed at 50 Hz, and the method of frequency conversion is used to change the motor power supply frequency. There are two common methods: direct AC–AC frequency conversion and indirect AC–DC–AC frequency conversion. The AC–AC frequency conversion uses a thyristor rectifier to directly convert the power frequency AC into the pulsating AC with a lower frequency, with the positive group outputting a positive voltage pulse, and the negative group outputting a negative voltage pulse, and the fundamental voltage of this pulsating AC is the desired frequency conversion voltage. However, the AC current obtained by this method fluctuates greatly.

The AC–DC–AC frequency conversion (Fig. 5–5) first rectifies AC into DC, then turns the DC voltage into a rectangular pulse wave voltage, and the fundamental wave of this rectangular pulse wave is the desired frequency conversion voltage. Due to the small fluctuation of AC current, wide frequency modulation range and good adjustment linearity, the AC–DC–AC frequency conversion is often used on CNC machine tools. In the AC – DC – AC frequency conversion, there are the intermediate DC voltage adjustable PWM inverter and intermediate DC voltage fixed PWM inverter. According to the large capacitance or large inductance of the energy storage component on the intermediate DC circuit, there are the voltage type PWM inverter and current type PWM inverter. A fixed voltage PWM inverter is a typical AC–DC–AC inverter.

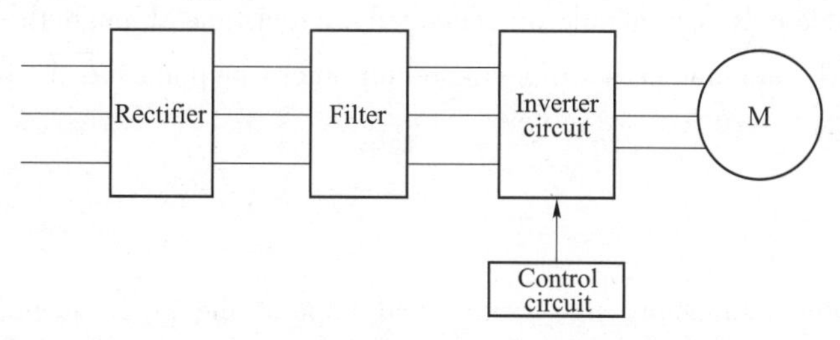

Fig. 5–5 Composition of AC–DC–AC Inverter

(2) SPWM principle.

Sine pulse width modulation is one of the PWM methods. The SPWM frequency converter is not only suitable for AC permanent magnet servo motors, but also for AC induction servo motors. SPWM adopts the principle of sine regular pulse width modulation, and is characterized by a high power factor and good output waveform. Therefore, it is widely used in AC speed regulation systems.

The analog single-phase SPWM principle is shown in Fig. 5–6. In SPWM, the output voltage is obtained by a sine voltage modulated by a triangular carrier. The output voltage U_0 of the SPWM is a square wave signal with equal amplitude and different widths. The area of each pulse is proportional to the area under the sine wave, so the pulse width is basically sinusoidal, and the fundamental wave is the equivalent sine wave. The output pulse signal is used as the phase voltage (current) of AC servo motor after power amplification, and the frequency of motor phase voltage (current) can be changed by changing the frequency of sine fundamental wave to achieve the purpose of frequency modulation and speed regulation.

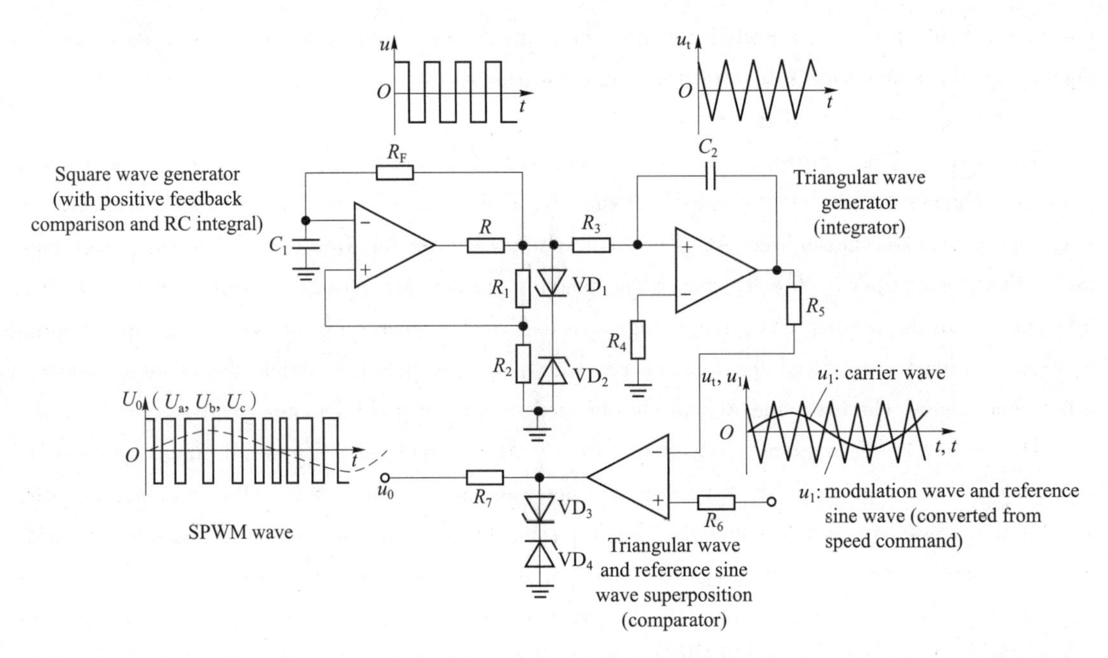

Fig. 5-6 Single-Phase SPWM Principle

In addition to the analog SPWM method, digital control SPWM is being used more and more widely, and common methods are as follows: ① The microcomputer stores the pre-calculated SPWM data, which is called out by commands or generated in real-time through the system; ② Special integrated chip; ③ The microprocessor of a single-chip microcomputer has the function of generating SPWM signals directly.

Task Implementation

(1) Divide students randomly into groups and appoint the group leaders. Fill the personal information of the qroup members in Table 5-1.

(2) According to the observation and discussion results, record the names, classification and composition, main functions, applications and etc. of servo system of the CNC machine tool in Table 5-1.

(3) Fully discuss the problems found during the observation, and fill the problems and solutions in Table 5-1.

Table 5-1 Task Work Order for Servo System of the CNC Machine Tool

Task Work Order				
Name:	Class:		Student ID:	
List of materials, tools, measuring instruments, and equipment required				
S/N	Name	Quantity	Model or specification	Remark

Continued

Operating process(Observation)				
S/N	Name	Classification and composition	Main function	Application
Problems			Solutions	

Task Evaluation

S/N	Evaluation contents	Total point	Score
1	Be proficient in the classification and composition of the CNC machine tool servo systems	25	
2	Can master the basic principles of position control in the feed servo systems of CNC machine tools	25	
3	Can interpret the control working principles of servo systems	25	
4	Have a professional spirit of effective communication, expression and solidarity	25	

Task II Common Fault Diagnosis and Maintenance of Spindle Drive Systems

Learning Objectives

(1) To master the operation of the FANUC CNC system spindle monitoring page.

(2) To understand the meaning of serial spindle alarm messages of FANUC CNC systems.

(3) To master the fundamentals of maintenance of FANUC spindle drive systems.

(4) To master the diagnosis numbers and their meanings of the spindle in the system diagnosis page.

(5) To be able to independently analyze and troubleshoot spindle drive system faults.

(6) To master the engineering working methods, cultivate a rigorous work style, and abide by the 5S management system.

Task Description

Observe the FANUC CNC system spindle diagnosis and maintenance page of the CNC machine tool in groups, and fill the names, signal symbols, main functions, applications, and etc. of each message on the FANUC CNC system spindle diagnosis and maintenance page in the task work order. Make sure to be careful and pay attention to the relevant operation specifications of the practical training site during observation.

Knowledge Link

I. Spindle diagnosis and maintenance pages of FANUC CNC systems

1. FANUC serial spindle provides maintenance and repair means

The FANUC serial spindle provides various maintenance and repair means for convenience. Diagnostic messages related to the spindle has been provided form the start of the system diagnosis 400; The LED seven-segment display of the spindle amplifier also shows the operating state. The spindle monitoring page is shown in Fig. 5-7.

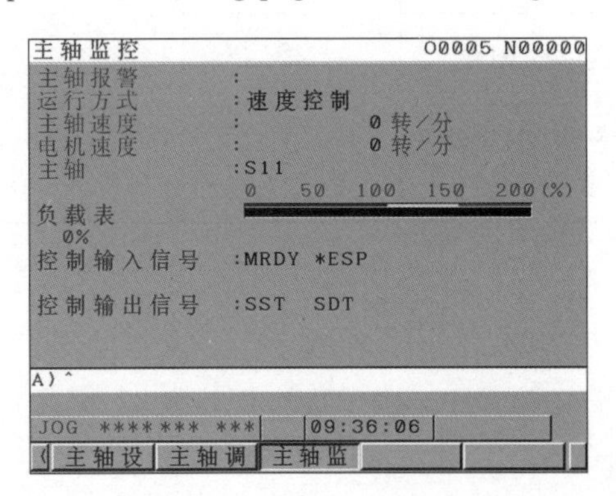

Fig. 5-7 Spindle Monitoring Page

In the page shown in Fig. 5-7, you can select the spindle setting page, spindle adjustment page, and spindle monitoring page. The spindle monitoring page provides rich maintenance and repair messages, which offer great convenience. Modern CNC machine tools should make full use of the abundant messages provided by the CNC system for fault diagnosis and maintenance.

2. FANUC spindle monitoring page

There are monitoring messages on the FANUC spindle monitoring page, as shown in Fig. 5-7. Different operation modes have different parameter adjustments and monitoring contents.

(1) Spindle alarm.

The "Spindle Alarm" message bar provides the spindle alarm messages of the spindle and spindle motor displayed immediately when the spindle alarms. There are 63 different spindle alarm messages, and some are listed in Table 5-2.

Table 5-2 List of Some Spindle Alarm Messages

Alarm No.: Alarm message	Alarm No.: Alarm message	Alarm No.: Alarm message
1: Motor overheating	29: Transient overload	61: Half-side and full-side position return error alarm
2: Excessive speed deviation	30: Input circuit overcurrent	65: Abnormal movement amount of magnetic pole determination action
3: Blown DC link fuse	31: Limited motor	66: Alarm of communication between spindle amplifiers
4: Blown input fuse	32: Abnormal RAM for transmission	72: Inconsistent motor speed judgment
6: Temperature sensor disconnection	33: Abnormal DC link charging	73: Motor sensor disconnection
7: Overspeed	34: Abnormal parameter setting	80: Abnormal next spindle amplifier for communication
9: Main circuit overload	41: One-rotation signal detection error of position encoder	82: No detection of one-rotation signal of motor sensor
11: DC link overvoltage	42: No detection of one-rotation signal of position encoder	83: Abnormal signal of motor sensor
12: DC link overcurrent	43: Signal disconnection of position encoder for differential control	84: Spindle sensor disconnection
15: Output switching alarm	46: One-rotation signal detection error of position sensor for thread cutting	85: One-rotation signal detection error of spindle sensor
16: Abnormal RAM	47: Abnormal signal of position encoder	87: Abnormal signal of spindle sensor
19: Excessive U phase current bias	51: Frequency converter DC link overvoltage	110: Abnormal communications between amplifiers
20: Excessive V phase current bias	52: Abnormal ITP signal I	111: Low voltage of frequency converter control power supply
21: Position sensor polarity setting error	56: Internal cooling fan stopped	112: Excessive regenerative current of frequency converter
24: Abnormal or stopped data transmission	57: Excessive deceleration power of frequency converter	120: Communication data alarm
27: Position encoder disconnection	58: Overload of the main circuit of the frequency converter	137: Abnormal equipment communication

During repair, you can easily and intuitively understand relevant fault diagnosis message, such as feedback from the spindle amplifier, spindle motor, and spindle sensor, through the spindle

parameter adjustment monitoring page. It is necessary to make full use of the fault diagnosis messages provided on the spindle monitoring page.

(2) Operation mode.

The "Operation Mode" message bar provides the operation mode of the current spindle. The FANUC spindle has rich and flexible operation modes, mainly including ①speed control; ②spindle orientation; ③synchronous control; ④rigid tapping; ⑤spindle CS contour control; ⑥spindle positioning control (T series). Not every spindle has six operation modes, which mainly depends on whether the machine tool manufacturer conducts the second development of operation modes required by users. In addition, for some operation modes, the CNC system shall have corresponding software options, and the spindle motor shall have hardware to realize functions.

(3) Spindle control input signal.

When programming PMC to enable relevant functions of the spindle, the logical processing results are often output to the G address of PMC to finally realize the spindle functions. For example, to make the 1st spindle rotate forward, it is necessary to write a machining program containing M03 and output it to G70.5 through LAD logic processing. FANUC specifies that the symbol representing the G70.5 address is SFRA, i.e. as long as the 1st spindle is rotating forward, "SFR" is displayed in the "Control Input Signal" bar, and "S1" is displayed in the "Spindle" bar. The list of common spindle control input signals is shown in Table 5-3.

Table 5-3 List of Common Spindle Control Input Signals

Signal symbol: signal meaning	Signal symbol: signal meaning
TLML: torque limit signal (low)	* ESP: emergency stop (negative logic) signal
TLMH: torque limit signal (high)	SOCN: soft start/stop signal
CTH1: gear signal 1	RSL: output switching request signal
CTH2: gear signal 2	RCH: confirmation signal of power line state
SRV: spindle reverse rotating signal	INDX: oriented stop position change signal
SFR: spindle forward rotating signal	ROTA: rotation direction signal of oriented stop position
ORCM: spindle orientation signal	NRRO: oriented stop position shortcut signal
MRDY: mechanical ready signal	INTG: speed integral control signal
ARST: alarm reset signal	DEFM: differential mode command signal

If a certain function of the spindle is not realized, check whether there is signal display in the "Control Input Signal" bar on the spindle monitoring page as shown in Fig. 5-7. If yes, it is not necessary to analyze the program in LAD. If a signal for realizing functions is not displayed, the logical relationship must be analyzed with the help of LAD.

(4) Spindle control output signal.

The ideas for learning the spindle control output signal are similar to those of the spindle control input signal. When the spindle control is in a certain state, CNC outputs relevant state to the F

storage area of PMC, so that maintenance personnel can intuitively understand the current control state of the spindle. For example, when the speed of the 1st spindle reaches the operating speed, CNC outputs a speed arrival signal, and the signal address is F45.3, with the symbol SARA defined by FANUC. At this point, "SAR" is shown in the "Control Output Signal" bar. The list of common spindle control output signals in the "Spindle" bar is shown in Table 5-4.

Table 5-4 List of Common Spindle Control Output Signals

Signal symbol: signal meaning	Signal symbol: signal meaning
ALM: alarm signal	LDT2: load detection signal 2
SST: speed zero signal	TLM5: torque limit signal
SDT: speed detection signal	ORAR: orientation ending signal
SAR: speed arrival signal	SRCHP: output switching signal
LDT1: load detection signal 1	RCFN: output switching ending signal

When repairing the spindle, you can learn the current operating state of the spindle from the "control output signal" bar on the spindle monitoring page.

3. FANUC spindle maintenance page

To facilitate the maintenance and replacement of spare parts by users, the FANUC CNC system automatically reads out and records ID information from the connected spindle amplifier when the spindle amplifier is started for the first time. Due to the different spindle motors connected to the spindle amplifier, the spindle motor information is not automatically read out. The spindle information page is shown in Fig. 5-8.

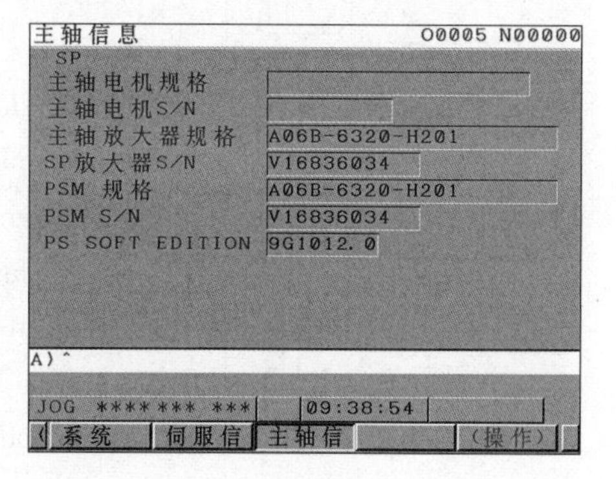

Fig. 5-8 Spindle Information page

It can be seen from Fig. 5-8 that the serial numbers in "Spindle Amplifier Specification" and "PSM Specification" are the order number of each component ordered from FANUC, which is convenient for maintenance personnel to order spare parts. After the CNC machine tool is started, if the stored spindle amplifier and other information are inconsistent with the real object, the difference shall be marked with " * ".

II. Fault diagnosis and maintenance of FANUC serial spindle drive systems

1. FANUC spindle drive systems

The FANUC spindle drive system is divided into the serial spindle drive system and the analog spindle drive system.

(1) Serial spindle drive system.

The FANUC spindle amplifier and FANUC spindle motor are selected for the serial spindle drive system. The CNC and spindle amplifier exchange data through a serial bus.

(2) Analog spindle drive system.

In the analog spindle drive system, the FANUC CNC system outputs $0\sim\pm10$ V analog voltage and inputs it into a spindle amplifier with an analog interface to control the spindle motor speed. At present, the spindle amplifiers of FANUC are equipped with serial interfaces, and there is no spindle amplifier with an analog interface.

2. Spindle alarm and error code display on spindle amplifier

The FANUC CNC system not only provides more than 130 alarm messages involving serial spindle alarms on the CNC but also provides alarm and error messages on the spindle amplifier. On the βi series spindle amplifier, FANUC uses the LED seven-segment display and indicators (Fig. 5-9) to reflect the operation and fault state of the spindle amplifier and spindle motor. When repairing the spindle amplifier and spindle motor, make full use of the alarm and error messages provided on the system and spindle amplifier for fault diagnosis.

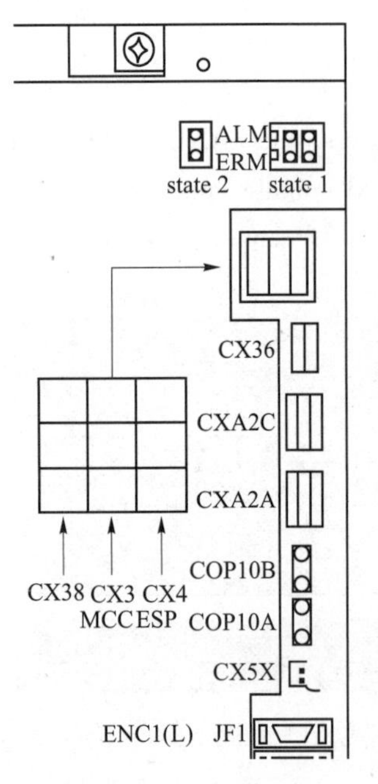

Fig. 5-9 Seven-Segment Display and Indicators on βi Spindle Amplifier Body

Spindle alarm and error messages display on βi spindle amplifier are introduced below.

Although the βi spindle amplifier and the βi servo amplifier are integrally designed, alarms and error messages related to the spindle are separated from the servo amplifier. As shown in Fig. 5-9, "State 1" is used to display spindle alarms and error messages, and "State 2" is used to display servo amplifier alarms and error messages. Like the αi spindle amplifier, the βi spindle amplifier also has LED seven-segment display and state indicators that display spindle alarms and error messages.

(1) Alarm messages display on βi spindle amplifier.

When the βi spindle amplifier gives an alarm, the red indicator next to "State 1" is on, and two LED seven-segment displays show the alarm code. The maintenance method is the same as that of the αi spindle amplifier module. The spindle alarm code and error message on the display screen are basically the same as those of the αi spindle amplifier module.

(2) Error messages display on βi spindle amplifier.

When there is an error message on the βi spindle amplifier, the yellow indicator next to "State 1" is on, and two LED seven-segment displays prompt the error message. The main contents still include parameter setting errors or improper logical sequence of CNC control. The maintenance method is the same as that of the error message prompt of the αi spindle amplifier module. The spindle alarm messages and the error messages on the display screen

are basically the same as those of the αi spindle amplifier module.

3. Spindle adjustment and monitoring pages

The spindle adjustment and monitoring pages provides maintenance diagnosis information, as shown in Fig. 5-10 and Fig. 5-7. During maintenance, fault diagnosis and analysis can be carried out in combination with the maintenance diagnosis information on the spindle adjustment and monitoring pages.

4. The system diagnosis pages provides spindle maintenance information

In addition to providing more than 130 spindle-related alarm information and spindle monitoring pages, the FANUC system also provides spindle fault diagnosis information on the system diagnosis page. The relevant diagnosis numbers on the system diagnosis page specifically indicate the actual operating data of the spindle, including spindle speed, spindle load display, rigid tapping position command and error, online spindle position, and encoder information, as well as error and alarm message and causes.

The system diagnosis interface involving spindle alarms (starting from the system diagnosis number 400) is shown in Fig. 5-11. During maintenance, when the spindle has a fault or error, the fault cause can be judged according to the specific diagnosis information provided by CNC with reference to the diagnosis page.

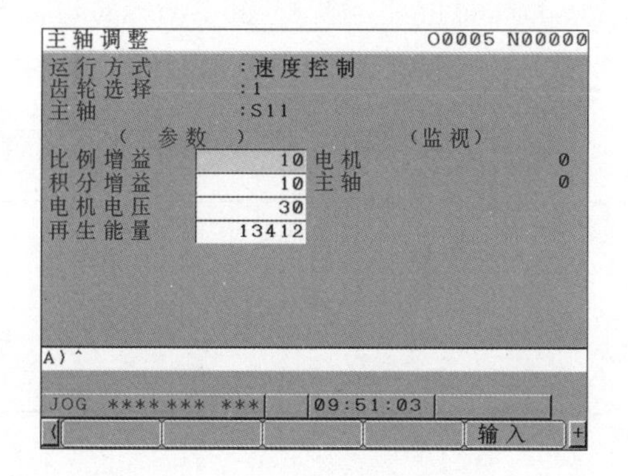

Fig. 5-10 Spindle Adjustment page

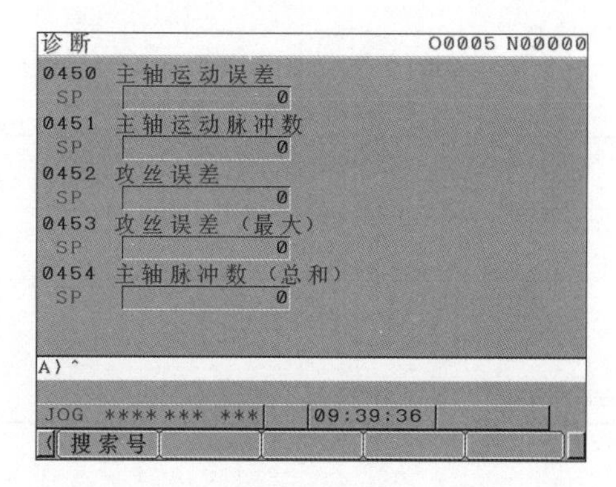

Fig. 5-11 Spindle System Diagnosis page

Specific steps: In the MDI mode, click [SYSTEM] several times, click [Diagnosis], enter "400", click [Search] to display the system 400 diagnosis pages, and then check other diagnosis numbers in turn.

Task Implementation

(1) Divide students randomly into groups and appoint the group leaders. Fill the personal information of the group members in Table 5-5.

(2) According to the observation and discussion results, record the names, signal symbols, main functions, application and etc of various messages on the spindle diagnosis and maintenance page of the FANUC CNC system in Table 5-5.

(3) Fully discuss the problems found during observation, and fill the problems and solutions in Table 5-5.

Table 5-5 Task Work Order for Observation of Spindle Diagnosis and Maintenance pages of FANUC CNC System

<table>
<tr><td colspan="5">Task Work Order</td></tr>
<tr><td colspan="2">Name:</td><td colspan="2">Class:</td><td>Student ID:</td></tr>
<tr><td colspan="5">List of materials, tools, measuring instruments, and equipment required</td></tr>
<tr><td>S/N</td><td>Name</td><td>Quantity</td><td>Model or specification</td><td>Remarks</td></tr>
<tr><td></td><td></td><td></td><td></td><td></td></tr>
<tr><td></td><td></td><td></td><td></td><td></td></tr>
<tr><td></td><td></td><td></td><td></td><td></td></tr>
<tr><td colspan="5">Operating process(Observation)</td></tr>
<tr><td>S/N</td><td>Name</td><td>Signal symbol</td><td>Main function</td><td>Application</td></tr>
<tr><td></td><td></td><td></td><td></td><td></td></tr>
<tr><td></td><td></td><td></td><td></td><td></td></tr>
<tr><td></td><td></td><td></td><td></td><td></td></tr>
<tr><td></td><td></td><td></td><td></td><td></td></tr>
<tr><td colspan="3">Problems</td><td colspan="2">Solutions</td></tr>
<tr><td colspan="3"></td><td colspan="2"></td></tr>
</table>

Task Evaluation

S/N	Evaluation content	Total point	Score
1	Be proficient in the operation of the FANUC CNC system spindle monitoring pages	25	
2	Can master the fundamentals of maintenance of FANUC spindle drive systems	25	
3	Be able to independently analyze and troubleshoot spindle drive system faults	25	
4	Have a professional spirit of effective communication, expression and solidarity	25	

Task III Common Fault Diagnosis and Maintenance of Feed Servo Systems

Learning Objectives

(1) To understand the operation of the servo adjustment and diagnosis pages of CNC system.

(2) To understand the meanings in the servo parameter adjustment page.

(3) To master the adjustment of common servo parameters;

(4) To understand the symptoms of servo system faults.

(5) To master the types and nodes of servo amplifier faults.

(6) To be able to independently analyze and troubleshoot feed servo system faults.

(7) To master the engineering working methods, cultivate a rigorous work style, and abide by the 5S management system.

Task Description

Observe the CNC system servo adjustment and diagnosis pages in groups, and fill in the names, numbers, meanings, main functions of each parameter and etc. on the system servo adjustment and diagnosis pages in the task work order. Make sure to be careful and pay attention to the relevant operation specifications of the practical training site during observation.

Knowledge Link

I. Servo parameter adjustment and diagnosis pages

1. Significance of servo parameter adjustment

After the CNC system is powered on properly through the connection of hardware, servo parameter initialization and optical cable initialization must be conducted. However, there are still many parameters that need to be adjusted to make sure that the machine tool truly meets users' normal operation requirements and usage requirements. For example, whether the position loop gain and the servo response time meet the operation needs of the machine tool, whether there is vibration or creeping during the operation of the machine tool, and whether the operation speed of the machine tool is satisfactory, etc. When machining workpieces, the requirements for the operating speed of the machine tool and each axis are different, so servo parameters need to be adjusted. Due to the different feed speeds of servo axes, the following errors of corresponding axes are different, and the lead screw clearance of each piece of equipment also varies. All of these parameters must be adjusted

to meet the actual needs of the machine tool.

2. Meanings in the servo parameter adjustment page

The FANUC CNC system has a specially designed servo parameter adjustment page (Fig. 5-12), so that users can intuitively understand the operation of the servo motor and whether there is an alarm during operation.

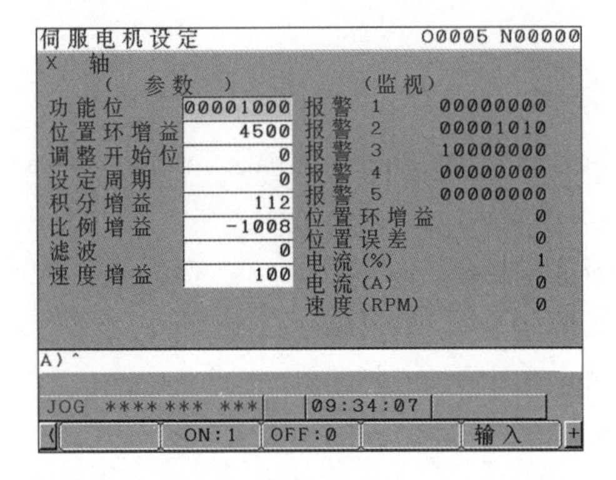

Fig. 5-12 Servo Parameter Adjustment page

Meanings of the parameters are as follows:

1) Function bit: corresponds to parameter 2003.

2) Position loop gain: Generally, the corresponding parameter 1825 is set to 3000 by default, and it can also be adjusted according to the machine tool conditions.

3) Adjustment start bit: used in the servo automatic adjustment function.

4) Set cycle: used in the servo automatic adjustment function.

5) Integral gain: corresponds to parameter 2043.

6) Proportional gain: corresponds to parameter 2044.

7) Filtering: corresponds to parameter 2067.

8) Speed gain. Set value = [(parameter 2021/256) + 1]×100, where parameter 2021 is the load ratio.

9) Alarm 1. It corresponds to diagnosis number 0200. See below for the specific meaning.

10) Alarm 2. It corresponds to diagnosis number 0201. See below for the specific meaning.

11) Alarm 3. It corresponds to diagnosis number 0202. See below for the specific meaning.

12) Alarm 4. It corresponds to diagnosis number 0203. See below for the specific meaning.

13) Alarm 5. It corresponds to diagnosis number 0204. See below for the specific meaning.

14) Position loop gain. It indicates the actual loop gain. The actual values after servo initialization and adjustment can be seen here.

15) Position error. It indicates the actual position error value (corresponding to diagnosis number 0300).

16) Current (%). The current is expressed as a percentage of the rated value of the servo motor.

17) Current (A). The peak value indicates the actual current.

18) Speed (RPM). It indicates the actual speed of servo motor of feed axis shown on this page.

On the page shown in Fig. 5-12, click [+] to view the values of the same parameters for other axes. The servo parameter adjustment page is very important for routine maintenance. Items 1)-8) in the page can be modified by users as needed. Generally, no modification is needed from the perspective of maintenance. The meanings of alarms 1-5 of alarm items 9)-13) in the page can only be understood by referring to maintenance data. In view of the refinement of system diagnosis

information, the alarm cause is already displayed on the interface when there is a fault alarm, so this alarm message can be used as a comprehensive reference factor for maintenance. Items 14) -18) on the interface can display the operating state of the servo motor in real time.

3. Detection contents for maintenance in the servo parameter adjustment page

In the servo parameter adjustment page, items 9) -13) are five groups of alarms. When there is a fault alarm in the FANUC CNC system and the specific alarm cause is unclear, you can first check whether items 9) -13) on the servo parameter adjustment page have an alarm with a value of 1, and then check the state change of each bit from diagnosis number 0200 to diagnosis number 0204. The meanings of diagnosis numbers from 0200 to 0204 are shown in Table 5-6.

Table 5-6 Meanings of Diagnosis Numbers from 0200 to 0204

Diagnosis number	#7	#6	#5	#4	#3	#2	#1	#0
0200	OVL	LV	OVC	HCA	HVA	DCA	FBA	OFA
0201	ALD	PCR	—	EXP	—	—	—	—
0202	—	CSA	BLA	PHA	RCA	BZA	CKA	SPH
0203	DTE	CRC	STB	PRM	—	—	—	—
0204	—	OFS	MCC	LDA	PMS	—	—	—

In Table 5-6, each bit of the diagnosis number is represented by the English abbreviation of the alarm. For specific meanings, please refer to *FANUC Series 0i-MODEL F maintenance manual (B-64605CM/01)* or the following introduction.

(1) Diagnosis number 0200.

The meaning of each bit of diagnosis number 0200 is as follows.

0200#0: OFA, overflow alarm;

0200#1: FBA, feedback cable disconnection alarm;

0200#2: DCA, discharging alarm;

0200#3: HVA, overvoltage alarm;

0200#4: HCA, abnormal current alarm;

0200#5: OVC, overcurrent alarm;

0200#6: LV, insufficient voltage alarm;

0200#7: OVL, overload alarm.

(2) Diagnosis number 0201.

The meaning of each bit of diagnosis number 0201 is as follows.

1) The meanings of state bits of diagnosis numbers 0201#4 and 0201#7 are shown in Table 5-7.

2) Diagnosis number 0201#6: PCR. When manually returning to the reference point, the one-rotation signal of the position detector is captured. As the grid for manually returning to reference point has been established, it can return to the reference point manually. This position is

meaningless when there is no action for manually returning to the reference point.

(3) Diagnosis number 0202.

The meaning of each bit of diagnosis number 0202 is as follows.

0202#0: SPH, abnormal serial pulse encoder or feedback cable or incorrect counting of feedback pulse signal;

0202#1: CKA, abnormal serial pulse encoder, and inoperation of the internal block;

0202#2: BZA, the battery voltage drops to 0. The battery will be replaced and the reference point shall be set;

0202#3: RCA, abnormal serial pulse encoder, and incorrect speed counting;

0202#4: PHA, abnormal serial pulse encoder or feedback cable, incorrect counting of feedback pulse signal;

0202#5: BLA, battery voltage drop (warning);

0202#6: CSA, abnormal hardware of serial pulse encoder.

Table 5-7 Meaning of State Bits of Diagnosis Numbers 0201#4 and 0201#7

Type	0201#7 (ALD)	0201#4 (EXP)	Meaning
Overload alarm	0	—	Servo motor overheating
	1	—	Servo amplifier overheating
Disconnection alarm	1	0	Built-in pulse encoder feedback cable disconnected (hardware)
	1	1	External pulse encoder feedback cable disconnected (hardware)
	0	0	Pulse encoder feedback cable disconnected (software)

(4) Diagnosis number 0203.

The meaning of each bit of diagnosis number 0203 is as follows.

0203#4: PRM, illegal parameters detected on the digital servo side. Then refer to the causes and countermeasures described in diagnosis number 0352.

0203#5: STB, abnormal communication of serial pulse encoder and incorrect data transmitted.

0203#6: CRC, abnormal communication of serial pulse encoder and incorrect data transmitted.

0203#7: DTE, abnormal communication of serial pulse encoder and no communication response.

(5) Diagnosis number 0204.

The meaning of each bit of diagnosis number 0204 is as follows.

0204#3: PMS, incorrect feedback due to abnormal serial pulse encoder C or feedback cable.

0204#4: LDA, abnormal LED of serial pulse encoder;

0204#5: MCC, contacting of electromagnetic switch in servo amplifier;

0204#6: OFS, abnormal A/D conversion of current value of digital servo.

Click [SYSTEM] several times, click [Diagnosis] to enter the diagnosis page, input "200",

and click [Search] to enter the diagnosis page of diagnosis number 0200.

II. Fault diagnosis and maintenance of servo systems

1. Overview

When the servo system is faulty, there are usually three symptoms: ① Alarm contents or alarm messages are displayed on the display screen or operation panel. ② The fault is displayed by an alarm lamp or segment display on the feed servo drive unit. ③ The movement is abnormal, but there is no alarm. For symptoms 1 and 2, because the meaning of some alarms is relatively explicit, please refer to the *Maintenance Manual for βi Series AC Servo Motor/Spindle Motor/Servo Amplifier (B-65325CM)*. For symptom 3, according to the control process of each link of the feed servo system, refer to the fault node of βi series servo amplifier unit shown in Fig. 5-13, consult the *Maintenance Manual for βi Series AC Servo Motor/Spindle Motor/Servo Amplifier (B-65325CM)*, and use the servo diagnosis information provided by the CNC system to check and troubleshoot step by step until the real cause is found out.

2. Fundamentals of routine repairs of servo amplifiers

(1) Repair category.

The repair of servo amplifier circuits is divided into board-level repair and chip-level repair. Correspondingly, the fault diagnosis of servo amplifiers is divided into board-level diagnosis and maintenance and chip-level diagnosis and maintenance. In fact, module or circuit board-level repair and chip-level repair are respectively conducted for end users and servo amplifier manufacturers. Currently, board-level repair is gradually adopted for high-density circuit boards. Module or circuit board-level repair is generally conducted on the site, that is rapid fault diagnosis and then replacement of circuit boards or modules.

(2) Maintenance conception.

A servo amplifier is a control component with high hardware integration, including both hardware circuits and software algorithms. Therefore, the maintenance of servo amplifiers is not carried out simply by a multimeter or ordinary instrument, but by the maintenance help provided by the servo amplifier manufacturer to determine where the fault exists. Nowadays, servo amplifiers have fault display and abundant fault diagnosis software to help users locate faults. More than that, the FANUC servo amplifier has abundant fault alarm function, which provides great maintenance convenience for users.

3. Types and nodes of servo amplifier faults

Although the servo amplifier is a technology-intensive product involving many automatic control theories and electronic circuits, as long as users and maintenance personnel are familiar with the circuit composition of the servo amplifier, unit hardware connection, and interface functions of the servo amplifier body and make full use of the rich fault diagnosis means provided by the system and servo amplifier itself, they can quickly locate and repair faults. According to the composition of the servo unit, faults can be divided into CNC (circlip) faults, servo amplifier module faults, servo

motor faults, encoder faults (including battery), peripheral connection faults, etc.

The fault nodes of a βi series servo amplifier is shown in Fig. 5-13, and the specific contents are shown in Table 5-8. It is also necessary to use FANUC CNC system software and servo diagnosis software for comprehensive judgment to diagnose and judge the fault nodes of βi series servo amplifiers.

Table 5-8 Fault Nodes of βi Series Servo Amplifiers

S/N	Fault nodes	Detailed possible fault points	Main measures
1	CNC (circlip) fault	CNC or circlip	Replace the CNC or circlip
2	Optical cable communication fault	CNC optical cable communication interface, optical cable connection, (the first) optical cable communication interface of the servo module	Replace the circlip, optical cable, servo module, or optical cable communication board
3	Servo amplifier fault	Various parts of the servo amplifier body (control and power)	Replace the servo module, power or control printed circuit board
4	Next optical cable communication fault	Previous optical cable communication interface, optical cable, and next optical cable communication interface	Replace the previous optical cable communication board, optical cable, next optical cable communication board, or next servo module
5	Control signal interconnection line fault	Output interface, control line, and servo module control line interface of power module	Replace the power module, power module components, control lines and servo module components
6	Dynamic braking module fault	Dynamic braking module resistor or overheating detection	Replace the dynamic braking resistor
7	Motor power cable fault	Short circuit and open circuit of power cable	Replace the power cable
8	Servo motor fault	Servo motor body fault	Replace the servo motor
9	Encoder fault (including battery)	Encoder body, battery	Replace the encoder and battery
10	Feedback cable fault	Servo module control PCB, feedback cable open/short circuit, encoder	Replace the servo module control printed circuit board, feedback cable, encoder, etc
11	Peripheral physical connection of the module	Peripheral control power supply (single-phase, 200 V), power supply (three-phase, 200 V), emergency stop circuit, MCC control circuit, etc	Replace the peripheral electrical components

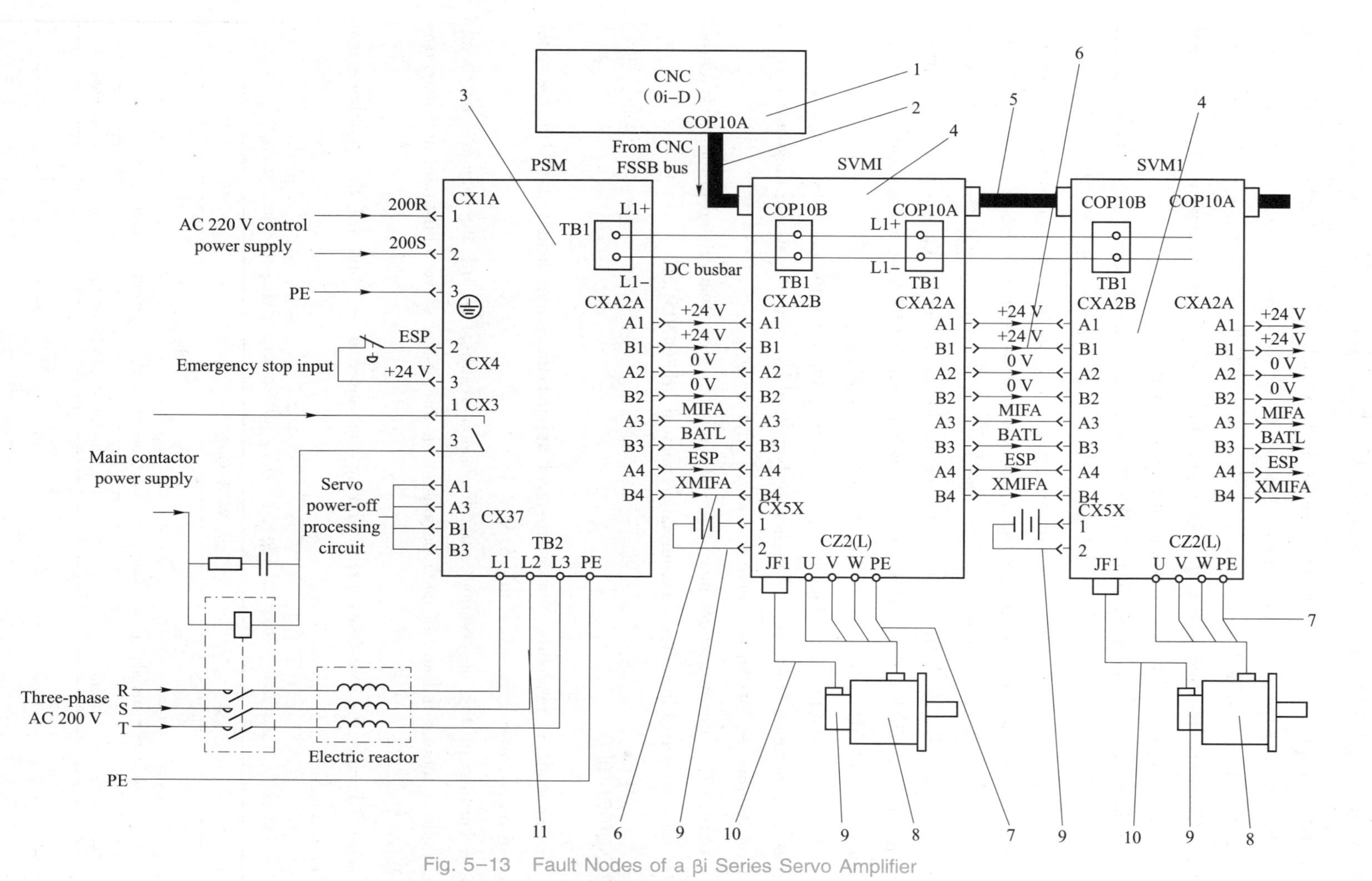

Fig. 5–13 Fault Nodes of a βi Series Servo Amplifier

1—CNC system; 2—FSSB bus; 3—Power supply module; 4—Spindle servo module; 5—FSSB bus;
6—Servo module interfaces; 7—Motor power supply cable; 8—Servo motor; 9—Motor encoder feedback cable; 10—Cable for power supply module

4. The system provides maintenance information of the servo amplifier and servo motor

Modern servo amplifiers are all-digital systems with a high level of integration, which can basically not be checked by conventional tools. Modern CNC maintenance mainly depends on the rich diagnostic functions provided by the servo amplifier itself. CNC maintenance personnel should have modern CNC maintenance concepts and make full use of the diagnostic functions and information provided by the CNC system to facilitate maintenance.

In case of fault of the components in the FANUC feed servo drive system, in addition to timely indication of fault alarm information on the display interface of CNC system, FANUC CNC system also provides 40-bit alarm information bits (alarm 1-alarm 5) on the servo parameter adjustment interface, with detailed introduction in the servo parameter adjustment interface (click[SYSTEM] several times—[+]—[SV adjustment] to display the interface as shown in Fig. 5-12). At the same time, there are servo-related alarm information contents in the system diagnosis interface starting from diagnosis number 0200 (click [SYSTEM], click [+] and [Check and diagnose], enter the diagnosis number, and click [Search] to display the system page).

Since the CNC system provides abundant alarm information, when the fault alarm information is not very specific and clear, it can be comprehensively judged by using the system diagnosis number and the alarm information bits on the servo parameter adjustment page.

The FANUC CNC system provides more than 70 kinds of information related to the servo alarm. For details, please refer to the relvant maintenance manual (B-64305CM).

Task Implementation

(1) Divide students randomly into groups and appoint the group leaders. Fill the personal information of the group members in Table 5-9.

(2) According to the observation and discussion results, record the names, numbers, meanings, main functions and etc. of each parameter on the system servo adjustment and diagnosis pages in Table 5-9.

(3) Fully discuss the problems found during the observation, and fill the problems and solutions in Table 5-9.

Table 5-9 Task Work Order for System Servo Adjustment and Diagnosis Interface

Task Work Order				
Name:	Class:		Student ID:	
List of materials, tools, measuring instruments, and equipment required				
S/N	Name	Quantity	Model or specification	Remark

Continued

Operating process(Observation)				
S/N	Name	No.	Meaning	Main function
Problems			Solutions	

Task Evaluation

S/N	Evaluation contents	Total point	Score
1	Be proficient in the operation of the CNC system servo adjustment page	25	
2	Master the meanings in the servo parameter adjustment page	25	
3	Be able to independently analyze the feed servo system fault and troubleshoot	25	
4	Have a professional spirit of effective communication, expression and solidarity	25	

(1) Briefly describe the classification and composition of CNC machine tool servo systems.

(2) What are the requirements for the servo system of CNC machine tools?

(3) Briefly describe the differences between serial spindle drive systems and analog spindle drive systems.

(4) Discuss how to enter the spindle adjustment page.

(5) What are the common servo parameters?

(6) What are the symptoms of servo system fault?

Module VI Mechanical Installation, Commissioning and Testing of CNC Machine Tools

The mechanical part of the CNC machine tool is its main part, which is a mechanical structure to complete various cutting operations. All movement and action commands from the CNC machine tools must be converted into real and accurate mechanical movement and actions by the machine tool body to realize the functions of CNC machine tools. Its accuracy directly affects the quality of machined parts.

The mechanical structure of CNC machine tool is always in a process of gradual development and change. The feed system of a common machine tool is mainly renovated and transformed for the early CNC machine tools, and its shape and structure are basically the same as those of common machine tools. At present, there are many similarities in the mechanical structure between simple CNC machine tools and early CNC machine tools as these products have also been developed through local improvement on the basis of the overall structure of common machine tools. With the rapid development of CNC, computer, automatic control, information processing, sensing and detection, power components, hydraulics and pneumatics and new materials, as well as continuous adaptation to the needs of high accuracy and efficiency, the mechanical structure of CNC machine tools has gradually developed from the improvement of the local structure of common machine tools to the formation of unique mechanical structure of CNC machine tools.

In addition to analysis on the mechanical structure of CNC machine tools, this module includes comprehensive analysis of the main transmission system, feed system and tool change system of CNC machine tools, and explanation of the specific contents of geometric accuracy detection of CNC machine tool in detail.

Task I Composition of CNC Machine Tool Mechanical Systems

Learning Objectives

(1) To understand the mechanical structure of CNC machine tools.

(2) To master the structure and requirements of the main transmission system of CNC machine

tools.

(3) To master the structure of the feed system of CNC machine tools.

(4) To master the structure of the tool change system for CNC machine tools.

(5) To be able to analyze and solve problems independently.

(6) To master the engineering working methods, cultivate a rigorous work style, and abide by the 5S management system.

Task Description

Observe the mechanical system unit of YL569 CNC milling machine in groups, and fill in the task work order with the names, characteristics, main functions and applications of each mechanical system. Make sure to be careful and pay attention to the relevant operation specifications of the practical training site during observation.

Knowledge Link

I. Basic composition of the mechanical system of CNC machine tools

The mechanical system of CNC machine tools is mainly composed of the following parts.

(1) Main transmission system, which is used to realize the main motion.

(2) Feed system, which is used to realize the feed motion.

(3) Basic parts of the machine tool (also known as large parts of the machine tool), which usually refer to the bed, base, pillar, slide carriage, and workbench. They are the foundation and frame of the whole machine tool, used to support other parts of the machine tool body and ensure that these parts are fixed on the basic parts during operation and move on their guide rails.

(4) Systems and devices used to realize the actions and auxiliary functions of some parts, such as hydraulic, pneumatic, lubrication, cooling, protective, and chip removal devices.

(5) Tool magazine, tool rest, and automatic tool changer.

(6) Automatic pallet exchange device, such as dual-station automatic pallet exchange devices.

(7) Devices and accessories used for workpiece rotation and indexing, such as rotary workbench.

(8) Special functional devices, such as tool breakage detection, accuracy detection, and monitoring devices.

II. Main transmission system and spindle components of CNC machine tools

1. Requirements for main transmission system of CNC machine tools

The main transmission system of CNC machine tools is the spindle motion transmission system of CNC machine tools. The spindle motion of CNC machine tools is one of the forming motions of

machine tools, whose accuracy determines the machining accuracy of parts. The spindle transmission system of CNC machine tools must meet the following requirements.

(1) The system shall have a large speed regulation range and can realize stepless speed regulation. In order to ensure a reasonable cutting amount during the machining of CNC machine tools, and achieve higher productivity and better machining accuracy and surface quality, a large speed regulation range is required. For machining centers, in order to meet the requirements of various processes and processing materials, the speed regulation range of the spindle system shall be further expanded.

(2) The system shall have relatively high accuracy and rigidity, stable transmission, and low noise. The improvement of the machining accuracy of CNC machine tools is closely related to the accuracy of the spindle system. In order to improve the manufacturing accuracy and rigidity of transmission parts, high-frequency induction heating quenching process shall be adopted for gear surfaces to increase wear resistance performance. The last stage shall be helical gear transmission to ensure smooth transmission. High-precision bearings and reasonable support spans shall be used, so as to improve the rigidity of the spindle assembly.

(3) The system shall have good vibration resistance and thermal stability. During machining of CNC machine tools, due to factors such as intermittent cutting, uneven machining allowance, imbalance of moving parts and natural vibration during cutting, impact and alternating forces may be caused to have the spindle vibrated. This will further affect machining accuracy and surface roughness, and even damage tools and the parts in the spindle system in severe cases, making them unable to work. The heating of the spindle system may cause thermal deformation of all parts, reduce transmission efficiency, and compromise the relative position accuracy between parts and motion accuracy, resulting in machining errors. Therefore, the spindle assembly should have a high natural frequency, good dynamic balance and proper fit clearance, and should be subjected to circulation lubrication.

2. Configuration modes of spindle transmission

There are three configuration modes for the spindle transmission of CNC machine tools, as shown in Fig. 6-1.

(1) Spindle transmission with variable gear (Fig. 6-1 (a)). This mode is widely used in large and medium-sized machine tools. The sliding gear is mostly realized by a hydraulic shift fork or directly driven by a hydraulic cylinder.

(2) Spindle transmission with belt drive (Fig. 6-1 (b)). It is mainly used in small machine tools and can avoid vibration and noise caused by gear transmission. It can only be used at the low torque.

(3) Spindle transmission directly driven by adjustable speed motor (Fig. 6-1 (c)). This mode is excellent. It simplifies the mechanism and improves the rigidity of the spindle, while the output torque is relatively small.

3. Main configuration forms of spindle bearing

There are three main configuration forms of spindle bearings for CNC machine tools, as shown in Fig. 6-2.

(1) The front support is a combination of the double-row short cylindrical roller bearing and the 60° angular contact double-row radial thrust ball bearing, and the rear support is a paired radial thrust ball bearing (Fig. 6-2 (a)). This configuration can improve the overall rigidity of the spindle and meet the requirements of strong cutting.

(2) A high-precision radial thrust ball bearing is used for the front support (Fig. 6-2 (b)). The radial thrust ball bearing has good speed performance, but its carrying capacity is small. It is suitable for high-speed, light-load and precise CNC machine tool spindles.

(3) The double-row and single-row tapered roller bearings are adopted (Fig. 6-2 (c)). This bearing has high radial and axial rigidity, and can bear heavy loads, especially strong dynamic loads. It has good installation and adjustment performance, but it will restrict the spindle speed and accuracy, so it is used for the spindle of CNC machine tools with medium accuracy, at a low speed, and under heavy loads.

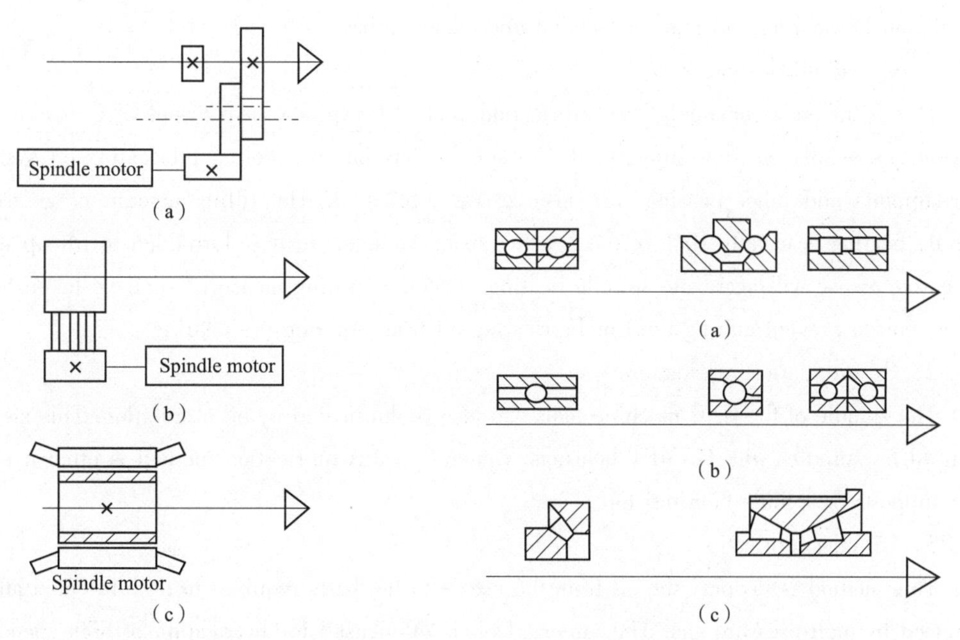

Fig. 6-1 Three Configuration Modes for Spindle Transmission of CNC Machine Tools

Fig. 6-2 Configuration Forms of Spindle Bearings for CNC Machine Tools

4. Spindle orientation devices (spindle exact stop devices)

Spindle orientation devices are special devices required for the tool change in the machining center, also known as spindle exact device. Since the tool is installed on the spindle, the cutting torque during cutting cannot be completely transmitted by the friction force of the taper. Therefore, a convex key is usually set at the front end of the spindle, and the keyway on the tool shank must be

aligned with this convex key when the tool is installed in the spindle. In order to ensure the smoothtool change, the spindle must stop in a fixed direction, for which the spindle orientation device is designed.

The spindle orientation device in the machining center usually uses a spindle encoder (connected with the spindle through synchronous toothed belt or gear transmission) to detect orientation and carry out initial positioning. Then, the locating pin (hydraulic or pneumatic) is inserted into the pin hole or pin slot on the spindle to achieve accurate position. After tool change, the locating pin is withdrawn and the spindle can rotate. This method is more reliable and accurate.

5. Lubrication and sealing of spindle assembly

Lubrication and sealing of the spindle assembly are two important aspects in the use and maintenance of machine tools. Good lubrication effect can reduce the working temperature of bearings and prolong their service life. The sealing shall avoid ingress of dust, chips and cutting fluid, as well as leakage of the lubricating oil.

(1) Lubrication methods of spindle bearings.

Lubrication methods of spindle bearings on CNC machine tools include grease lubrication, oil circulation lubrication, oil mist lubrication and oil-gas lubrication.

1) Grease lubrication.

This is the most commonly used lubrication method for spindle bearings of CNC machine tools at present, especially on front supports. If the spindle box has no cooling lubricating oil system, the rear supports and other bearings are also grease lubricated. The filling amount of grease in the spindle bearing is usually 10% of the bearing space volume, so it is forbidden to fill up it at will. Excessive grease will aggravate spindle heating. Effective sealing measures shall be taken for grease lubrication to prevent cutting fluid or lubricating oil from entering the bearing.

2) Oil circulation lubrication.

The spindle of the CNC machine tools can also be lubricated by oil circulation. This method can be used for spindles with GAMET bearings. Generally, this lubrication method is often used on the rear supports of spindle bearings too.

3) Oil mist lubrication.

This method is to spray the oil from the nozzle to the parts required to be lubricated after being atomized by high-pressure gas. The atomized oil is often used for lubrication of high-speed spindle bearings due to its good heat absorption and no need for oil stirring. However, the oil mist can be easily blown out and pollute the environment.

4) Oil-gas lubrication.

This method is a new type of lubrication developed for high-speed spindles. It is intended to lubricate the bearing with a small amount of oil (about 0.03 cm^3 within 8-16 min) to suppress bearing heating, with the lubrication principle shown in Fig. 6-3. The oil level switch in the oil tank and the pressure switch in the pipeline can ensure that the power supply of the main motor can be automatically cut off in case of no oil in the oil tank or insufficient oil pressure.

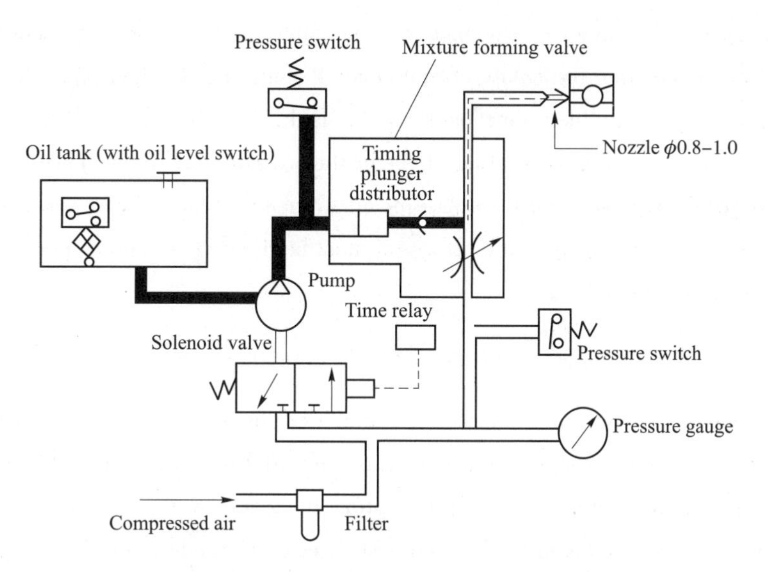

Fig. 6–3 Principle of Oil-Gas Lubrication

In addition, for lubrication of the angular contact bearing with oil, attention should be paid to the pumping effect of the angular contact bearing. The oil must flow from a small port.

(2) Spindle seal.

There are two types of seals for the spindle: contact and non-contact. There are mainly two types of contact seals, oil felt ring and oil-resistant rubber ring. The form of non-contact seal is shown in Fig. 6–4.

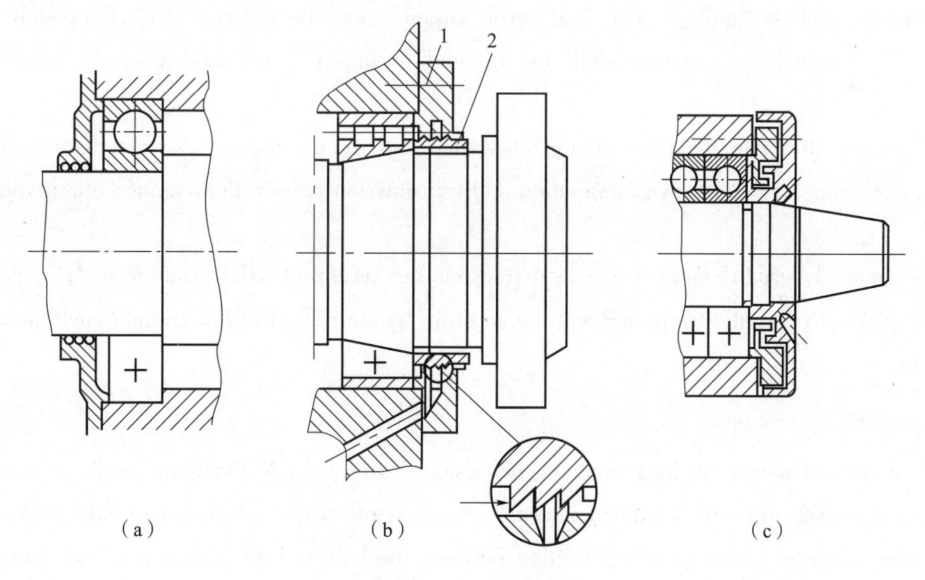

Fig. 6–4 Non-contact Seal

1—End cover; 2—Sealing ring

As shown in Fig. 6–4(a), the sealing is realized through the clearance between the bearing cap and the shaft, and grooves are made in the hole of the bearing cap to improve the sealing effect. This seal is used in grease lubrication where the working environment is relatively clean. As shown in

Fig. 6-4(b), a zigzag ring groove is opened on the outer circle of the nut. When the oil flows outward, it will be thrown into the cavity of end cover 1 along the inclined plane by the centrifugal force of spindle rotation, and then flow back to the tank. Fig. 6-4(c) shows a labyrinth seal structure, which can obtain reliable sealing effect in the working environment with many chips and heavy dust. This structure is suitable for grease or oil lubricated seals. When non-contact oil seal is adopted, in order to prevent leakage, the return oil must be drained as soon as possible to ensure the smoothness of the oil return hole.

III. Feed transmission system of CNC machine tools

In order to ensure the transmission accuracy and working stability of the feed system of CNC machine tools, the feed system should have no gap, low friction, low inertia, high rigidity, high resonance frequency and appropriate damping ratio. In order to meet these requirements, the following measures are mainly taken in the feed system of CNC machine tools.

(1) Adopt low-friction transmissions to reduce friction, such as hydrostatic guide rails, rolling guide rails, and ball screws.

(2) Select the best transmission ratio. This can not only improve the resolution of the machine tool, but also make the workbench track instructions faster and reduce the transmission inertia converted by the system to the drive shaft.

(3) Shorten the transmission chain, and improve the rigidity of the transmission system through pre-tightening. If a motor is used to directly drive the lead screw, preloaded rolling guide rails and ball screw pair shall be applied. The lead screw support shall be designed with both ends axially fixed, and pre-stretching structure shall be adopted to improve the rigidity of the transmission system.

(4) Try to eliminate the transmission clearance and reduce the reverse deadband error. For example, couplings with clearance elimination and transmission pairs with clearance elimination shall be used.

The commonly used devices in the feed transmission system of CNC machine tools include ball screw nut pairs, hydrostatic worm and worm gear strip, guide rails with low friction coefficient, gear racks, etc.

1. Ball screw nut pair

Ball screw nut pair is an ideal motion conversion device for CNC machine tools. It consists of lead screw 1, nut 2 and ball 3 located between threaded raceways, as shown in Fig. 6-5. During transmission, there is basically rolling friction between the ball and the lead screw and between the ball and the nut. Therefore, compared with ordinary rolling screw nut pairs, this ball screw nut pair has many advantages, such as high transmission efficiency (η: 0.92 - 0.96), sensitive transmission, less liability to creep, high transmission accuracy and positioning accuracy, and good synchronization, small rolling friction and wear, and long service life. It is reversible, i.e. mutual conversion between rotary motion and linear motion. When proper pre-tightening force is applied, the axial clearance can be eliminated and there is no idle stroke in reverse direction. The accuracy of axial motion, rigidity and repeated positioning accuracy can also be improved. Its disadvantages are high cost and self-locking failure. The braking devices are required for vertical installation.

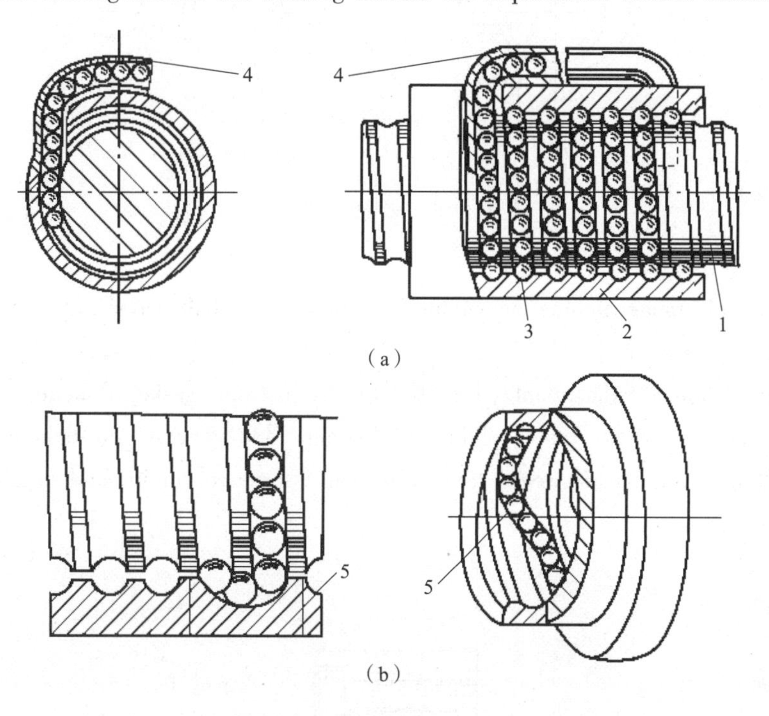

Fig. 6-5 Structure of Ball Screw Nut Pair

1—Screw rod; 2—Nut; 3—Ball; 4—Elbow; 5—Reverser

(1) Structure of ball screw nut pair.

The structure of ball screw nut pair has two types, internal circulation and external circulation, both of which are used in production. Fig. 6-5(a) shows the external circulation, and Fig. 6-5(b) shows the internal circulation. Their difference lies in the different return modes of balls. One is guided by elbow 4, and the returned ball is not in contact with the outer circle of the lead screw, while the other one is guided by reverser 5, and the returned ball returns through the reverser and the outer circle of the lead screw.

From the perspective of lead screw raceway thread profiles, there are mainly two types, circular arc profile (Fig. 6-6(a)) and double-arc profile (Fig. 6-6(b)). The former is easy to machine and the latter has better performance.

(2) Methods for the clearance adjustment of ball screw nut pair.

The transmission clearance of the ball screw nut pair is the axial clearance. To ensure reverse transmission accuracy and axial rigidity, the axial clearance must be eliminated. When the double-nut ball screw nut pair is pre-tightened to eliminate the axial clearance, it should be noted that the pre-tightening force should not be too large. The excessive pre-tightening force will increase the no-load torque, resulting in the reduction of the transmission efficiency as well as the shortening of the service life. In addition, The clearance of the mounting part and the drive part of the screw nut pair

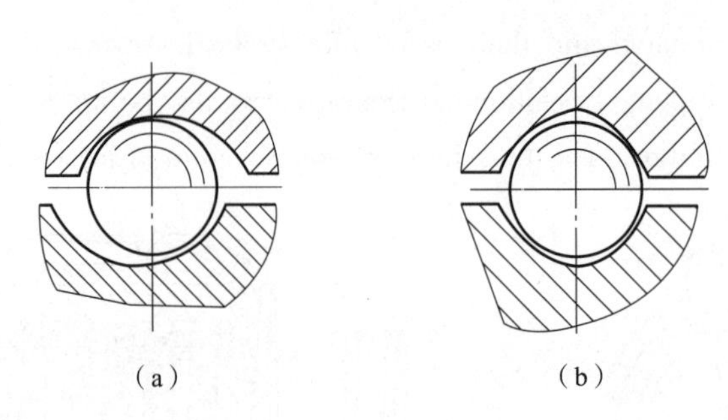

Fig. 6-6 Thread Profiles

shall be eliminated as well.

The common clearance elimination methods for double-nut ball screw nut pairs are shown as below.

1) Gasket clearance adjustment (Fig. 6-7). Adjust the gasket thickness to make axial displacement of left and right nuts to eliminate clearance and generate pre-tightening force. This method is simple and reliable. However, the adjustment is time-consuming and suitable for machine tools with general accuracy.

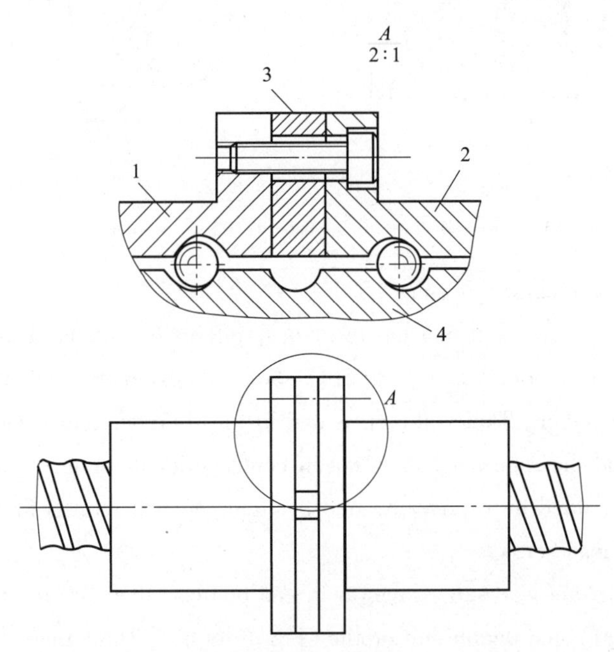

Fig. 6-7 Gasket Clearance Adjustment

1—Left nut; 2—Right nut; 3—Gasket; 4—Lead screw

2) Thread clearance adjustment (Fig. 6-8). The left nut 1 has threads on the outer circle and is fixed with two round nuts 2 and 3. By rotating the round nut 3, the axial clearance of nut 1 can be adjusted and it can be pre-tightened. Then lock it with round nut 2. This structure is compact and reliable. And it is adjustable at any time when the raceway is worn, but the pre-tightening amount is inaccurate.

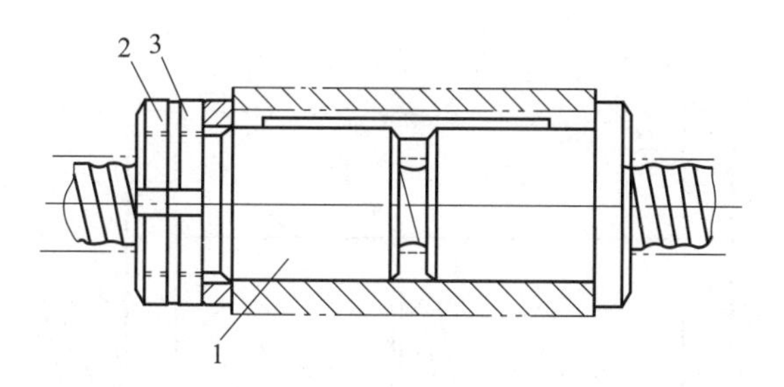

Fig. 6-8 Thread Clearance Adjustment

1—Left nut; 2, 3—Round nut

3) Tooth difference clearance adjustment (Fig. 6-9). In Fig. 6-9(a), the flanges of two nuts 2 and 5 are cylindrical gears with only one tooth difference, which are respectively installed in inner ring gears 1 and 4. The inner ring gear is fastened on nut seat 3 with screws and locating pins. During adjustment, the inner ring gear shall be removed first. When two nuts rotate one tooth in the same direction, their axial displacement is $s=(1/z_1-1/z_2)t$. For example, if the number of teeth of two cylindrical gears is $z_1=80$ and $z_2=81$ respectively, and the lead of ball screw is $t=6$ mm, then $s=6/6\ 480\approx 0.001$ mm, so the clearance and pre-tightening force can be adjusted accurately. However, the structure is complex and the size is large, so it is suitable for high accuracy transmission. Another tooth difference clearance adjustment structure is shown in Fig. 6-9(b), which is characterized by compact structure and pre-adjustment.

4) Displacement pitch preloading (Fig. 6-10). It has a variable pitch in the middle of its thread, which is simple in structure, but the load capacity needs to be preset and cannot be changed.

2. Guide rail with low friction coefficient

(1) Plastic sliding guide rail.

A plastic sliding guide rail is a cast iron-plastic or steel-plastic sliding guide rail. The guide rail plastic is commonly used as PTFE guide rail flexible belt, which adopts the bonding method and is customarily called "stick-on plastic guide rail". It is used for small and medium-sized CNC machine tools with a feed rate less than 15 m/min. Another guide rail plastic is a resin-type wear-resistant coating, which takes epoxy resin and molybdenum disulfide as the substrate, adds plasticizer to mix into liquid or paste as one component, and adds curing agent to mix into another component to form a two-component plastic coating.

(2) Rolling guide rail.

The rolling guide rail has a low friction coefficient (about 0.003), small dynamic and static friction coefficients, hardly affected by the change of motion speed, high positioning accuracy and sensitivity, and is also widely used.

It has two types. The first is a rolling guide block. The rolling elements are balls or rollers, and the rolling elements make circular motion in the rolling guide rail. It is suitable for guide rails with

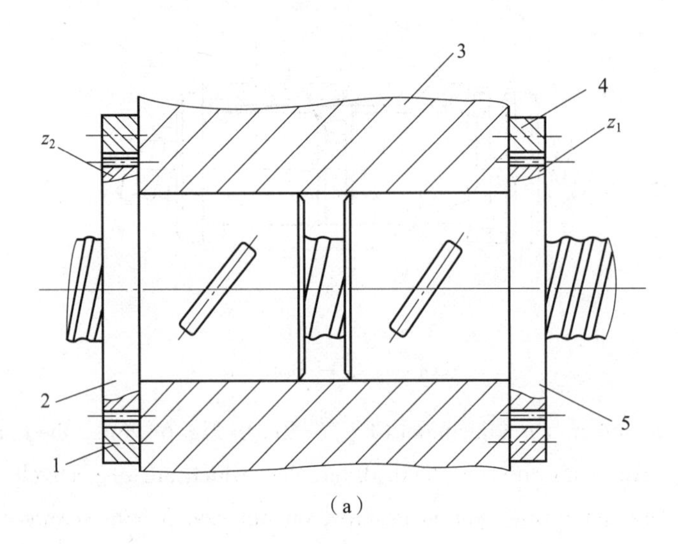

(a)

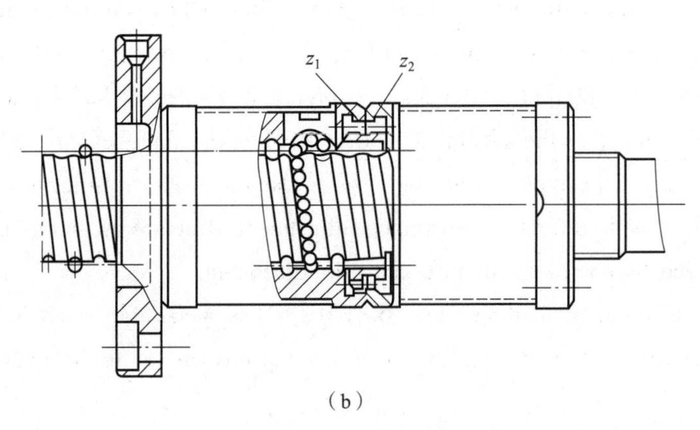

(b)

Fig. 6-9 Tooth Difference Clearance Adjustment

1, 4—Internal ring gear; 2, 5—Nut; 3—Nut seat

medium load. Guide rail blocks are installed on moving parts, and at least two or more pieces shall be used for each guide rail. The matching guide rails are mostly steel-inlaid quenched guide rails, as shown in Fig. 6-11.

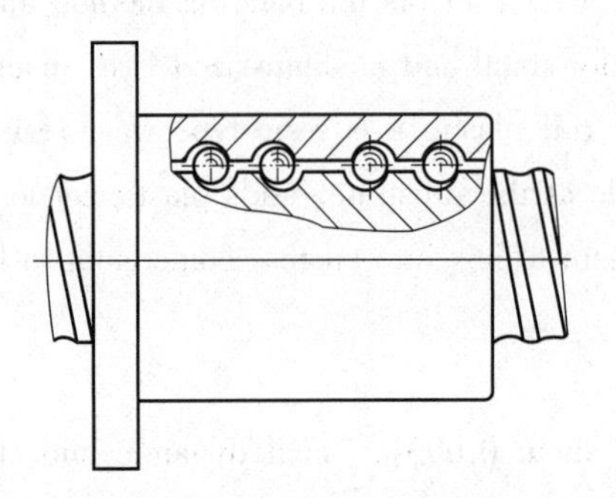

Fig. 6-10 Displacement Pitch Preloading

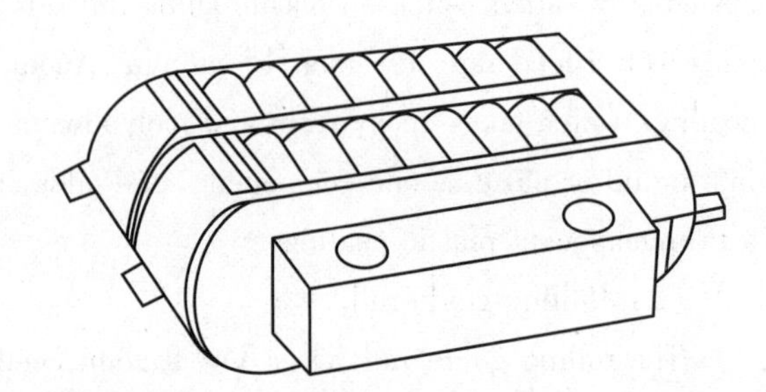

Fig. 6-11 Roller-type Rolling Guide Block

The second is the linear rolling guide rail. Its outstanding advantage is that it has no clearance and can apply pre-tightening force. Its outline and structure are shown in Fig. 6-12 and Fig. 6-13. It consists of guide rail body 1, slider 7, ball 4, retainer 3, end cover 6, etc. In use, the guide rail body is fixed on a stationary part and the slider is fixed on a moving part. When the slider moves along the guide rail body, the ball rolls in the arc straight groove between the guide rail body and the slider, passes through the raceway in the end cover, from the working load area to the non-working load area, then roll back to the working load area, and continuously circulate, thus changing the movement between the guide rail body and the slider into rolling of balls. In order to prevent dust and dirt from entering the guide rail raceway, end gasket 5 and side gasket 2 are installed at both ends and lower part of the slider. There is also a lubricating oil filling cup 8 on the slider.

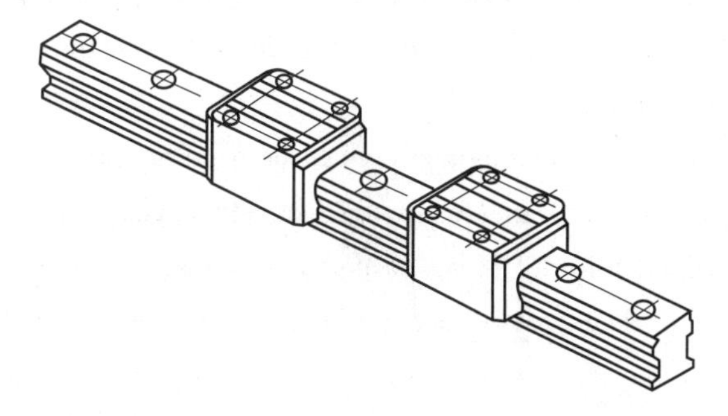

Fig. 6-12 Appearance of Linear Rolling Guide Rail (Unit Type)

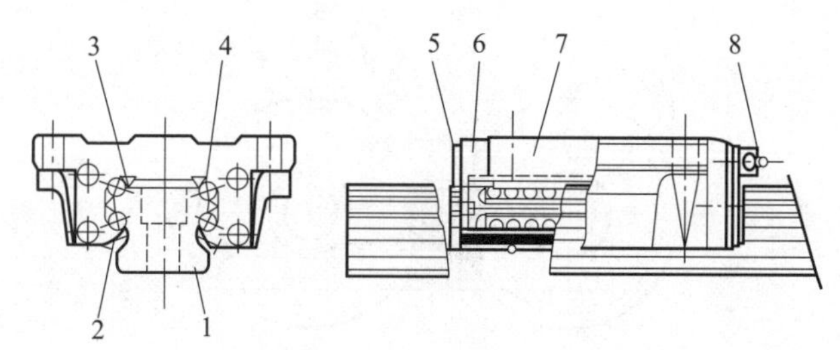

Fig. 6-13 Structure of Linear Rolling Guide Rail (HSR Type)

1—Guide rail body; 2—Side gasket; 3—Retainer; 4—Ball; 5—End gasket; 6—End cover; 7—Slider; 8—Lubricating oil filler cup

(3) Hydrostatic guide rail.

The hydrostatic guide rail is usually filled with pressure oil between two relatively moving guide rail surfaces to float the moving parts. During operation, the oil pressure in the oil chamber on the guide rail surface can be automatically adjusted with the change of applied load to ensure that there is always pure liquid friction between the guide rail surfaces. Therefore, the friction coefficient of the hydrostatic guide rail is extremely small (about 0.000 5) and the power consumption is low. This kind of guide rail will not be worn, so the accuracy of the guide rail is well maintained and its life is long. Its oil film thickness is hardly affected by the speed, and it has large bearing capacity, high rigidity and good vibration absorption. This kind of guide rail runs smoothly, without crawling or vibration. However, the hydrostatic guide rail has a complex structure and requires a set of hydraulic devices with good filtering effects, which results in relatively high manufacturing costs. At present, hydrostatic guide rails are generally used in large and heavy CNC machine tools.

The hydrostatic guide rail can be divided into open type and closed type according to the form of

guide rail. Closed hydrostatic guide rail is often used on CNC machine tools. The hydrostatic guide rail can also be divided into constant pressure (set pressure) type and constant flow (set quantity) type in terms of the oil supply mode.

3. Gear and rack

If the stroke of CNC machine tool is very long, gear and rack transmission is generally used to realize its feed motion. However, the transmission of gear and rack has clearance like that of gear transmission, so it should be eliminated.

When the load is small, the gear and rack can be adjusted by staggered teeth of double thin gears, which are closely attached to the left and right sides of the tooth groove of the rack respectively to eliminate the clearance.

When the load is large, radial loading method can be used to eliminate the clearance. As shown in Fig. 6-14, pinion gear pinion 1 and 6 are respectively engaged with rack 7, and preload is applied on gear 3 by pre-tightening device 4, so that gear 3 makes the meshed bull gear 2 and bull gear 5 extend outward, while their coaxial gear pinion 1 and 6 extend outward at the same time, so as to be closely attached to the left and right sides of the tooth groove on rack 7 respectively without clearance.

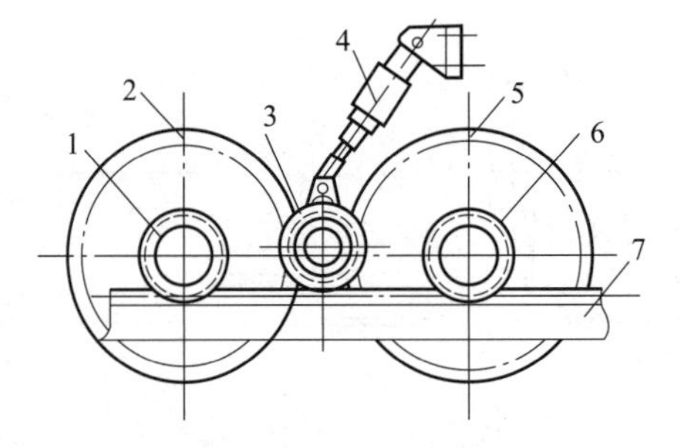

Fig. 6-14 Elimination of Clearance in Gear and Rack Transmission

1, 6—Gear pinion; 2, 5—Bull gear; 3—Gear; 4—Pre-tightening device; 7—Rack

IV. Automatic tool changer

The automatic tool changer has the function of automatically changing the required tools according to process requirements. The automatic tool changer shall meet the requirements of short tool change time, high repeated positioning accuracy of tools, large storage capacity of tools, small floor area of tool magazine, safety and reliability.

1. Classification of automatic tool changer

According to its composition and structure, the automatic tool changer can be divided into rotary tool rest type, tool turret type and automatic tool changer with tool magazine.

(1) Rotary tool rest type automatic tool changer.

The rotary tool rest is a simple automatic tool changer, which is often used in CNC lathes. According to the machining requirements, it is designed into square tool rest and hexagonal tool rest, and 4 and 6 tools are installed accordingly. Fig. 6-15 shows the rotary tool rest of CNC lathe (i.e. square tool rest). The working principles of this model are as follows.

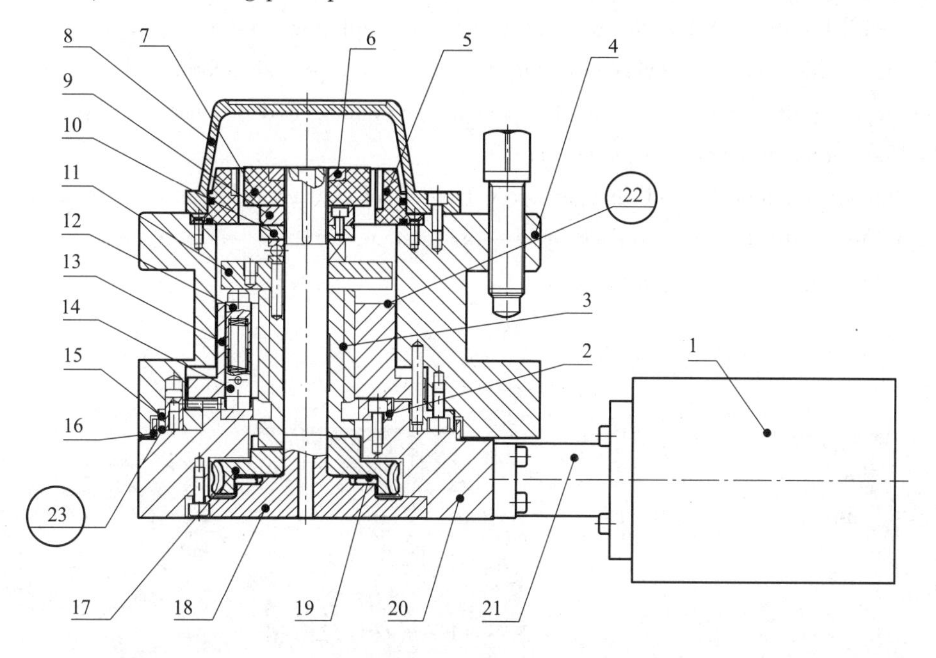

Fig. 6-15 Rotary Tool Rest

1—Motor; 2—Backing disc; 3—Screw; 4—Upper tool body; 5—Cover seat; 6—Small nut; 7—Transmitting disc; 8—Aluminum hood; 9—Large nut; 10—Retaining ring; 11—Clutch disc; 12—Clutch pin; 13—Nut; 14—Backing pin; 15—Outer end gear disc; 16—Protective cover; 17—Worm gear; 18—Spindle; 19—Needle bearing; 20—Lower tool body; 21—Connecting seat; 22—Plane F; 23—Base surface of tool rest

The system sends a tool change signal to control the forward rotation relay to act, and the motor rotates forward. Through the coupling, the worm drives the worm gear and the worm wheel drives the screw to rotate, and the nut starts to rise. At the same time, the screw drives the clutch disc to rotate, and the clutch pin slides on the plane of the clutch disc. When the nut rises to a certain height (the rising height has been adjusted during factory assembly), the three-end teeth are disengaged, and the clutch pin enters the clutch disc groove. At this time, the screw drives the clutch disc, clutch pin, nut, upper tool body, outer end tooth and backing pin to start indexing. The backing pin climbs out of the groove of the backing disc, that is, the upper tool body starts to tool change action.

When the upper tool body rotates to the required position, the hall element circuit sends out an in-place signal, the forward rotation relay is released, the reverse rotation relay is closed, the motor starts to rotate reversely, and the screw drives the clutch disc, clutch pin, backing pin and upper tool body to rotate reversely. When the backing pin moves past the backing groove on the plane of the backing disc, the backing pin is spring-loaded into the backing groove. As the backing pin enters the backing groove, the right angle surface of the backing pin and the right angle surface of the backing groove push against each other (the tool rest completes coarse positioning), preventing the rotation of

the backing pin, clutch pin and upper tool body. At this time, the clutch disc continues to rotate reversely driven by the screw, the clutch pin climbs out of the clutch disc groove, and the nut starts to descend until the three-end teeth are fully engaged, completing accurate positioning and locking the tool rest. At this time, the reverse rotation time is up, the reverse rotation relay is released and the motor stops rotating. A response signal is sent to the system, and the machining program starts.

(2) Turret-type automatic tool changer.

The turret-type automatic tool changer is a common tool changer in the CNC lathe with tilt bed, as shown in Fig. 6-16. It is characterized by two-way rotation, arbitrary tool position nearby selection, compact structure and high positioning accuracy.

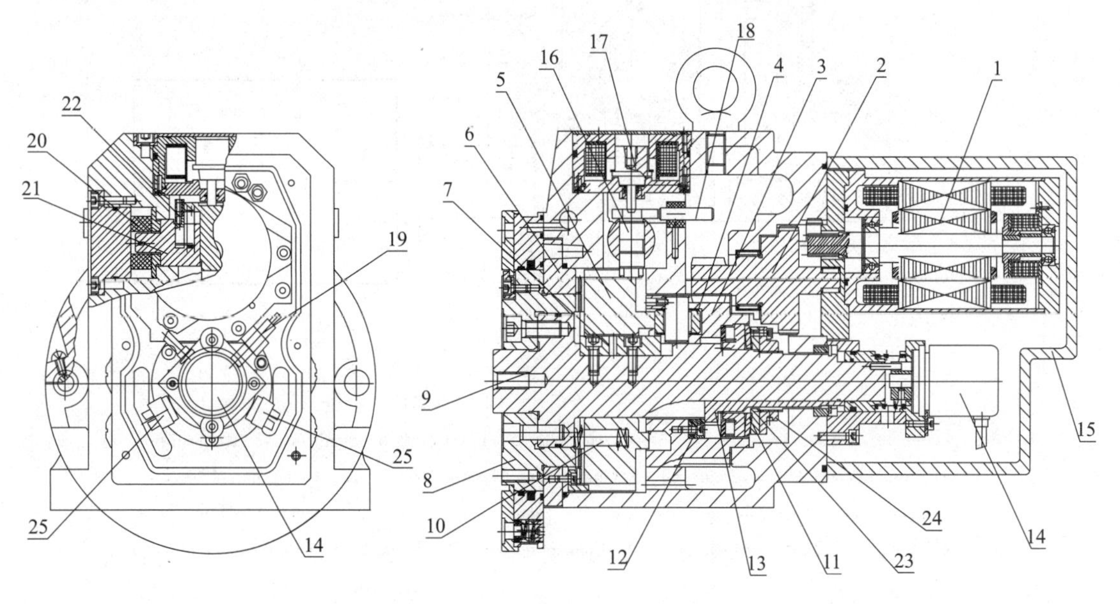

Fig. 6-16 Turret-type Automatic Tool Changer

1—Motor; 2—Secondary duplex gear; 3—Roller seat; 4—Roller; 5—Pin disc; 6—Outer end gear disc; 7—Central shaft key; 8—Inner end tooth; 9—Central shaft; 10—Loosening spring; 11—Belleville spring; 12—Transmission head; 13—Transmission rod; 14—Encoder; 15—Rear cover; 16—Locating pin; 17—Electromagnetic magnet; 18—Pin proximity switch; 19—Clamping proximity switch; 20—Damping pad; 21—Damping rod; 22—Small pin; 23—Check ring; 24—Nut; 25—Wiring terminal

After the turret is driven by the motor, it completes accurate positioning through gear reduction, electromagnetic magnet pin acting and cam clamping. The working procedure is that the system sends out a tool change command, the brake in the tool turret motor is released, the system judges the rotation direction of the motor according to the position of the required tool and the current tool number position, the motor rotates, the cam is released, the tool turret indexes. When the tool turret rotates to the lower edge of the gating signal at the previous station before the required tool position, the pre-positioning electromagnetic magnet supplies power, the pin acts, the pin proximity switch receives the insertion signal, the motor stops and stops for 50 ms, the motor reverses rotation, the clamping proximity switch receives the clamping signal, the motor stops rotating, the brake in the motor supplies power, the pin electromagnet loses power after a delay of 200 ms, the pin proximity switch detects whether the pin is loosened, and the tool change program ends.

(3) Automatic tool changer with tool magazine.

At present, automatic tool changers with tool magazines are widely used. Due to the tool magazine, a machining center only needs one spindle holding tools for cutting. When it is necessary to cut with a certain tool, the tool will be automatically exchanged from the tool magazine to the spindle, and after cutting, the used tool will be automatically put back into the magazine from the spindle. As the tool change process is carried out between components, it is required that the actions of each component involved in tool change must be accurately coordinated. The tool changer can improve not only the rigidity of spindle but also machining accuracy and efficiency. Moreover, the storage capacity of tools increases, which is conducive to machining complex parts. The tool magazine can also leave the machining area, eliminating many unnecessary disturbances. The automatic tool changer with tool magazine can be divided into two types, tool changer without manipulator and tool changer with manipulator.

The tool changer without manipulator generally places the tool magazine in a position that can be moved by the spindle box, i.e. the whole tool magazine or a certain tool position of the tool magazine can be moved to a position that can be reached by the spindle box. The storage direction of tools in the tool magazine is generally consistent with the tool loading direction of the spindle box. During tool change, the relative motion of spindle and tool magazine is used to perform tool change action, and the spindle is used to take out or put back the tool. Fig. 6-17 shows the tool changer for the machining center.

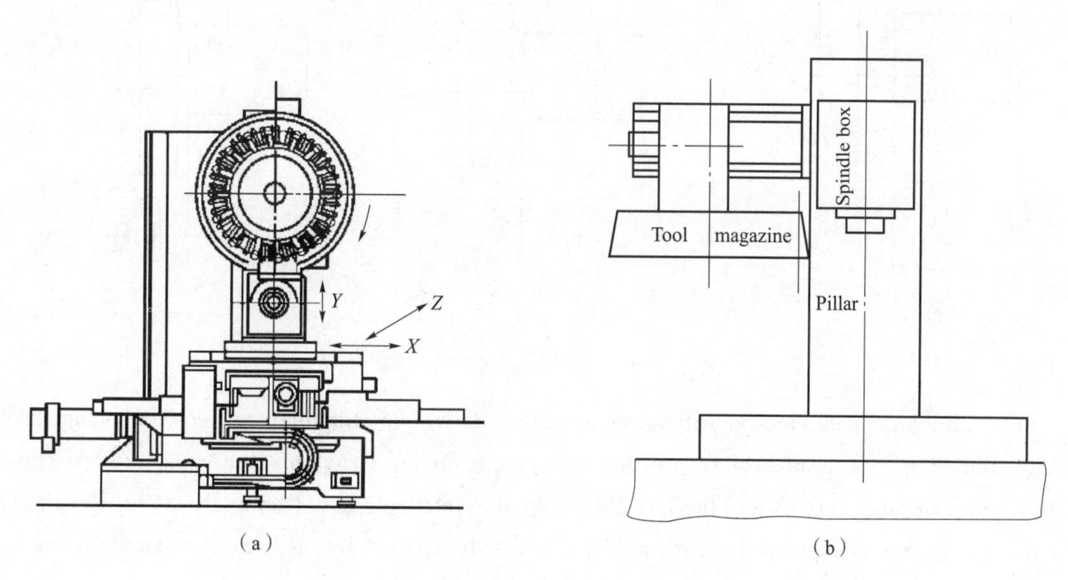

Fig. 6-17 Tool Changer for the Machining Center

(a) Horizoutal machining center; (b) Vertical machining center

Fig. 6-18 shows the tool change process of horizontal machining center tool changer without manipulator as shown in Fig. 6-17(a). Fig. 6-18(a) shows that after the previous step is completed, the spindle stops exactly for positioning and the spindle box rises. Fig. 6-18(b) shows

that the spindle box rises to the top tool change position, and the tool enters the empty exchange position of the tool magazine. The tool is fixed by the clamping jaw on the tool magazine, and the automatic tool clamping device on the spindle is released. Fig. 6-18(c) shows that the tool magazine moves forward and pulls out the tool to be replaced from the spindle hole. Fig. 6-18(d) shows the tool magazine indexing. According to the program, the tool to be used in the next process is switched to the tool change position, and the spindle hole cleaning device cleans the tool hole on the spindle. Fig. 6-18(e) shows that the tool magazine moves backward, inserts the required tool into the spindle hole, and the tool clamping device on the spindle clamps the tool. Fig. 6-18(f) shows that the spindle box is lowered to the working position and starts the next step of work.

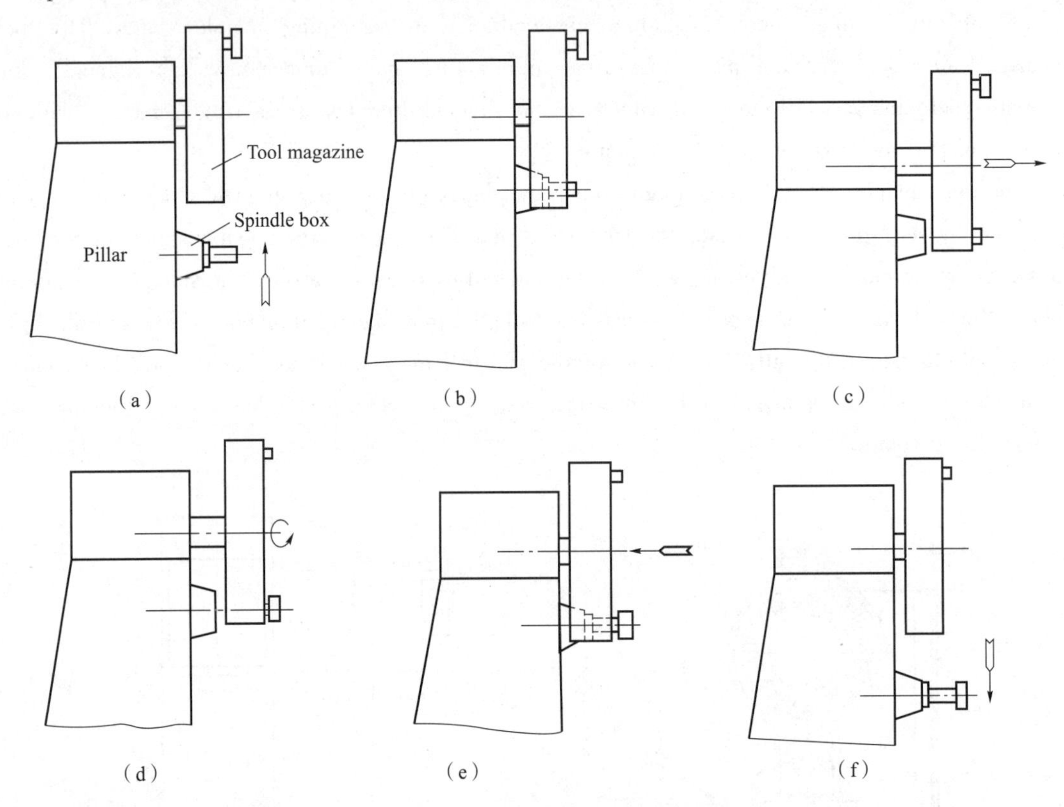

Fig. 6-18 Tool Change Process

The advantages tool changer without manipulator are simple structure, low cost, and high tool change reliability. Its disadvantages are that due to the limited structure, the capacity of the tool magazine is not large enough and the tool change time is relatively long (generally taking 10-20 s). Therefore, the tool changer without manipulator is mostly adopted by medium and small machining centers.

2. Tool magazine

The tool magazine is used to store machining tools and auxiliary instruments. Since the pick-up and feed positions of most machining centers are at a fixed tool position in the tool magazine, the tool magazine also needs to have a mechanism for moving and positioning the tools to ensure reliable tool change. The power for moving the tool may be a servo motor or a hydraulic motor. The tool positioning mechanism is used to ensure that each tool and tool sleeve to be replaced can accurately stop at the tool change position. Its control part can adopt a simple position controller or servo position control similar to semi-closed loop feed system, or a pin positioning mode combining electrical and mechanical methods. Generally, its comprehensive positioning accuracy is required to reach 0.1–0.5 mm.

According to the structural forms, tool magazines can be divided into disc-type tool magazines, chain-type tool magazines and box-type tool magazines. Fig. 6–19 shows several typical tool magazines. Disc-type tool magazine has simple structures and many applications. However, due to the annular arrangement of tools, space utilization is low. Therefore, the tools are arranged in double or multiple rings in the disc to increase space utilization. However, in this way, the outer diameter of the tool magazine expands, the rotational inertia also increases, and the tool selection time is longer. Therefore, the disc-type tool magazine is generally used for tool magazines with small tool capacity. The chain-type tool magazine has a compact structure and large capacity. The shape of the chain can be configured according to the layout of the machine tool, and the tool change position can be highlighted to facilitate tool change. When it is necessary to increase the tool capacity of the chain-type tool magazine, only the length of the chain needs to be increased. Within a certain range, there is no need to change the linear speed and inertia of the tool magazine. Generally, chain-type tool

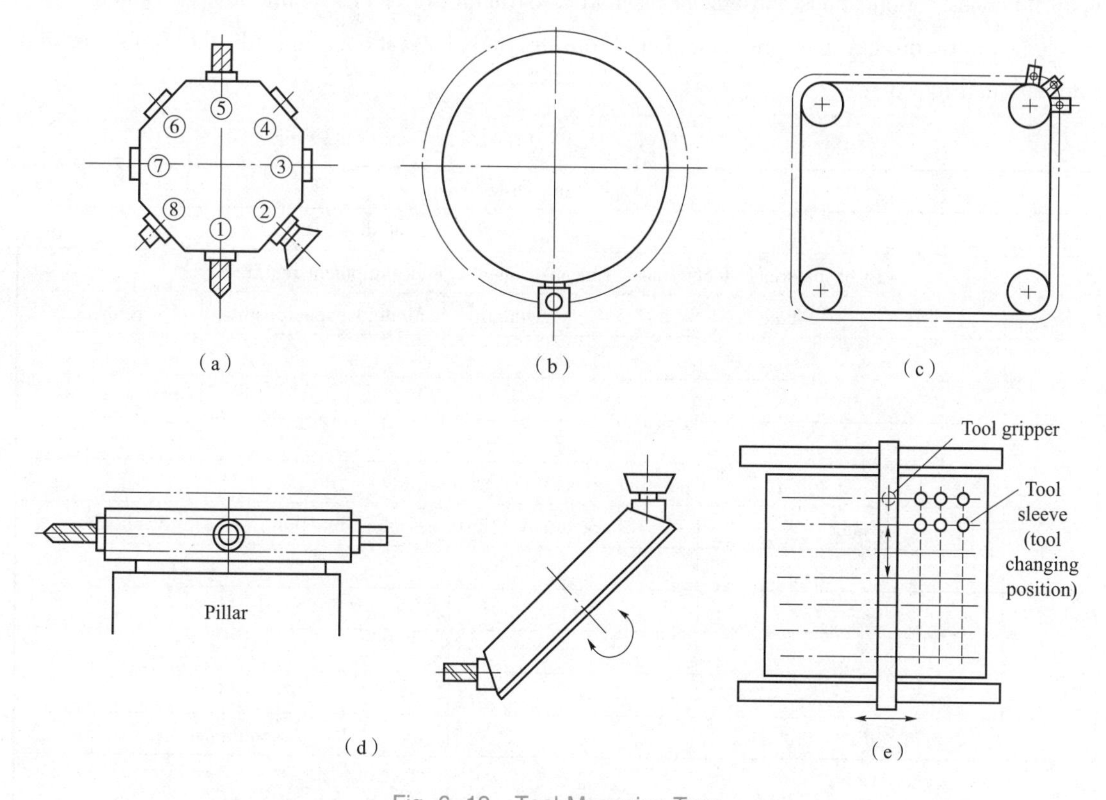

Fig. 6–19 Tool Magazine Type

(a) Turret type; (b) Disc type (side type); (c) Chain type; (d) Disc type (overhead type); (e) Box type

magazine is adopted when the number of tools is 30–120. The structure of the box-type tool magazine is also simple, with two types: linear and box-shaped. Linear tool magazines are generally used for tool changers without manipulators, while box-shaped tool magazines with large capacity are often used in unit-type machining centers.

In addition, the tool magazine can be divided into top-mounted type, side-mounted type, suspended type and floor-mounted type according to different installation positions. According to tool exchange or spindle exchange, the tool magazine can be divided into two types, ordinary tool magazine (referred to as tool magazine) and spindle box tool magazine.

The principle of "shortcut movement" is generally adopted for tool magazine selection, that is, no matter which tool selection method is adopted, when the tools to be used in the next process are moved to the tool change position according to the program command, they shall be moved in the direction less than half a circle of the tool magazine to save the time for tool selection.

Task Implementation

(1) Divide students randomly into groups and appoint the group leaders. Fill the personal information of the group members in Table 6–1.

(2) According to the observation and discussion results, record the names, characteristics, main functions, applications and etc. of mechanical structure of each machine tool in Table 6–1.

(3) Fully discuss the problems found during the observation, and fill the problems and solutions in Table 6–1.

Table 6–1 Task Work Order for Machine Tool Observation

Task Work Order				
Name:		Class:		Student ID:
List of materials, tools, measuring instruments, and equipment required				
S/N	Name	Quantity	Model or specification	Remark
Operating process(Observation)				
S/N	Name	Characteristic	Main function	Application
Problems			Solutions	

Task Evaluation

S/N	Evaluation contents	Total point	Score
1	Be proficient in the mechanical structure of CNC machine tools	25	
2	Master the feed system structure of CNC machine tools	25	
3	Master the tool change system structure of CNC machine tools	25	
4	Have a professional spirit of effective communication, expression and solidarity	25	

Task II Spindle Installation and Commissioning of CNC Machine Tools

Learning Objectives

(1) Understand the spindle structure of CNC machine tools.

(2) Be proficient in using the tools and measuring instruments used for spindle installation of CNC machine tools.

(3) Read and master the spindle assembly drawing of CNC machine tools.

(4) Master the installation process of CNC machine tool spindle.

(5) Safe and civilized operation.

(6) To master the engineering working methods, cultivate a rigorous work style, and abide by the 5S management system.

Task Description

Observe the spindle structure of YL569 CNC milling machine in groups, and fill the steps, methods, main functions and accuracy, and etc. of spindle installation and commissioning of CNC machine tools in the task work order. Make sure to be careful and pay attention to the relevant operation specifications of the practical training site during observation.

Knowledge Link

The spindle is an important component of the main transmission system of CNC machine tools, which is used to install tools and rotate at high speeds. Its assembly quality directly affects the machining quality of workpieces. This task takes the spindle of YaLong YL-1506B CNC milling machine as an example to specifically introduce the spindle assembly process. The general assembly drawing of YL-1506B spindle is shown in Fig. 6-20.

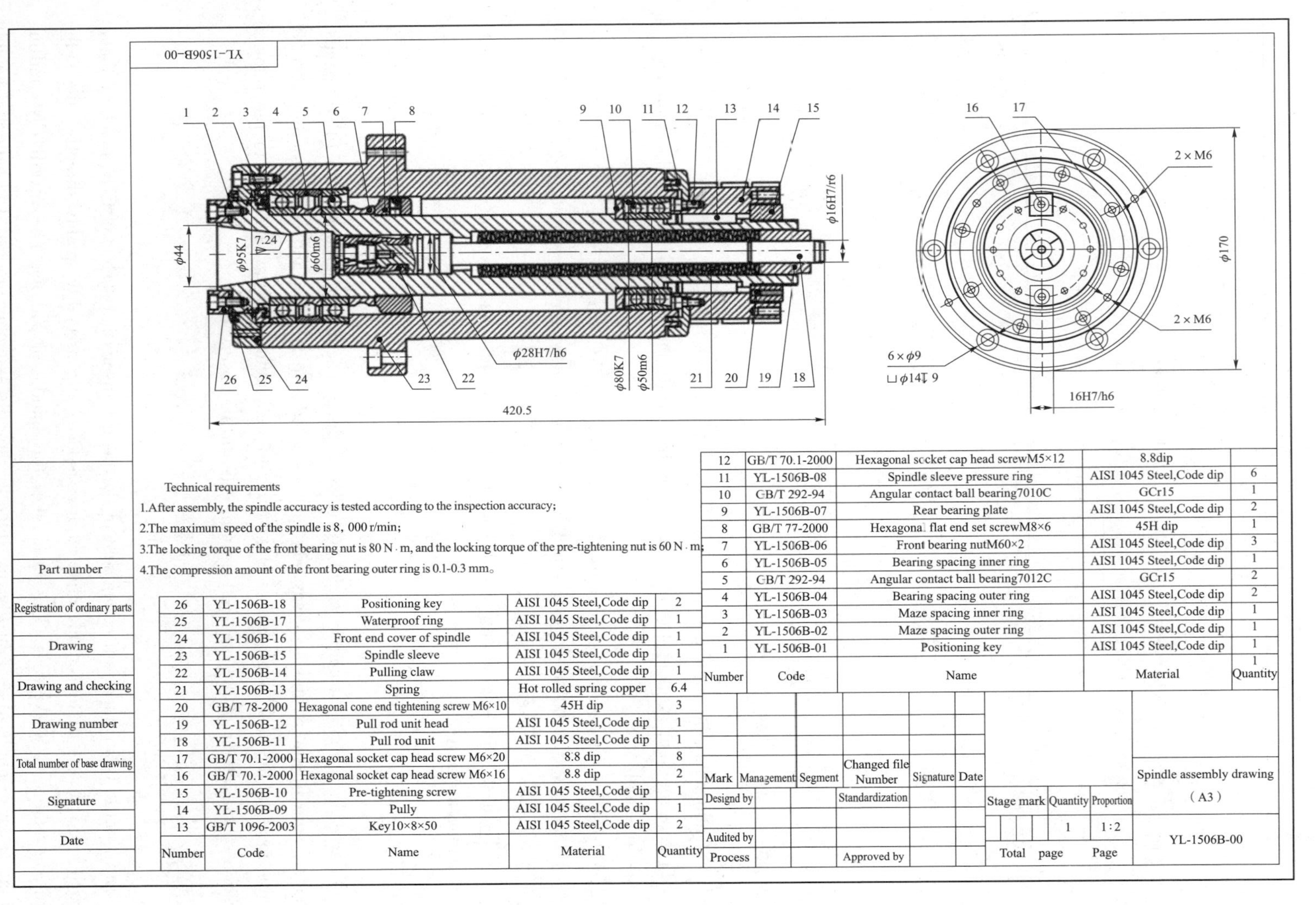

Fig. 6-20 Final Assembly Drawing of YaLong YL-1506B Spindle

I. Preparation before spindle assembling

(1) All parts in the spindle assembly need to be cleaned, especially the contact surface with the bearing which should be wiped with alcohol and checked for stains.

(2) Check that the reference surface of parts is free of scars, scratches, and rust spots, and lay more emphasis on checking the contact step surface and the bearing mating outer surface.

(3) Check that each sharp edge chamfer is free of burrs, and ensure that it is smooth without edges and corners when touched by hands during the assembly.

(4) Check the debris and depth of the fastening threaded hole, remove the debris using the screw tap, and blow it clean.

(5) Cleaned parts shall be placed on dust-free clean oil paper or cloth, while those that are not used temporarily shall be provided with dust covers.

(6) The parts shall be placed at a distance of ≥800 mm from the working area.

(7) Clean the bearing.

1) The bearing cleaning fluid is contained in two containers, one for cleaning and the other for rinsing the bearing.

2) In the process of initial bearing cleaning, it is not allowed to rotate the bearing relatively. It can be shaken up, down, left and right in the liquid.

3) After cleaning, place the bearing in a rinsing pool and rotate the inner and outer rings of the bearing while brushing.

4) The bearing shall not be placed at the bottom of the pool during cleaning, and must be removed from the pool after cleaning.

5) After cleaning the bearing, throw off the liquid beads on the bearing before leaving the liquid pool, and repeat this operation after rotating the bearing.

6) Put the bearing in a clean place for drying and cover it with oil paper or wiping paper. In order to shorten the drying time, dry it with an electric hairdryer. It is strictly prohibited to blow the bearing with air compressor dust.

II. Accuracy detection and parts inspection between spindle fitting parts

(1) Trial assembly of angular contact ball bearings (7010C) and (7012C) with the spindle respectively.

(2) Wipe the inspection platform clean with a rag before measuring, place the fittings on the inspection platform, and check the accuracy.

Check the high accuracy of front/rear bearings (7010C) and (7012C), which shall be ≤0.002 mm, and check the inner and outer rings one by one. The detection process is shown in Fig. 6-21.

(3) The pulley balance jackscrew (M6×10; GB70) shall be weighed with a balance and then grouped, with the difference mass of each group ≤0.2 g.

(4) The fastening screws (M5×12; GB70) on the reverse buckling disc are weighed with a balance and then grouped, with the difference mass of each group ≤0.2 g.

(5) The key screws (M5×12; GB70) of the spindle front end cover are weighed with a balance and grouped.

(6) After the pulley tensioning fixing screw is weighed with a balance, the mass difference of each group is ≤0.2 g.

(7) Inject lubricating grease into the spindle bearing.

1) Ensure that the bearing is dried before lubricating grease injection. Fill the cleaned syringe barrel with lubricating greases, and then push to exhaust air so that the syringe has a specified capacity of 3.6 cm^3 for the front bearing and 2.6 cm^3 for the rear bearing.

2) Each rolling body of the bearing shall be evenly injected and distributed on both sides.

(8) Date measurement related to spindle front bearing assembly.

1) Measure the value $K1$ from the end face of spindle sleeve to the spacer of spindle sleeve with a depth gauge, as shown in Fig. 6-22.

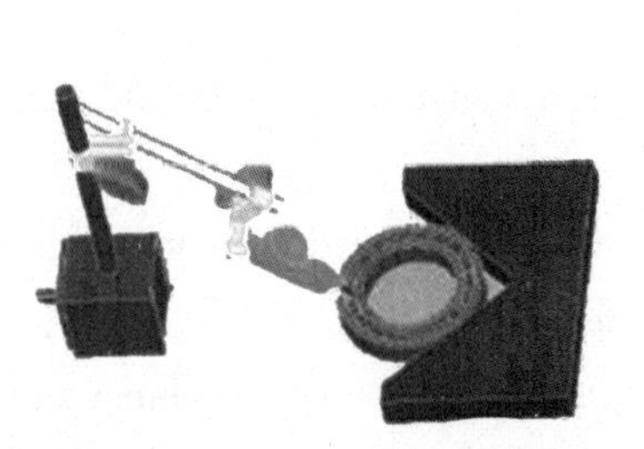

Fig. 6-21 High Accuracy Detection of Front/Rear Bearings

Fig. 6-22 Depth Gauge Measurement $K1$

2) The cleaned bearings shall be stacked together. They are angular contact ball bearing (7012C), inner ring of bearing spacer sleeve, outer ring of bearing spacer sleeve, angular contact ball bearing (7012C) and inner and outer rings of labyrinth spacer ring respectively. The measured superposition height is $K2$, as shown in Fig. 6-23.

3) Measure the depth of concave recess on front end cover of spindle as H, as shown in Fig. 6-24. (Measure once at two groups of positions perpendicular to each other, and the resulting values are weighted to calculate the average)

Fig. 6-23 Depth Gauge Measurement $K2$

Fig. 6-24 Depth Gauge Measurement H

4) During factory installation, the front end cover of spindle shall be repaired and adjusted

according to the deviation between $K=K2-K1+0.2$ mm and H value.

When measuring each value, ensure that each workpiece is clean and free of stains. Wipe the contour platform clean with alcohol and wiping paper before measurement.

III. Assembly of spindle components

(1) The spindle shall be erected on the workbench with its front end facing downwards.

(2) Put in the outer ring of labyrinth spacer ring, and install it into the spindle with its annular groove facing upward.

(3) Put in the inner ring of labyrinth spacer ring, and install it into the spindle with its annular groove facing downward, as shown in Fig. 6-25.

(4) Place the angular contact ball bearing (7012C) on the inner ring of the spindle labyrinth ring, and install it into the spindle with the wide end face side of the bearing outer ring facing upwards, as shown in Fig. 6-26.

Fig. 6-25 Installation of the Inner and Outer Rings of Labyrinth Spacer Ring

Fig. 6-26 Installation of Angular Contact Ball Bearing

(5) Install the bearing spacer inner ring into the spindle, then place the bearing spacer outer ring, and install the second angular contact ball bearing (7012C) into the spindle with its wide end face side facing downward, as shown in Fig. 6-27.

(6) Install the other bearing spacer inner ring into the spindle, as shown in Fig. 6-28.

Fig. 6-27 Installation of Bearing Spacer Inner Ring and Second Angular Contact Ball Bearing

Fig. 6-28 Installation of the Other Bearing Spacer Inner Ring

(7) Install the front bearing nut (M60×2) into the spindle, the required tightening torque is 80 N · m. Tighten the front bearing nut with a hook wrench, and then tighten its three M8×6 jackscrews with a 4 mm Allen wrench, as shown in Fig. 6-29.

(8) Use a magnetic gauge stand to suck it on the spindle, with the gauge head contacting the outer ring of the angular contact ball bearing (7012C). Rotate, measure and adjust the excircle to be concentric with the spindle, with a tolerance ≤0.05 mm, as shown in Fig. 6-30.

Fig. 6-29 Tightening of the Front Bearing Nut

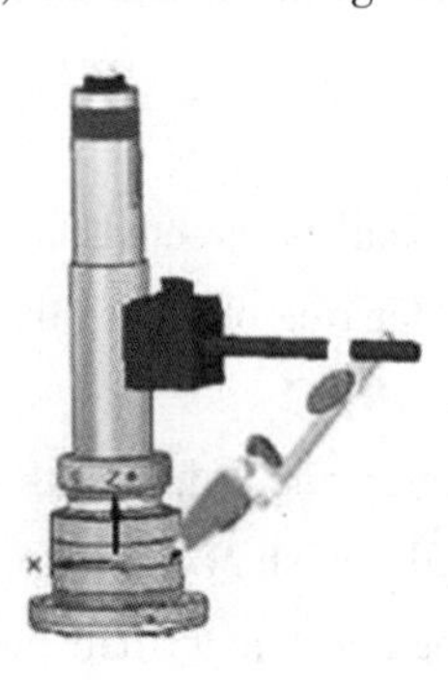

Fig. 6-30 Measure and Adjust the Excircle to Be Concentric with the Spindle

(9) Use a magnetic gauge stand to suck on the journal of the rear angular contact ball bearing, with the gauge head in contact with the spindle, and check its rotation runout, as shown in Fig.6-31. The tolerance is ≤0.04 mm.

(10) Suck the magnetic gauge stand on the spindle, keep the magnetic gauge stand still, let the gauge head contact the end face of the outer ring of the angular contact ball bearing, rotate the outer ring, and check that the tolerance of end face runout is ≤ 0.02 mm, as shown in Fig. 6-32.

Fig. 6-31 Check the Rear Angular Contact Ball Bearing Rotation Runout

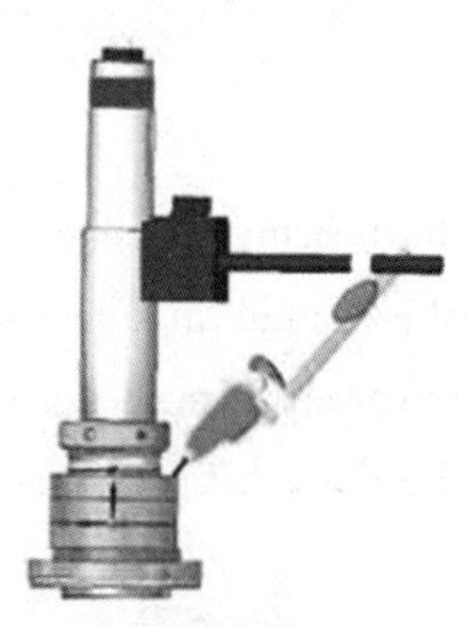

Fig. 6-32 Check the End Face Runout Tolerance of Angular Contact Ball Bearing

(11) In Processes "9" and "10", if the runout is out of tolerance, it can be adjusted by adjusting 3 jackscrews on the front bearing nut (M60×2) or tapping the corresponding direction of the nut to meet the requirements.

(12) Install the rear bearing baffle (convex side upward).

(13) Install the angular contact ball bearing (7010C) combination DB and place it on the rear bearing baffle.

(14) Sleeve the spindle sleeve into the spindle.

(15) Assemble the spindle sleeve pressing ring and pulley with (M5×12) screws.

(16) Install the key (C10×8×50).

(17) Install the assembled spindle sleeve pressing ring and pulley into the spindle.

(18) Install the pre-tightening nut on the spindle with a tightening torque of 60 N · m. Use an adjustable round nut wrench to install it in place, and adjust the three jackscrews (M6×10) on the pre-tightening nut.

(19) Install the spindle front end cover and waterproof ring, and lock them with 8 hexagon socket head cap screws M6× 20. The calculated pressing amount A of the front bearing outer ring is within the tolerance range of technical requirements, where $A=K2-K1-H$.

(20) Install the locating key and lock it with 2 hexagon socket head cap screws M6×16.

IV. Detect the accuracy of spindle components

Place the spindle on the test bench to detect its runout, which shall be ≤0.01 mm.

Task Implementation

(1) Divide students randomly into groups and appoint the group leaders. Fill the personal information of the group members in Table 6-2.

(2) According to the observation and discussion results, record the steps, methods, main functions, accuracy and etc. of spindle installation and commissioning of CNC machine tools in Table 6-2.

(3) Fully discuss the problems found during the observation, and fill the problems and solutions in Table 6-2.

Table 6-2 Task Work Order for Spindle Installation and Commissioning of CNC Machine Tools

<table>
<tr><td colspan="6">Task Work Order</td></tr>
<tr><td colspan="2">Name:</td><td colspan="2">Class:</td><td colspan="2">Student ID:</td></tr>
<tr><td colspan="6">List of materials, tools, measuring instruments, and equipment required</td></tr>
<tr><td>S/N</td><td colspan="2">Name</td><td>Quantity</td><td>Model or specification</td><td>Remark</td></tr>
<tr><td></td><td colspan="2"></td><td></td><td></td><td></td></tr>
<tr><td></td><td colspan="2"></td><td></td><td></td><td></td></tr>
<tr><td></td><td colspan="2"></td><td></td><td></td><td></td></tr>
<tr><td colspan="6">Operating process(Observation)</td></tr>
<tr><td>S/N</td><td>Step</td><td colspan="2">Method</td><td>Main function</td><td>Accuracy</td></tr>
<tr><td></td><td></td><td colspan="2"></td><td></td><td></td></tr>
<tr><td></td><td></td><td colspan="2"></td><td></td><td></td></tr>
<tr><td></td><td></td><td colspan="2"></td><td></td><td></td></tr>
<tr><td></td><td></td><td colspan="2"></td><td></td><td></td></tr>
<tr><td colspan="4">Problems</td><td colspan="2">Solutions</td></tr>
<tr><td colspan="4"></td><td colspan="2"></td></tr>
</table>

Task Evaluation

S/N	Evaluation contents	Total point	Score
1	Be able to state the spindle structure of CNC machine tools	25	
2	Be able to correctly use the tools and measuring instruments used for spindle installation of CNC machine tools	25	
3	Be able to master the spindle installation process of CNC machine tools	25	
4	Have a professional spirit of effective communication, expression and solidarity	25	

Task III Understand the Geometric Accuracy Detection of CNC Machine Tools

Learning Objectives

(1) To understand the contents of geometric accuracy detection for CNC machine tools and corresponding tools and measuring instruments.

(2) To master the straightness detection method.

(3) To master the parallelism detection method.

(4) To master the verticality detection method.

(5) To master the spindle runout detection method.

(6) To master the flatness detection method.

(7) To master the engineering working methods, cultivate a rigorous work style, and abide by the 5S management system.

Task Description

Observe the geometric accuracy of YL569 CNC machine tool in groups, and fill the project name, inspection tools, inspection methods, tolerance ranges and etc. of various geometric accuracy detection of CNC machine tools, into the task work order. Make sure to be careful and pay attention to the relevant operation specifications of the practical training site during observation.

Knowledge Link

I. Geometric accuracy detection contents of CNC machine tools

The geometric accuracy of the CNC machine tool comprehensively reflects the key parts and components of the machine tool and their errors in geometrical form after assembly. Because some items in geometric accuracy are interrelated and mutually affected, the geometric accuracy detection of the machine tool must be completed at one time after the fine commissioning of the machine tool, and it is not allowed to adjust and detect by item.

The detection projects of geometric accuracy generally include straightness, flatness and parallelism. For example, the geometric accuracy detection of machining center usually includes the following items.

(1) Flatness of workbench.

(2) Mutual perpendicularity of movement in each coordinate axis direction.

(3) Parallelism of workbench when moving in the coordinate X and Y axes direction.

(4) Axial movement of the spindle.

(5) Radial runout of the spindle hole.

(6) Parallelism of spindle axis when the spindle box moves along Z coordinate direction.

(7) Perpendicularity of spindle rotation axis to workbench.

(8) Straightness of spindle box movement in Z coordinate direction.

II. Accuracy detection experiment of CNC milling machines

1. Experimental instruments

(1) CNC milling machines;

(2) Two leveling rulers (400 mm, 1,000 mm, Grade 0);

(3) One angle square (300 mm×200 mm, Grade 0);

(4) One direct inspection bar (80 mm×500 mm);

(5) One BT40 taper inspection bar;

(6) Two dial gauges;

(7) Two magnetic gauge stands;

(8) Two level gauges (200 mm, 0.02/1,000);

(9) Three contour blocks;

(10) Two adjustable gauge blocks.

2. Experiment contents: Accept of geometric accuracy

Acceptance of geometric accuracy.

1) Machine tool leveling.

Inspection tools: Precision level gauge.

Inspection method: As shown in Fig. 6–33, place the workbench in the middle of the guide rail stroke, place two level gauges in the center of the workbench along the *X* and *Y* coordinate axes respectively, and adjust the height of the machine tool shim plate to make the level gauge bubble in the middle of the reading; Move the workbench along *X* and *Y* coordinate axes respectively in full stroke, observe the change of level gauge reading, adjust the height of machine tool shim plate, so that the change range of level gauge reading is less than 2 grids when the workbench moves along *X* and *Y* coordinate axes in full stroke, and the reading is in the middle position.

2) Detect the flatness of workbench.

Inspection tools: Dial gauge, leveling ruler, adjustable gauge block, contour block and precision level gauge.

Principle of detecting the flatness error of the workbench with a leveling ruler: Within the specified measurement range, if all points are included in two planes parallel to the general direction of the plane and spaced apart by a given value, the plane is considered flat.

Inspection method: As shown in Fig. 6–34, first select three points *A*, *B* and *C* on the inspection surface as zero marks, and place three equal-height gauge blocks on these three points to determine the datum plane for comparison with the inspected surface. Place the leveling ruler on points *A* and *C*, and place an adjustable gauge block at point *E* on the inspection surface to make it contact with the small surface of the leveling ruler. At this time, the upper surfaces of the gage blocks *A*, *B*, *C* and *E* are all on the same surface. Then place the leveling ruler on point *B* and *E* to find the deviation of point *D*. Place an adjustable gauge block at point *D* and adjust its upper surface to the plane defined by the upper surface of the gauge block already in place. Place the leveling ruler on points *A* and *D*, *B* and *C* respectively to find out the deviation between points *A* and *D* and between points *B* and *C* on the inspected surface. Deviations between the remaining points can be found in the same way.

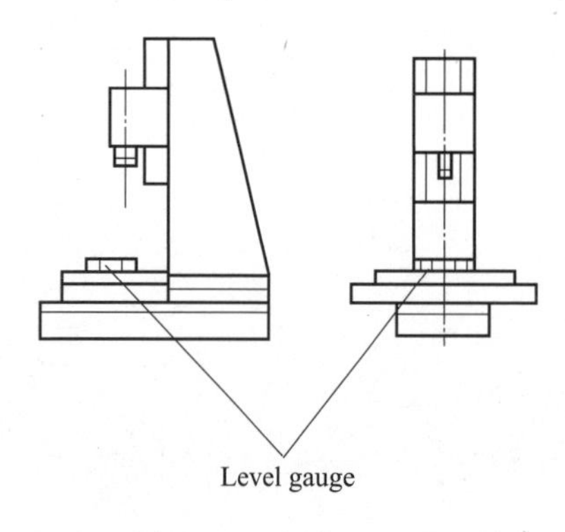

Fig. 6–33 Diagram of Machine Tool Leveling

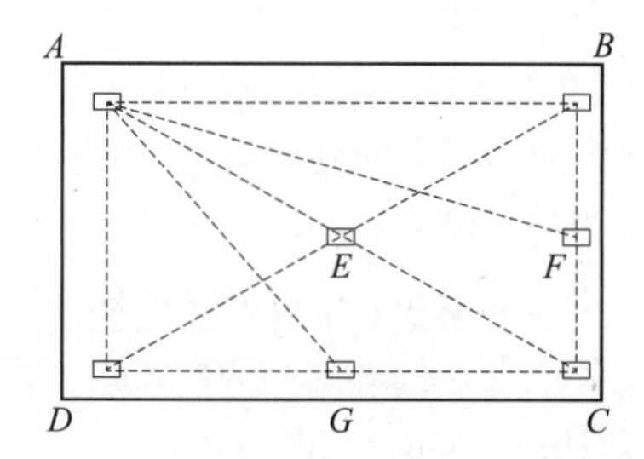

Fig. 6–34 Diagram of Detecting Flatness

3) Radial runout of spindle taper axis, deflection of spindle end face and deflection of outer wall of spindle sleeve.

Inspection tools: Inspection bar and dial gauge.

Inspection method: As shown in Fig. 6–35, insert the inspection bar into the spindle taper, install the dial gauge on the fixed part of the machine tool with its probe vertically touching the measured surface, rotate the spindle, record the maximum reading difference of the dial gauge, and measure at *a* and *b* respectively. Mark the relative position between the inspection bar and the spindle in the circumferential direction, remove the inspection bar, rotate the inspection bar by 90°, 180° and 270° respectively in the same direction, re-insert it into the spindle taper, and detect each position separately. Take the average value of 4 detections as the radial runout error of spindle taper axis.

The deflection of spindle end face and the outer wall of spindle sleeve are detected as shown in Fig. 6-36.

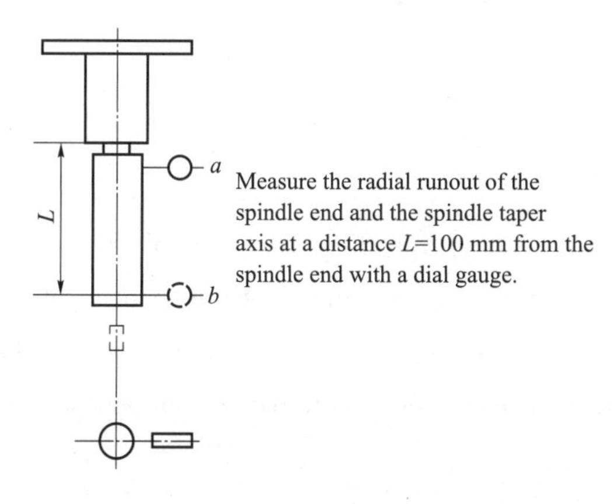

Fig. 6-35 Diagram of Detecting Radial Runout of Spindle Taper Axis

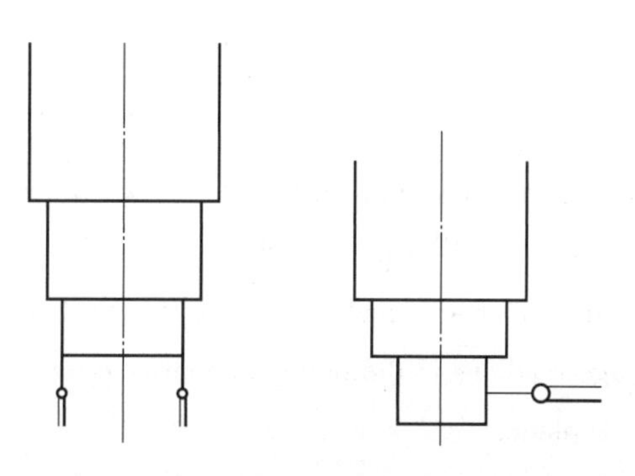

Fig. 6-36 Diagram of Detecting Spindle End Face Deflection and Outer Wall Deflection of Spindle Sleeve

4) Perpendicularity of spindle axis to workbench.

Inspection tools: Leveling ruler, contour block, dial gauge and gauge bracket.

Inspection method: As shown in Fig. 6-37, install the indicator bracket with a dial gauge on the spindle and adjust the probe of the dial gauge to be parallel to the spindle axis. The parallelism deviation between the measured plane and the datum plane can be measured by the inspection method that the probe of the dial gauge swings on the measured plane. When the spindle rotates for one circle, the maximum difference between dial gauge readings is the perpendicularity deviation. Record the reading difference of dial gauge at two positions 180° apart in $X-Z$ and $Y-Z$ planes respectively. In order to eliminate the measurement error, the inspection instrument can be rotated 180° relative to the spindle after the first inspection and then inspect again.

5) Perpendicularity of vertical movement of spindle box to workbench.

Inspection tools: Contour block, leveling ruler, angle square, dial gauge.

Inspection method: As shown in Fig. 6-38, place the contour block on the workbench along the Y-axis direction, place the leveling ruler on the contour block, place the angle square on the leveling ruler (in the $Y-Z$ plane), fix the dial gauge on the spindle box with its probe vertically touching the angle square, move the spindle box, and record the reading and direction of the dial

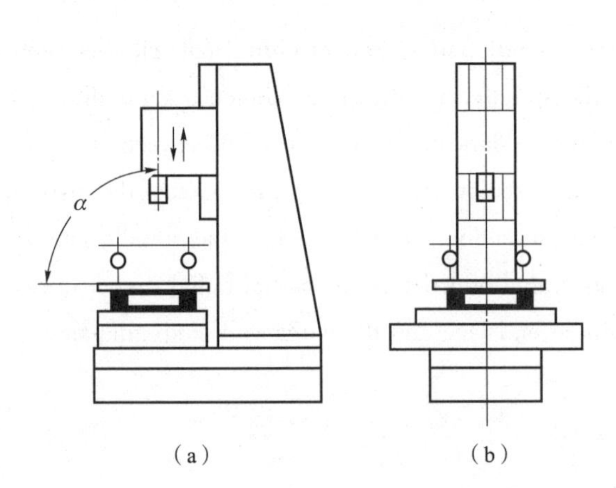

Fig. 6-37 Diagram of Perpendicularity Detection of Spindle Axis to Workbench

(a) $Y-Z$ plane; (b) $X-Z$ plane

gauge. The maximum difference between its readings is the perpendicularity error of the vertical movement of the spindle box to the workbench in the $Y-Z$ plane; Similarly, place the contour block, leveling ruler and angle square in the $X-Z$ plane for remeasurement. The maximum difference of dial gauge readings is the perpendicularity error of vertical movement of spindle box to workbench in the $X-Z$ plane.

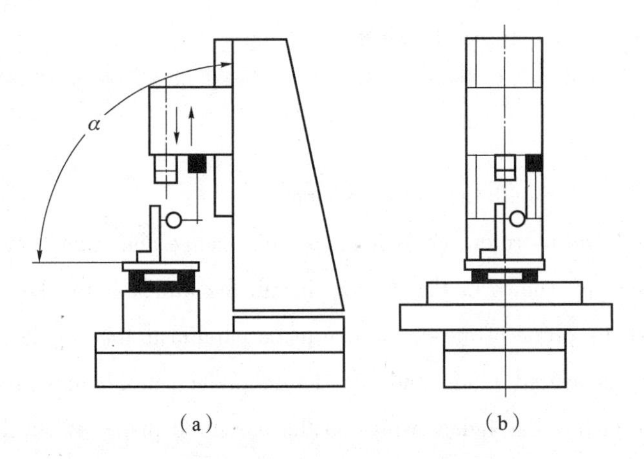

Fig. 6-38 Diagram of Detecting the Perpendicularity of Spindle Box Vertical Movement to Workbench

(a) $Y-Z$ plane; (b) $X-Z$ plane

6) Perpendicularity of vertical movement of spindle sleeve to workbench.

Inspection tools: Contour block, leveling ruler, angle square and dial gauge.

Inspection method: As shown in Fig. 6-39, place the contour block on the workbench along the Y-axis direction, place the leveling ruler on the contour block, place the angle square on the leveling ruler, and adjust the position of the angle square to make its axis coincide with the spindle axis; fix the dial gauge on the spindle box, and its probe vertically touches the angle square in the $Y-Z$ plane. Move the spindle and record the reading and direction of the dial gauge. The maximum difference between the readings is the perpendicularity error of the spindle sleeve vertical movement to the workbench in the $Y-Z$ plane; Similarly, the dial indicator probe vertically touches the angle square gauge in the $X-Z$ plane for remeasurement. The maximum difference of dial gauge readings is the perpendicularity error of the spindle sleeve vertical movement to the workbench in the $X-Z$ plane.

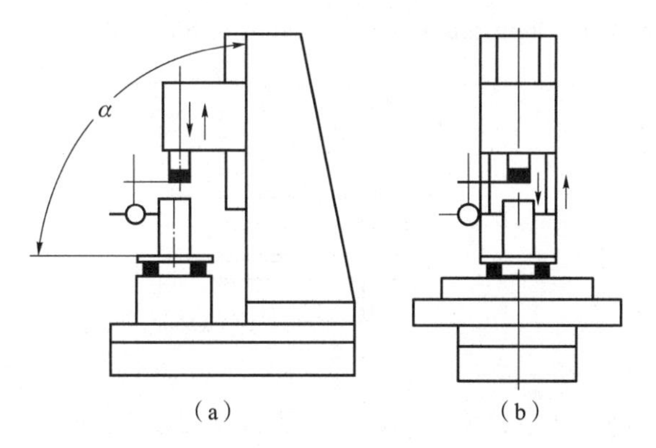

Fig. 6-39 Diagram of Perpendicularity of Spindle Sleeve Vertical Movement to Workbench

(a) $Y-Z$ plane; (b) $X-Z$ plane

7) Parallelism of workbench movement in X-axis direction or Y-axis direction to workbench.

Inspection tools: Contour block, leveling ruler and dial gauge.

Inspection method: As shown in Fig. 6-40, place the contour block on the workbench along the Y-axis direction and the leveling ruler on the contour block, make the dial gauge probe vertically touch the leveling ruler, move the workbench along the Y-axis direction, and record the reading of the dial gauge. The maximum difference between the readings is the parallelism error of workbench movement in Y-axis direction to workbench; Place the contour block on the workbench along the X-axis direction, move the workbench in the X-axis direction for remeasurement. The maximum difference of readings is the parallelism error of workbench movement in X-axis direction to workbench.

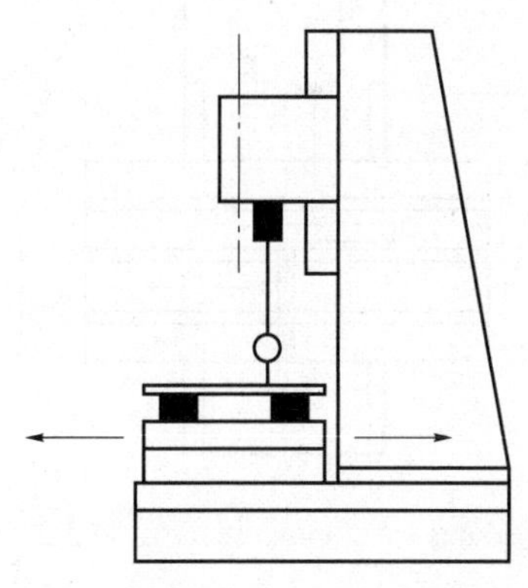

Fig. 6-40 Diagram of Parallelism Detecting of Workbench Movement in X-axis Direction or Y-axis Direction to Workbench

8) Parallelism of workbench movement in X-axis direction to workbench reference (T-groove).

Inspection tool: Dial gauge.

Inspection method: As shown in Fig. 6-41, fix the dial gauge on the spindle box so that its probe vertically touches the reference (T-groove), move the workbench in the X-axis direction and record the reading of the dial gauge. The maximum difference between the readings is the parallelism error of the workbench moving along the X-axis direction to the workbench reference (T-groove).

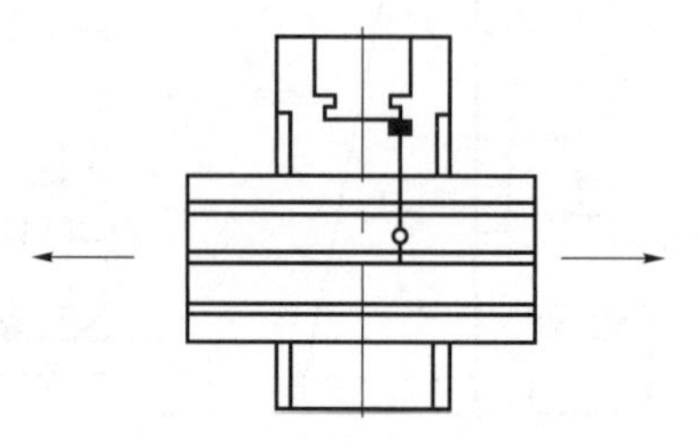

Fig. 6-41 Diagram of Parallelism Detecting of Workbench Movement in X-axis Direction to Workbench Reference (T-groove)

9) Working perpendicularity of workbench movement in X-axis direction to that in Y-axis direction.

Inspection tools: Angle square and dial gauge.

Inspection method: As shown in Fig. 6-42, the workbench is in the middle of the stroke. Place the angle square on the workbench, fix the dial gauge on the spindle box so that the dial gauge probe vertically touches the angle square (Y-axis direction), move the workbench in Y-axis direction, adjust the position of the angle square so that one side of the angle square is parallel to the Y-axis axis, then vertically touch the other side of the angle square (X-axis direction), move the workbench in X-axis direction, and record the readings of the dial gauge. The maximum difference of readings is the working perpendicularity error of workbench movement in X-axis direction to that in Y-axis direction.

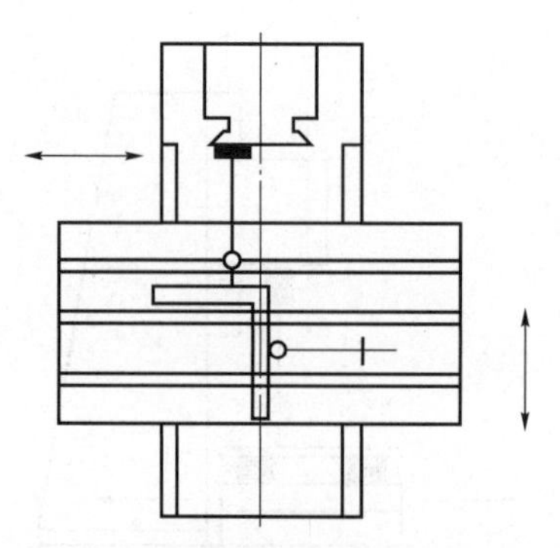

Fig. 6-42 Diagram of Working Perpendicularity of Detecting Workbench Movement in X-axis Direction to That in Y-axis Direction

Record the measurement results of the above test items in Table 6-3.

Table 6-3 Data Record of Accuracy Detection of CNC Milling Machine

Machine tool model	Machine tool No.	Ambient temperature	Inspector	Test date

S/N	Inspection item	Tolerance range/mm	Inspection tool	Measured/mm
1	Machine tool leveling	0.06/1,000		
2	Flatness of workbench	0.08/Overall length		
3	Radial runout of spindle taper axis near the spindle end	0.01		
	Radial runout of spindle taper axis at L (L = 100 mm) from the spindle end	0.02		
4	Perpendicularity of spindle axis to workbench in Y–Z plane	0.05/300($\alpha \leqslant 90°$)		
	Perpendicularity of spindle axis to workbench in X–Z plane			
	Deflection of spindle end face	0.01		
	Deflection of outer wall of spindle sleeve	0.01		
5	Perpendicularity of vertical movement of spindle box to workbench in Y–Z plane	0.05/300($\alpha \leqslant 90°$)		
	Perpendicularity of vertical movement of spindle box to workbench in X–Z plane			
6	Perpendicularity of spindle sleeve movement to workbench surface in Y–Z plane	0.05/300($\alpha \leqslant 90°$)		
	Perpendicularity of spindle sleeve movement to workbench surface in X–Z plane			
7	Parallelism of workbench movement in X-axis direction to workbench	0.056($\alpha \leqslant 90°$)		
	Parallelism of workbench movement in Y-axis direction to workbench	0.04($\alpha \leqslant 90°$)		
8	Parallelism of workbench movement along X-axis direction to workbench reference (T-groove)	0.03/300		
9	Working perpendicularity of workbench movement in X-axis direction to that in Y-axis direction	0.04/300		

Task Implementation

(1) Divide students randomly into groups and appoint the group leaders. Fill the personal information of the group members in Table 6-4.

(2) According to the observation and discussion results, record the names of various geometric accuracy inspection items, inspection tools, inspection methods tolerance ranges and etc. of the CNC machine tool in Table 6-4.

(3) Fully discuss the problems found during the observation, and fill the problems and solutions in Table 6-4.

Table 6-4 Task Work Order for Geometric Accuracy Detection of CNC Machine Tool

Task Work Order					
Name:		Class:		Student ID:	
List of materials, tools, measuring instruments, and equipment required					
S/N	Name		Quantity	Model or specification	Remark
Operating process(Observation)					
S/N	Project name	Inspection tool	Inspection method		Tolerance range
Problems			Solutions		

Task Evaluation

S/N	Evaluation contents	Total point	Score
1	Be familiar with and know the names of machine tool mechanical devices; know the CNC machine tool geometric accuracy related tools and measuring devices	30	

Continued

S/N	Evaluation contents	Total point	Score
2	Be able to master the contents of CNC machine tool geometric accuracy measurement	30	
3	Be able to use tools and measuring devices to measure the CNC machine tool geometric accuracy	30	
4	Have a professional spirit of effective communication, expression and solidarity	10	

(1) What are the main parts of the mechanical structure of CNC machine tools?
(2) What are the requirements of CNC machine tools for main transmission systems?
(3) What are the commonly used mechanisms of CNC machine tool feed systems?
(4) What are the precautions during spindle installation of CNC milling machines?
(5) What is the significance of geometric accuracy inspection of CNC machine tools?
(6) How to measure the radial runout of spindle taper axis of CNC machine tools?

References

[1] YANG ZHONGLI, HU ZONGZHENG, ZUO WEI. Fault Diagnosis and Maintenance of CNC Machine Tools [M]. Dalian: Dalian University of Technology Press, 2019.

[2] ZHOU LAN, CHEN SHAOAI. Connection Commissioning and PMC Programming of FANUC 0i-D/0i Mate-D CNC System [M]. Beijing: China Machine Press, 2022.

[3] JIANG HONGPING, WANG BEI, LIU CAIXIA. Fault Diagnosis and Maintenance of CNC Equipment [M]. Beijing: Beijing Institute of Technology Press, 2018.

[4] YU TAO, WU HONGEN. CNC Technology and Machine Tools [M]. Beijing: Tsinghua University Press, 2019.